Emerging Concepts of Tumor Exosome–Mediated Cell–Cell Communication

Huang-Ge Zhang
Editor

Emerging Concepts of Tumor Exosome–Mediated Cell–Cell Communication

Springer

Editor
Huang-Ge Zhang
University of Louisville
Louisville, KY
USA

ISBN 978-1-4614-3696-6 ISBN 978-1-4614-3697-3 (eBook)
DOI 10.1007/978-1-4614-3697-3
Springer New York Dordrecht Heidelberg London

Library of Congress Control Number: 2012948293

Printed on acid-free paper

Springer is part of Springer Science+Business Media (www.springer.com)

Preface

Over the past decades, secretion of proteins that bind to receptors on neighboring cells was considered a primary mode for communication between cells. Recently, the release of membrane vesicles including exosomes/microvesicles has become recognized as an important mode of intercellular communication. Unlike a single soluble factor secreted by cells, exosomes can carry a group of functional molecules enriched with particular proteins, lipids, and microRNAs (miRNAs) and provide an increased stability of the transported molecules. Therefore, exosomes can serve as intercellular communicators not only locally but also systemically. Exosomes are released by many different cell types, and their biological effects on the recipient cells are dependent on the composition of the exosomes and the microenvironment where the recipient cells are residents.

This book will focus on exosomes as they relate to tumor cells and cancer. The first four chapters will cover the basic biology of exosomes and the roles of specific components of exosomes, primarily focusing on proteins, mRNAs/miRNA, and lipids. In the first chapter, Dr. Graça Raposo provides an introduction to exosomes, describes the composition, biogenesis, function of exosomes in general, and the methodology for isolation and characterization of exosomes/microvesicles. Chapter 2 written by Dr. Jan Lötvall further elucidates the role of exosomal shuttle RNA in cell-to-cell communication. Dr. Lötvall's group discovered that miRNAs are encapsulated in exosomes and that they play a critical role in cell-cell communication and also as potential markers for prodiagnosis of diseases. Dr. Michel Record wrote Chapter 3 and he focuses on the effects of exosomal lipids in cell-cell communication. Development of cutting-edge lipidomic technology has provided a means to recognize the biological and pathobiological effects of lipids on many different cell types and diseases. Exosomes carrying lipids can function in exosomal lipid mediated targeting as well as carrying lipids to recipient cells. The final chapter in this section by Dr. Margot Zöller will cover the mechanisms underlying the selective recruitment of molecules into exosomes. Dr. Zöller uses tetraspanin as an excellent example to demonstrate how tetraspanins regulate protein assembly of exosomes and the possible recruitment of miRNA.

Book Chapters 5–9 focus on the effects of exosomes at the cellular and molecular levels, on tumor cell proliferation, on metastasis, and on immunosuppression. Topics in these chapters deal with the role of tumor exosomes in the progression of tumor development (Drs. Huang-Ge Zhang, William Grizzle, and Douglas Taylor);

the mechanisms underlying tumor exosome-mediated immunosuppression including induction of regulatory T cells and myeloid cells, immune modulation of T cells, and natural killer cells by tumor-derived exosomes (Dr. Theresa Whiteside), tumor-stroma interactions (Dr. Johan K. O. Skog), and microenvironment regulation of metastasis by exosomes (Dr. David Lyden).

The final chapter deals with the future of exosome biology and more specifically the induction of stem cells in regenerative medicine (Dr. Mariusz Ratajczak).

There are many individuals I wish to thank, without whose efforts this book would not have become a reality. First and foremost, my thanks go to all the authors who contributed material for publication in this book. Their expertise and scientifically informed viewpoints, coupled with a collaborative spirit under tight time constrains, has made working with them a pleasure. I also wish to thank our acquisitions editor, Fiona Sarne for her patience, encouragement, and support. This book marked a lot of firsts for me, so in the process of learning the ins and outs of publishing a book of this nature, I also developed a very strong friendship with her. I look forward to working with her again in the future.

Last, but not least, I thank my wife Lifeng Zhang, and my twin daughters Shuangqin Zhang and Shuangyin Zhang for their support and understanding during the many times that book-writing took precedence over family matters.

Contents

Chapter 1
The Cell Biology of Exosomes: Historical and Perspectives

Alessandra Lo Cicero and Graça Raposo

Abstract Cells release into the environment not only soluble but also membrane-associated factors. Whereas membranous microvesicles up to 1 μm are generated by outward budding of portions of the plasma membrane, membrane vesicles of 30–100 nm, called exosomes are secreted during fusion of endosomal multivesicular bodies with the cell surface. Exosomes contain membrane and cytosolic components that include proteins, lipids and RNAs. Exosomes from different cell types not only harbor a common set of components but unique subsets of proteins are also associated with cell type-associated functions. The composition of exosomes is related to their biogenesis within the endosome, a process that requires the inward invagination of the endosomal membrane to form small membrane vesicles. The mechanisms involved in their formation and secretion only begin to be elucidated. Intervention on their function represents certainly ways to understand their involvement in essential cellular processes that extend from immune regulation to neuronal communication but also pathogen transmission. Exosomes are present in biological fluids and are potential candidate biomarkers for disease that could also possibly be exploited for therapeutical delivery of drugs.

Abbreviations

ILV	Intraluminal vesicle
MVE	Multivesicular endosomes

G. Raposo (✉)
Institut Curie, CNRS-UMR144, 26 Rue d'Ulm,
75248 Paris Cedex, France
e-mail: graca.raposo@curie.fr

A. Lo. Cicero
Institut Curie, Centre de Recherche, 75248 Paris Cedex, France

H.-G. Zhang (ed.), *Emerging Concepts of Tumor Exosome–Mediated Cell–Cell Communication*, DOI 10.1007/978-1-4614-3697-3_1,

1.1 On the Discovery of Exosomes

1.1.1 The Reticulocyte Unravels a Novel Pathway of Secretion

Differentiation of the reticulocyte into a mature erythrocyte requires that one-third of the surface area is lost accompanied by the eradication of many membrane-associated proteins [1]. One of the main mechanisms for the loss of plasma membrane proteins appeared to be through the selective intracellular budding of vesicles into a so-called multivesicular endosome (MVE) [2]. The groups of R. Johnstone in Montreal and P. Stahl in Saint Louis reported in the mid-1980s that immature sheep reticulocytes released the bulk of their transferrin receptor in association with small membrane vesicles released upon fusion of MVEs with the cell surface of the reticulocyte during its maturation process [2–5] (Fig. 1.1). Reticulocytes cultured with radioactive transferrin, with radioactive antibody bound to the cell surface exposed transferrin receptor or with gold-labeled antibodies released rapidly the majority of their transferrin receptor. Electron microscopical observations have been fundamental to uncover exocytic profiles of MVEs fusing with the cell surface and to characterize the secreted membrane vesicles [2]. The exocytosed vesicles were recovered after ultracentrifugation at $100.000 \times g$ for 90 min and were named exosomes (exocytosed endosomal contents) [6]. Red blood cells of different species secrete exosomes and the Johnstone's group further showed that, in addition to the transferrin receptor other proteins such as acetylcholinesterase, cytochalasin B-binding (glucose transporter) nucleoside binding (i.e., nucleoside transporter), Na^{+}-independent amino acid transport, can be lost from the reticulocyte by such unconventional secretory process [6, 7]. Dr Rose Johnstone was a scientist with an incredible strength and enthusiasm and who believed that the process of exosome secretion that her group described in reticulocytes would be a key in additional cellular processes. As she commented with humor after the first workshop on exosomes in Montreal in 2005: An "Alice in Blunderland" approach led to the discovery of exosomes and their involvement in protein extrusion from maturing reticulocytes. Current studies have shown that the process of exosome formation extends to many mammalian cells. Full knowledge of the contribution of exosomes to intercellular information transmission and the potential medical application of this knowledge will depend on the ingenuity of future investigators and their insight into biological processes (Citation from [8]).

1.1.2 Exosomes in the Immune System

While studies on reticulocytes uncovered exosome secretion as a process of eradication of unwanted molecules required for cell differentiation, only 10 years later a similar process has been described for immune cells granting exosomes extracellular functions [9]. Immunoelectron microscopy of B lymphoblastoid cells demonstrated that the limiting membrane of major histocompatibility complex (MHC) class II compartments that have all the hallmarks of MVEs can fuse directly with the plasma

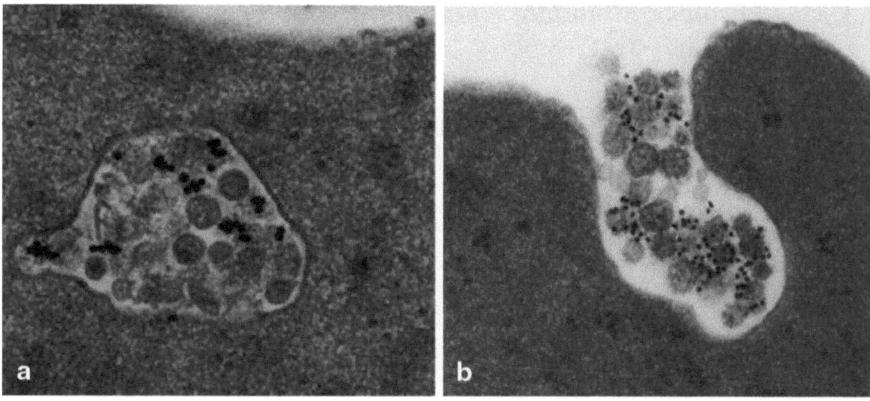

Fig. 1.1 Release of exosomes by reticulocytes. Intracellular localization of MVEs and their fusion with the plasma membrane. **a** In the MVEs, the gold label for the transferrin receptor is associated with the small dense bodies of ~50 nm, which are present inside larger vesicles of 300–500 nm. Note the absence of surface label. Colloidal gold of 20 nm was used. Incubation was for 60 min at 37 °C. **b** Fusion of MVEs with the plasma membrane and release of the so-called round bodies: exosomes. Note, exocytosis into the medium of small dense bodies labeled with gold after a 3-h incubation at 37 °C. The gold label is only present on the 50-nm bodies, which are inside vesicles of 300–800 nm in diameter. The limiting membrane of the MVE is devoid of label. Magnification × 61,500. (Copyright permission 1985 Rockefeller University Press. Originally published in [2])

membrane. Similar to the process described for reticulocytes, fusion resulted in release from the cells of internal MHC class II-containing vesicles [9]. By analogy, these secreted vesicles were also named exosomes. They were isolated from the cell culture media by differential centrifugation followed by flotation on sucrose density gradients in which exosomes and the majority of MHC class II floated to a mean density of 1.13 revealing their vesicular nature. Importantly, the overall surface protein composition of exosomes differed significantly from that of the plasma membrane. Exosome-bound MHC class II were in a compact, peptide-bound conformation and metabolically labeled MHC class II were released into the extracellular medium with relatively slow kinetics, 10 ± 4 % in 24 h, indicating that direct fusion of MIICs with the plasma membrane is not the major pathway by which MHC class II reaches the plasma membrane but by which it is released extracellularly. Importantly and despite the initial hypothesis that exosomes could serve for eradicating nonfunctional MHC class II exosomes derived from both human and murine B lymphocytes- induced antigen-specific MHC class II-restricted T cell responses [9]. These data suggested a role for exosomes in antigen presentation in vivo. From these studies it was speculated that in vivo, in the circulation, exosomes might function as transport vehicles for MHC class II peptide complexes responsible for the maintenance of long-term T cell memory or T cell tolerance [9]. Moreover, they certainly opened a new avenue to explore the usefulness of exosomes, in particular as biological vehicles, in immunotherapy.

1.1.3 Exosomes Stimulate Antitumoral Immune Responses In Vivo

Only 2 years after the findings that B lymphocytes secrete exosomes with antigen-presenting capacities, a first breakthrough on the way to transfer from the bench to the bedside has been the demonstration, in vivo, in mice bearing tumors, that exosomes can eradicate and/or suppress tumor growth [10]. Firstly, it was shown that the dendritic cell (DC), a professional antigen-presenting cell (APC) also has the capacity to secrete exosomes. Secondly, tumor peptide-pulsed DC-derived exosomes once injected into mice, prime specific cytotoxic T lymphocytes in vivo and eradicate or suppress growth of established murine tumors in a T cell-dependent manner [10]. These results provided the basis for the hypothesis that exosomes could play an active role in intercellular communication, at least in the immune system, and prompted the very first attempt at using them in the clinic, as a new type of exosome-based cell-free vaccines representing an alternative to DC-adoptive therapy for suppressing tumor growth [11, 12]. Phase I trials using the Dex strategy in melanoma and nonsmall-cell lung cancer patients have been held (1999–2002) [13, 14] and a phase II trial is currently ongoing at the Gustave Roussy Institute (with participation of the Curie Institute) in France [15]. Despite the accumulating evidence for the bioactivity of exosomes and their involvement in intercellular communication, however, and as detailed in the following sections of this chapter, there is still limited understanding on the mechanisms involved in their biogenesis and their secretion. This has largely precluded understanding their physiological relevance in vivo.

1.2 Definition of Exosomes

1.2.1 Exosomes, Microvesicles, and Other Secreted Vesicles

In different conditions, cells release different types of membrane vesicles. Not only endosome-derived exosomes but also vesicles generated by shedding or exfoliation of the plasma membrane [16, 17] (see also Fig. 1.2). One major ongoing challenge is to define methods that will allow separating both types of vesicles and comparing their respective properties. Several studies reporting on the composition and functions of secreted vesicles do not make a clear distinction between the different types, certainly because the purification and characterization of each type of vesicle is not an easy task and requires a multiplicity of complementary methods. The process of "ectocytosis" is proposed to describe triggered shedding of right-side-out membrane vesicles from the surface of eukaryotic cells but the term ectosome is more generally related to vesicle shedding by neutrophiles [18, 19]. Other vesicles are generally called microvesicles (MVs), or exosome-like if incompletely characterized. Interestingly, the release of bioactive membrane vesicles from the cell surface is not only a feature of mammalian cells but is also conserved across microbial life, in bacteria, archaea,

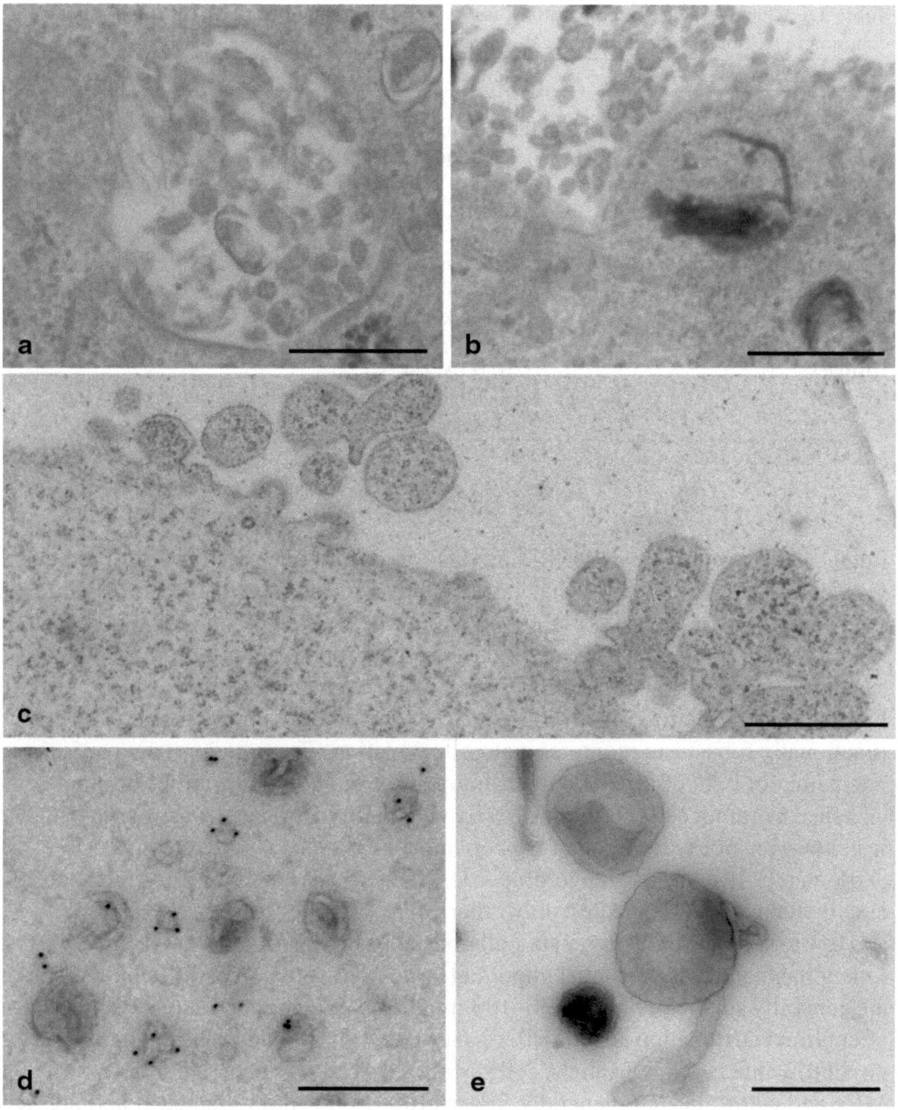

Fig. 1.2 Electron microscopy of exosomes and microvesiscles. **a** and **b** Melanoma cells were fixed and embedded in epon for conventional electron microscopy. **a** Multivesciular endosome. **b** Secretion of exosomes. Note that the secreted vesicles have the same morphology and size of the intraluminal vesicles of the MVE. **c** Breast cancer cells were fixed and embedded in epon for conventional electron microscopy. Large membrane vesicles bud from the plasma membrane. **d** and **e** Isolated exosomes and microvesicles were contrasted and embedded in a mixture of uranyl-acetate and methylcellulose. In **d**, exosomes were immunogold labeled with antibodies anti-CD63 and Protein A gold 10 nm. Magnification bars in **a**, **b**, **d**, and **e** are 200 nm. Magnification bar in **c** is 1,000 nm

Table 1.1 Characteristic protein composition of exosomes found from MS/MS analysis that are common to exosomes from different cell types

Classifications	Proteins
Antigen presentation	MHC class I
Adhesion molecules	Tetraspanins: CD63, CD81, CD9, CD37, CD53, CD82; Integrins: α3, α4, αM, αL, β1, β2
Membrane transport and fusion	Annexins: I, II, IV, V, VI, VII, XI; Flotillin-1 Rab 2, Rab 5c, Rab 10, Rab 7; Arf 3, Arf 6, Arf 5
ESCRT proteins	Alix, Tsg101
Heat-shock proteins	Hsc70, Hsp84/90
Cytoskeletal proteins	Actin, cofilin 1, moesin; tubulin: α1, α2, α6, β5, β3
Enzymes	Pyruvate kinase, alpha-enolase, GAPDH
Signal transduction	14-3-3 ξ, γ, ε; Gβ1, Gi2α
Histones	H2B, H2A, H4
Others	MFG-E8, lactadherin, clathrin, elongation factor 1α, C3, Lamp2

MS Mass spectrometry

fungi, and parasites in which MV production occurs not only in vitro, but also in vivo during infection [20]. It has been reported for decades that mammalian cells shed fragments of their plasma membrane in the form of MVs [21]. Several studies, in particular using cancer cell lines, correlate this feature with their metastatic capacities as inhibition of shedding decreases the appearance of metastasis [22, 23]. Further studies reported that protease cargo contained within the microvesicles is functionally robust and promotes matrix degradation. These findings are particularly interesting in light of reports demonstrating that proteolytic activities of microvesicles shed by tumor cells correlate directly with malignancy and invasiveness [24, 25]. It is however not excluded that cancer cells and other cells secrete both MVs and exosomes [24] with respective properties. This is also the case for platelets [26], endothelial cells [27], and certainly other cells.

How to separate the different populations of vesicles and to access their respective functional features is one of the important issues of discussion in the field. Exosomes are generally smaller within the 30–100-nm diameter ranges and can only be pelleted after ultracentrifugation at 100.000 × *g* (see below). Although we cannot rule out the possibility that budding from the cell surface [28] can also generate smaller vesicles, MVs and ectosomes are generally larger (100–1,000 nm in diameter) and are pelleted at lower *g* forces (Fig. 1.2). Because of their different origins, the lipid, protein, and maybe RNA composition is likely to be different [16, 29]. However, whereas several studies revealed that exosomes secreted from very different cell types depict a common signature (Table 1.1 and Fig. 1.1, see also [30]), still little is known on the composition of MVs. Efforts are currently being made to compare the composition of the different types of secreted vesicles and relate that to their functional features. It is also likely that the cellular machineries involved in the formation of MVs/ectosomes/exosomes are distinct. Interestingly, not only exosome formation requires the specific sorting of proteins and lipids, there are also sorting mechanisms that may operate during the formation of shed vesicles [31]. The mechanism of formation of MVs from the cell surface appears to require the actomyosin machinery and the action of small GTPases such as ARF6, as demonstrated for several cancer lines [24, 25]. This aspect will not be further developed in this chapter.

1.2.2 Isolation of Exosomes

Differently from cell fractionation that is generally used to purify membranous fractions from cells and tissues, exosomes are isolated directly from cell culture supernatants. For morphological and biochemical characterization, exosomes can be collected from the cell culture-conditioned medium of cultured cell lines by differential ultracentrifugation [9, 32]. The supernatant of viable cell cultures is centrifuged; cell debris and dead cells are removed with different centrifugation at increasing speeds. The final supernatant is ultracentrifuged and exosomes, similar to any small membrane vesicle, are pelleted at 100.000 × *g*. The pellet obtained is washed in PBS, to eliminate contaminating proteins, and ultracentrifuged one last time. One possible flow is the presence of exosomes in fetal calf serum. Therefore, the culture media should be depleted of exosomes by ultracentrifugation of the culture media 1–2 h at 100.000 × *g* [33]. It is also possible to ultracentrifuge directly the fetal calf serum though the viscosity of the serum requires longer ultracentrifugation. The same recommendations apply to body fluids (serum, ascites) that should also be diluted before ultracentrifugation [32]. Some protocols described in the literature bypass the first steps of ultracentrifugation by using a direct filtration of the cell medium using a 0.22 μm filter that is thought to eliminate dead cells and cell debris. However, this filtration step can also remove exosomal membranes if they are secreted as small aggregates. It should be noted that the classical differential ultracentrifugation as described above, ended by the 100.000 × *g*, is generally not enough to obtain a fraction of exosomes devoid of contaminants. Generally, and as evaluated by morphology, these fractions contain vesicles of exosomal nature. Nevertheless, many other secreted proteins can adsorb to exosomal membranes and heavily contribute to the amount of proteins found on exosomes as evaluated by protein assays. Several proteomic and lipidomic studies have been performed on exosomes collected by differential ultracentrifugation and it cannot be excluded that some of the proteins identified could be nonspecifically associated with exosomes. A better isolation can be achieved by performing continuous sucrose density gradients [9, 33–35]. Exosomes "float" to a density close to 1.13–1.15 g/ml, but this range of densities can vary from cell to cell depending on the protein content of exosomes as, for example, exosomes secreted by intestinal epithelial cells float even to a density of 1.19 [35]. The fractions corresponding to these densities are generally pooled for further analysis, although electron microcopical observations of the different fractions reveal that membrane vesicles with different morphologies and size are present in the fractions highlighting the heterogeneity of the exosomal population (see also below for additional discussion). In some studies, exosomes have been immmunoisolated using magnetic beads coated with antibodies directed enriched on exosomes, as, for example, tetraspannins or MHC class II when dealing with exosomes from APCs [32, 33, 36]. This step certainly leads to better yields of purified exosomes, however, and as noted above and further discussed below, the exosomal population is heterogeneous in their enrichment for specific proteins and a subpopulation could be lost if not using more than one antibody. To prepare clinical-grade purified exosomes derived from APCs, an alternative method based on ultrafiltration and centrifugation in sucrose deuterium oxide (D_2O) cushions has been developed [37, 38]. Because of the increasing interest for exosomes, their use in therapeutical protocols or as biomarkers in body fluids, commercially available methods to isolate exosomes start to be developed. These approaches that could allow to isolate exosomes "easily" should nevertheless be followed by a thorough characterization of the ending product because they do not allow to discriminate between exosomes or other types of secreted vesicles and do not exclude adsorbed, nonspecifically bound components.

1.2.3 Characterization of Exosomes

Once purified using the different methods detailed above, identification of membrane vesicles as exosomes requires the use of complementary imaging and biochemical methods. Given their small size (30–100 nm), exosomes can be visualized only with an electron microscope at a magnification of 17.000 and higher; thus, evaluation of the purity of exosome preparations as well as their characterization should be firstly evaluated by electron microscopy. Exosomes observed by whole mount microscopy, fixed, contrasted, and embedded in a mixture of uranyl-acetate and methylcellulose [9] depict a cup-shaped morphology (Fig. 1.2). It is a quick and easy approach to evaluate the presence of small membrane vesicles after differential ultracentrifugation and to analyze the enrichment for exosomal-associated components exposed at the outer membrane. Access to the cytosolic components sequestered in exosomes requires permeabilization and given their small size labeling with antibodies and gold probes is often difficult to perform. It should be noted however that the cup-shaped appearance is very likely a consequence of the fixation and contrasting procedure that induces shrinkage of membranes. Exosomal vesicles analyzed by high-resolution cryomicroscopy without fixation and staining appear perfectly round ([39] and our own unpublished observations). Separation of exosomes by SDS-PAGE, followed by Coomassie blue staining of all exosomal proteins also gives an idea of the quality of the exosome preparation: exosomes should contain a large number of proteins, i.e., at least a few dozen, which are distinct from the proteins detected in whole cell lysates and of protein composition by immunoblotting. As determined by western blot, they contain exosomal proteins (Tetraspanins, Hsc70, Tsg101, Alix, Flottilins) and they do not contain endoplasmic reticulum contaminants (calnexin, Hsp90) or proteins that are mostly present at the plasma membrane, or in other compartments of the endocytic pathway (NaK ATPase; Transferrin Receptor if not studying exosomes from reticulocytes!). As a complement, FACS analysis can also be used to analyze the protein composition, but exosomes are too small to be analyzed directly by cell sorting. They can be adsorbed to latex beads and labeled with fluorophore-conjugated antibodies directed against exosomal proteins. These protocols are largely detailed in [32]. Another reliable FACS method to analyze small amounts of exosomes has been recently published by Wauben and collaborators.

One limiting factor in the characterization of exosomes has been a reliable manner for quantifying the amounts of vesicles released. A possible approach for quantitation under development is to exploit the NanoSight LM10 instrument that is currently used to quantify small particles and liposomes. The NanoSight is an optical microscope that uses a laser light source to illuminate nanoscale particles within a 300 μl sample that is introduced to the viewing unit with a disposable syringe. Enhanced by a black background, particles appear individually as point scatterers moving under Brownian motion. The image analysis nanoparticle tracking analysis (NTA) software allows users to automatically track and size nanoparticles on an individual basis. Results can be displayed as a frequency size distribution graph and output to spreadsheet. These methods have certainly to be used in complement with the other approaches.

1.2.4 Exosome Composition

As a consequence of their endosomal origin, exosomes from different cell types contain protein involved in membrane transport and fusion (e.g., Rab GTPase, Annexins, flotillin), in MVEs biogenesis (e.g., Alix and Tsg101), and in adhesion such as integrins and tetraspanins (e.g., CD63, CD9, CD81, and CD82; reviewed in [30, 40, 41]). Several proteins that are generally part of the exosome "signature" are, for example, heat shock (hsc70) and cytoskeletal proteins (actin, moesin; summarized in Table 1.1). However, the particular role of the different proteins present in exosomes and whether they play a particular role in their formation, secretion, targeting to recipient cells is not fully understood. The lipid composition has been determined only for exosomes secreted from few cell types such as DCs and mast cells [42], basophilic leukemia cells [43], reticulocytes [44] and B cells [33]. The results taken together reveal that exosomes have a particular lipid organization and composition, as enrichment in sphingomyelins (1.3 times in erythrocytes and 2.3 times in B lymphocytes) [43]. It has been shown also the presence of lipid raft-like domains, enriched in cholesterol and containing high amount of sphingolipids and lower amount of phosphatidylcholine [33, 42]. Exosomes secreted by the mast cell line RBL-2H3 appeared to display a random unconventional transmembrane phospholipid organization, probably due to the presence of a phospholipid scramblase, which have the ability to mix the phospholipids in the two membrane sides [42]. Research in the field has been certainly reboosted by the finding that exosomes also contain genetic materials. In 2007, the group of J. Lotvall reported the presence of mRNA and miRNA in exosomes secreted by mast cells [45]. Microarray revealed mRNAs from approximately 1,300 genes and quality control RNA analysis of total RNA derived from exosomes also revealed the presence of small RNAs, including microRNAs. Importantly, RNAs from mast cell exosomes were transferable to other mouse and human mast cells, after which new mouse proteins were found in the recipient cells, indicating that transferred exosomal mRNA can be translated after entering another cell [45]. There are an increasing number of studies reporting the presence of RNAs in exosomes also from body fluids [46–48]. Recently, plant miRNAs have been identified in human cells and it has been suggested that epithelial cells in the gastrointestinal tract [49] could take up the plant microRNAs. Intestinal cells could, in turn, secrete the microRNAs into the body via exosomes or microvesicles. It is clear that vesicles of endosomal/exosomal origin contain RNAs that are sequestered within the endosome by mechanisms that are not totally understood. Nevertheless, most studies published so far reported the release of RNAs and short DNA sequences within a population of microvesicles that have been obtained by ultracentrifugation not preceded by differential centrifugation steps. It is unclear therefore whether those identified RNAs, mRNAs, and miRNAs can be clearly included into exosomal components. The approaches used to characterize exosome-associated miRNA are very different; therefore it is also uncertain whether specific miRNA sequences, rather than the whole set of intracellular miRNAs, are targeted to exosomes/microvesicles. In addition, in some studies it is unclear whether miRNA and mRNA are sequestered within the cytosolic moiety of exosomes. Determining the resistance to RNAse digestion and flotation on sucrose gradients would help resolving this issue.

Over the past years, efforts have been made to assemble together proteins, lipid and RNAs associated with exosomal vesicles [50]. Such compilation of components has the possible drawback that it reports components of both exosomes and microvesicles but it is certainly of use for the growing community. The modes of exosomal purification and characterization, the biophysical and molecular properties are listed in the database to help biomedical scientists in assessing the quality of the exosomal preparation and the corresponding data obtained. ExoCarta (http://www.exocarta.org) catalogs information from both published and unpublished exosomal studies. The current ExoCarta (Version 3.1) contains information on 11,261 protein, 2,375 mRNA, and 764 miRNA entries that were obtained from 134 exosomal studies. Lipids identified in exosomes will be added to ExoCarta.

1.3 The Biogenesis of Exosomes

1.3.1 Multivesicular Endosomes Mature from Early Endosomes

The endosomal system consists of morphologically, compositionally and functionally distinct domains classified as early and late as they are sequentially accessed by tracers internalized by receptor-mediated endocytosis and/or bulk flow [51, 52]. Early endosomes are generally subclassified into sorting and recycling endosomes based on differential kinetic access to recycling proteins, such as transferrin receptors, following internalization (see Fig. 1.3). At the ultrastructural level, sorting endosomes consist of electron-lucent vacuoles with few internal membranes and

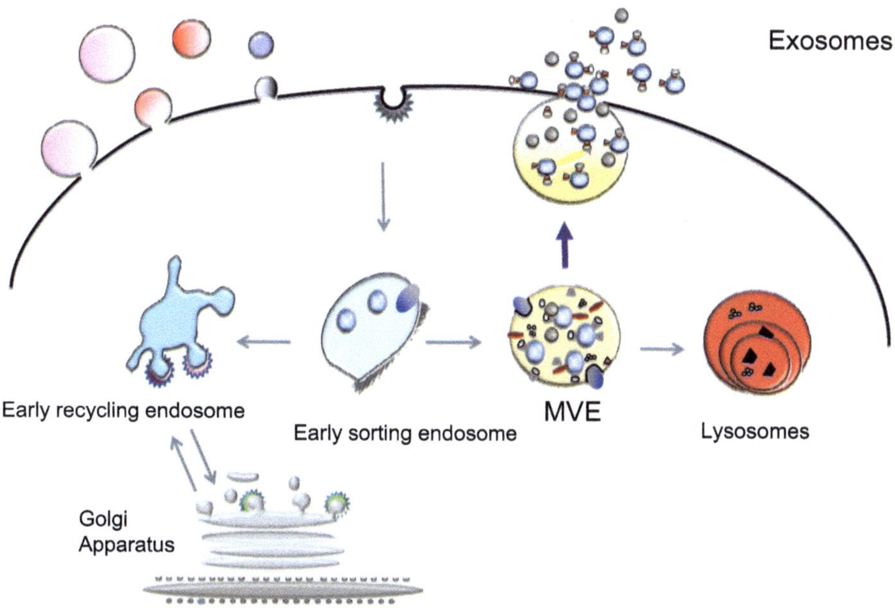

Fig. 1.3 The endocytic pathway, release of microvesicles and exosomes. Representation of organelles of the endosomal system, highlighting the best characterized compartments, found in all cell types. The clathrin-containing coat of the plasma membrane invagination and the coat at the cytosolic side of the early-sorting endosome (coated endosome) is represented by *gray shading*. Intralumenal vesicles present in early-sorting endosomes and multivescular endosomes (MVEs), most often assimilated to late endosomes, are represented by *small circles* of different sizes; irregularly shaped membranes and vesicles are generally present in lysosomes. *Arrows* represent proposed directions of protein and lipid transport between organelles and between MVEs and the plasma membrane for exosomes secretion. Budding of microvesicles from the plasma membrane is also represented

closely apposed tubules in continuity with the limiting membrane. The tubules likely mediate cargo transport both into and out of the vacuoles, including receptors that undergo rapid recycling to the plasma membrane. A certain proportion of recycling receptors accumulate in and recycle from recycling endosomes, which correspond to tubulovesicular elements generally concentrated in the perinuclear region apposed to the Golgi apparatus [53]. The vacuolar domains of early endosomes are partially coated on the cytosolic side with an electron-dense bilayered coat [54]. This coat contains clathrin, but is distinct from those present on buds at the plasma membrane, TGN or tubular endosomes, and functions at least partially in sorting cargo between the vacuolar-limiting membrane and membrane destined to invaginate to form the internal membranes. Endosome maturation is accompanied by the progressive accumulation of internal vesicles, often enriched in integral membrane proteins destined for lysosomal degradation, in the lumen of the vacuolar domains. Such maturing endosomes are called multivesicular bodies (MVBs) or multivesicular endosomes (MVEs) that have been described by Sotelo and Porter in the late 1950s [55].

MVB/MVE populations with variable amounts of intralumenal vesicles are observed within a given cell type at steady state. These MVEs are most often assimilated to late endosomes because of their content of newly synthesized lysosomal enzymes and accessed to endocytic cargo by 30 min. The main fate of MVEs is fusion with lysosomes to degrade their contents. However, and as described above, the findings that in some cell types compartments with the hallmarks of MVEs fuse with the plasma membrane conducted to the idea that MVEs have a dual fate, fusion with the lysosome, or fusion with the plasma membrane. Nevertheless, until recently it has been uncertain whether cells depict different MVE subpopulations that would explain such dual destiny.

1.3.2 Distinct Subpopulations of MVEs

One first line of evidence for different subclasses of MVEs comes from studies on the localization of cholesterol using the toxin perfringolysin O [56, 57]. Labeling on ultrathincryosections of a B-EBV lymphocytic cell line revealed cholesterol-positive and cholesterol-negative MVEs. Interestingly, those enriched in cholesterol showed exocytic fusion with the cell surface to release exosomes [57]. Additional evidence for the existence of MVE subpopulations are the findings that EGF and its receptor traffic through a subpopulation of MVEs that do not contain the lipid bis (monoacylglycero)phosphate/lysobisphosphatidic acid (BMP/LBPA) [58]. These LBPA-containing MVEs are probably not those that fuse with the plasma membrane as mass spectrometry (MS) analysis of the lipid composition of exosomes from human B cells revealed that BMP/LBPA is not particularly enriched in exosomes [58]. Studies on exosomes secreted by intestinal epithelial cells (IECs) largely support the actual thought that subpopulations of exosomes secreted from different types of MVEs coexist in cells [59]. Exosomes with different compositions are secreted from the apical and the basolateral sides of the polarized cell monolayers. Both populations shared common proteins such as MHC I, tetraspanins, and Hsp, but also contain specific sets of molecules. Whereas apical exosomes contain syntaxin 3, microsomal dipeptidase CD26 and are typically addressed to the apical membrane [60]. More recently, evidence for the generation of different types of MVEs in nonpolarized cells came from the analysis of DCs and their interactions with cognate $CD4^+$ T cells [60]. Whereas DCs alone depict MHC class II-containing MVEs that are mostly fated for degradation, DCs in the presence of T cells secrete an increased amount of exosomes derived from newly generated MVEs enriched in the tetraspannin CD9. Together these data indicate two distinct MVE-sorting mechanisms, one for lysosomal targeting and the other for exosome secretion. Interestingly, the mechanisms underlying the sorting of MHC II into MVEs destined for degradation or fated for exosome secretion appeared to be distinct. Incorporation of MHC II into exosomes was, in contrast to lysosomal targeting, independent of its ubiquitination but rather correlated with its incorporation into CD9-containing detergent-resistant membranes in agreement with the localization of CD9 to this "secondly generated" secretory MVE subpopulation [60]. These studies also revealed that the different types of MVEs have a slightly different morphology (size, electron density of the ILVs) and maybe, a different composition. A better characterization of these different subpopulations of MVEs is certainly required but methods allowing separating them are still not available and awaiting a better characterization of the mechanisms involved in their formation and fate. Only recently it appeared clear that different MVEs and ILVs are generated by several mechanisms that can be classified as endosomal sorting complex responsible for transport (ESCRT)-dependent and ESCRT-independent.

1.3.3 Mechanisms Involved in the Generation of Exosomes

1.3.3.1 ESCRT-Dependent Mechanisms

Exosomes correspond to the ILVs of MVEs, thereby they are formed within the endocytic pathway during the maturation of endosomes (see above). The generation of the ILVs into MVEs involves the lateral segregation of the cargo on the limiting membrane of an endosome, by the formation of an inward budding vesicles and the release of the membrane vesicle containing a small portion of cytosol into its lumen (Fig. 1.3). Over the past 10 years, a major breakthrough onto the mechanisms involved in the formation of MVEs and ILVs has been through studies on several yeast mutants defective in the transport to the vacuole (see for recent reviews [61, 62]). These so-called vacuolar protein sorting (Vps) mutants showed impaired formation of MVEs and the accumulation of abnormal endosomal membranes, the class E compartment [63]. These proteins assembled into four complexes named ESCRT [62, 64]. Just after these initial findings, it appeared evident that the proteins present in yeast had their orthologues in mammalian cells. Interestingly, while it was shown that these proteins were involved in the sorting of cargo proteins and formation of ILVs at the level of the endosome, it was also reported that they could be exploited by HIV-1 to bud in intracellular compartments with features of MVEs or at the plasma membrane, a mechanism that is topological similar to the formation of ILVs [65, 66].

The ESCRT machinery comprises approximately 20 proteins that assemble into four complexes (ESCRT-0, -I, -II, and -III) with associated proteins conserved from yeast to mammals [67, 68]. The ESCRT-0, -I, and -II complexes recognize and sequester ubiquitinated proteins in the endosomal membrane, whereas the ESCRT-III complex seems to be responsible for the membrane budding and vesicle scission [69].

ESCRT-0 contains five ubiquitin-binding domains (UBDs) and a subunit, Hrs, associates in a complex with STAM, Eps15, and clathrin that recognize the monoubiquitinated cargo proteins. Hrs recruits Tsg101 of the ESCRT-I complex and ESCRT-I is then involved in the recruitment of ESCRT-III, via ESCRT-II or Alix, an ESCRT accessory protein. Finally, the dissociation and recycling of the ESCRT machinery requires interaction with the AAA-ATPase Vps4. It has been unclear whether ESCRT-II has a direct role on biogenesis of ILVs or its sorting function could be just limited to a particular cargo [70, 71].

The discovery of a machinery involved in MVE and ILV biogenesis has been taken with great enthusiasm in the field of exosomes because it would certainly open an avenue to modulate their formation through inhibition of the components of the machinery. Proteomic analyses of purified exosomes, from various cell types, show the presence of ESCRT components (Tsg101, Alix) and ubiquitinated proteins [72]. The Transferrin Receptor in reticulocytes that is generally fated for exosome secretion, although not ubiquitinated, interacts with Alix for MVE sorting [73]. More recently, it has been reported that Hrs, an ESCRT-0 component, could be required for exosome formation/secretion by DCs and thereby impact in their antigen-presenting capacity [74]. However, as discussed above and elsewhere [75]. due to the existence in cells of different MVEs and ILVs subpopulations, we are still at an early stage of understanding exactly whether at least some ESCRT components could be involved in the formation of a subpopulation of exosomes.

1.3.3.2 ESCRT-Independent Mechanisms

Studies in oligodendrial cell lines secreting the proteolipid PLP in association with exosomes [76] and in melanocytic cells producing melanosomes [77] highlighted that mammalian cells developed additional pathways of MVE formation. Moreover, concomitant inactivation of ESCRT subunits belonging to the four ESCRT complexes does not totally impair the formation of MVEs indicating that other mechanisms are thought to operate in the formation of ILVs and thereby of exosomes [78]. One of these pathways is thought to require a type II sphingomyelinase that hydrolyzes sphingomyelin to ceramide [76]. Whereas depletion of different ESCRT components did not lead to a clear reduction in secretion of PLP via exosomes and in the formation of MVEs, inhibition of sphingomyelinase expression with siRNA or its activity with the drug GW4869 decreased exosome formation and release. Ceramides are known to induce lateral phase separation and domain formation in model membranes [79] and the cone-shaped structure may induce spontaneous negative curvature on the membranes, promoting ILV invagination. Exosomes produced by the shingomyelinase pathway appeared to be enriched in tetraspanins, such as CD63. Proteins of the tetraspanin family are also possibly involved as effectors of the endosomal-sorting system, as they are known to form oligomers with other testraspanins and also with different transmembrane and cytosolic proteins [80]. Our most recent studies in melanocytic cells show for the first time an implication for CD63 in an ESCRT-independent endosomal-sorting pathway. The sorting of the premelanosomal protein PMEL relies on a lumenal determinant [77] and our recent data reveal that CD63 directly participates to the ESCRT-independent sorting of its luminal domain, but not of conventional ESCRT-dependent cargoes, to ILVs [81]. However, interestingly, while CD63 is required for sorting the PMEL luminal domain, the disposal of the remaining PMEL transmembrane fragment requires functional ESCRTs but not CD63. In the absence of functional CD63, the PMEL luminal domain follows this fragment and is targeted for ESCRT-dependent degradation. These recent data put forward the concept that tetraspanins constitute essential regulators of ESCRT-independent ILV formation but also reveal a tight interplay between two distinct endosomal ILV-sorting processes for a single cargo [81]. This pathway leads to the formation of MVEs that are not fated to degradation but which destiny is to produce melanosomes, one class of secretory lysosomes [82]. Interestingly, PMEL is also an exosomal cargo in melanoma cells [83]. These studies further support the observations made in DCs revealing that MHC II-containing MVEs generated in an ubiquitin and maybe ESCRT-independent pathway are enriched in the tetraspanin CD9 [60]. Although purely speculative, such balance between tow mechanisms would allow on one hand ubiquitinated MHC class II to be targeted to degradation by an ESCRT-dependent pathway, while the ESCRT-independent machinery would drive MHC class II toward a secretory MVE during interaction with an activator T cell. It would be of interest to understand what are the signals that switch between the two pathways.

1.3.3.3 Other Mechanisms

The studies reported so far would suggest a model in which proteins sorted to ILVs in an ESCRT-independent manner are fated for secretion whereas those sorted via the ESCRT machinery are destined to degradation. Nevertheless, the general picture maybe more complex depending not only on the cargo but also on the cell system. For example, it was recently found using a combination of proteomic and lipidomic approaches that the composition of exosomes changes during reticulocyte maturation [84]. The data together support the model, whereby in reticulocytes but also maybe other cells, the biogenesis of exosomes involves several distinct mechanisms for the preferential recruitment of particular proteins and lipids (see Fig. 8 in [84]). The balance of these pathways changes over the course of the differentiation process. This could be also true for DCs during maturation after LPS treatment a process during which the amount and composition of exosomes also varies [85]. This may also be the basis for the different functional properties of exosomes, their targeting to different cells, and their fate (see below).

The presence of cytosolic protein into the ILVs can be due to cosorting with other proteins. This mechanism is suggested for chaperones, such as Hsp70 and Hsc70, found in exosomes from most cell types [73, 86]. GlycosylPhosphatidyInostol (GPI)-anchored proteins are sequestered in exosomes likely because of their sequestration within lipid domains and rafts. The lipid microdomains could be involved themselves in the generation of ILVs [87], or interacting with other proteins that have an affinity to the lipid raft domains. It has also been proposed that some proteins could associate to proteins modified by ubiquitination to be sequestered in ILVs of MVEs, but whether these are proteins thought to be fated for degradation rather than secretion is not known.

How RNAs, mRNAs, and miRNAs are sorted is enigmatic. The recent observations that miRNA, depending on their sequences, are incorporated into exosomes suggests a selectivity of miRNA targeting to vesicles. Although merely speculative, the findings that ESCRT-II is an RNA-binding complex [88] opens the possibility that it could function to select RNA at the endosomal level. Moreover, the observations that MVEs are sites of miRNA-loaded RISC (miRISC) [89] accumulation and that exosome-like vesicles are considerably enriched in GW182 and AGO2, proteins involved in miRNA-mediated postranslational silencing, is also in favor of a selective sorting.

1.4 Exosome Release

By analogy with the process of exocytic fusion of secretory lysosomes [82, 90], the release of the endosomally generated ILVs in the extracellular space is preceded by the transport of the formed MVE toward the cell periphery and its docking and fusion with the plasma membrane. Exosome release is sensitive to changes in intracellular calcium in a human erythroleukemia cell line [91] and in mast cells [92]; however, these later cells have well defined regulated secretory pathways induced by IgE binding during inflammatory and allergic reactions. Depolarization induced by K^+ appears to increase the secretion of neuronal exosomes [93]. Cross-linking of CD3 in T cells stimulates exosome release by T cells [94]. However, we are still far from deciphering the machinery involved in mobilization of secretory MVEs to the cell periphery, their docking and fusion with the cell surface.

These processes generally require the association of organelles with the cell cytoskeleton (actin, microtubules), associated molecular motors (kinesins, myosins) but also the fusion machinery (SNAREs, Tethering factors) and molecular switches (Small GTPases) [95, 96]. As for the machinery involved in exosome biogenesis, a detailed knowledge of the components and mechanisms involved in exosome release would be essential to get further insights into their functional relevance. As described below only recently we have started to get insights on the molecular switches that coordinate exosome release.

1.4.1 A Role for Small GTPases

Most, if not all, intracellular pathways are controlled by Small GTPases of the Rab superfamily, the largest family of monomeric small GTPases that are distributed to distinct intracellular compartments where they are involved in the maintenance of membrane subdomains, controlling membrane identity and vesicle budding, uncoating, motility and fusion commuting between a GTP/GDP bound state and through the recruitment of effector proteins, such as sorting adaptors, tethering factors, kinases, phosphatases and motors (reviewed in [97]). Crosstalk between multiple Rab GTPases through shared effectors, or through effectors that recruit selective Rab activators, ensures the spatiotemporal regulation of vesicle traffic. Impairment of the function of selected Rab GTPases either by expression of mutants or inactivation by siRNA and shRNA consequently affects the morphology and distribution of MVEs and exosome secretion [91, 98, 99].

The first indication for the possible involvement of Rab GTPases in exosome secretion came from studies on reticulocyte cell lines showing that exosome release may require the function of Rab11, a small GTPase generally associated with recycling endosomes [91]. However, it has been unclear at what step Rab11 acts and whether the effects on exosome secretion observed with Rab11 mutants are due to the involvement of this Rab protein in fusion events or a consequence of the alterations of the biogenesis of MVEs. An alternative possibility is that Rab11 associated with early endosomes could be required for the recruitment of effectors that are directly involved in fusion of MVEs with cell surface. On line with this hypothesis another class of secretory organelle, the Cytolytic granule of $CD8^+$ T lymphocytes only fuse with the cell surface once they have interacted with Rab11 exocytic vesicles [100].

A highthrouput approach using a selected sh library targeting the expression of small GTPases revealed Rab27a and Rab27b as essential effectors of the secretion of exosomes in Hela expressing the CIITA transactivator that induces expression of MHC class II molecules and other components of the antigen presentation machinery such as the Invariant chain and HLA-DM [98]. The data demonstrated that the decreased expression of Rab27a and Rab27b induced a 50 % reduction in exosome secretion, suggesting that most secreted exosomes originate from a Rab27-dependent secretion event. Since Rab27 is generally specifically associated with lysosomal related organelles [98], these findings also importantly supported that exosomes are secreted from an intracellular endosomal compartment that is secretory in nature [98]. By analogy with other cell systems hosting secretory endo/lysosomes, Rab27 could be involved directly or indirectly in the transport and tethering at the cell periphery of the secretory MVEs. On this line, silencing of two known Rab27 effectors, Slp4 (also known as SYTL4, synaptotagmin-like 4) and Slac2b (also known as EXPH5, exophilin 5), inhibited exosome secretion and phenocopied silencing of Rab27a and Rab27b, respectively.

Whereas silencing Rab27a increases the size of MVEs and reduces the docking at the cell surface, Rab27b is likely to be involved in the transfer of MVEs to the cell periphery, because its reduced expression by shRNA results in smaller sized MVEs [98].

Using a different screen-targeting Rab-activating proteins (Rab-GAP) proteins, studies in oligodendroglial cells revealed Rab35 as the main target of the Rab-GAP TBC1D10A–C. Rab35 with its catalytic domain, regulate exosome release possibly by controlling the docking/tethering of endocytic vesicles with the plasma membrane [99]. Silencing of Rab35 does not affect the size of the vesicles, whereas the total number of vesicles inside the cells increases by 35 %.

The association of Rab proteins with lipids present in MVEs could regulate the motility of these compartments. For example, the accumulation of cholesterol in endosomal membranes increases the amount of membrane-associated Rab7 and inhibits the motility of late endocytic structures [101]. Therefore, sequestering specific lipids away from the limiting membrane by sorting them in ILVs, could affect the motility of the late endocytic structures and thus the docking of secretory MVEs with the plasma membrane. As major coordinators of membrane-trafficking events Rab GTPases and their effectors are certainly good candidates to interfere with exosome release even if they are ubiquitously expressed and involved in additional trafficking events [97].

The Rab proteins mentioned so far although highlighted by their impact on exosomes secretion on selected cell types (Hela cells expressing MHC class II and oligodendrocytes) may also regulate exosome secretion in other cell types. In concern to Rab27, it should be noted that the levels of expression of the two isoforms of Rab27 are variable from cell to cell (some cells only express Rab27b), thereby validation of these findings to other cell types is certainly required.

1.4.2 The Fusion Machinery

If Rab proteins may play a role in the docking and tethering through their ability to organize membrane domains and recruit effectors, little is known about the regulation of MVE fusion with the plasma membrane. The fusion process is likely to be mediated by SNAREs proteins and synaptotagmin family members. SNAREs are key mediators of all intracellular membrane fusion events and mostly studied for exocytosis of synaptic vesicles in neurons [102]. The SNARE complex is known to be implicated on the exocytosis of conventional lysosomes and consist of VAMP7 on the lysosomes, Syntaxin 7 on the plasma membrane and the lysosomal regulatory protein Synaptotagmin 7 [103]. It is not known whether the exocytic fusion of MVEs is similarly modulated and/or controlled by the same fusion machinery. Our studies in MDCK cells expressing the Longin domain of VAMP7 revealed impaired lysosomal secretion whereas the release of exosomes did not seem to be consequently affected suggesting that maybe VAMP7 is not the major player [104]. Nevertheless further studies using other cell types are needed and a thorough screening for components of this family of proteins is certainly worthy.

Another potential candidate that has been often mentioned is the V0-subunit of the V-ATPase because of the findings in a *Caernorhabditis elegans* mutant defective for MVE secretion as analyzed in situ by electron microscopy of the worm [105]. The V-ATPase is a multisubunit protein complex that functions as a proton pump establishing a pH gradient in the secretory and endocytic pathways. This pump has two subcomplexes called V0 and V1 sectors, the cytoplasmic V1 sector is responsible for ATP Hydrolysis and the transmembrane V0 sector is responsible for proton translocation. Independently of the V0-V1 ATPase complex, the V0 sector plays a role in membrane fusion [106]. The V0 subunit played a role in the secretion to the apical membrane of cuticles in *Caernorhabditis elegans* and is required for the apical secretion of hedgehog-related peptides through a multivesicular compartment [105]. However, the inability to further isolate exosomes from the worm did not allowed further studies of the secreted vesicles. Validation of these findings in mammalian cells have been so far difficult to perform because of the existence of multiple isoforms of the V-ATPase.

1.5 The Fate of Released Exosomes

1.5.1 Interaction of Exosomes with Target Cells

Once released from cells the function of exosomes in physiological and pathological processes depends on their ability to deliver their content, proteins, lipids, and RNAs to recipient cells. The mode of interaction of exosomes with target cells and their intracellular trafficking pathway is still a subject of debate and certainly depends on the system and cellular process in which they are involved. As described, exosomes express adhesion molecules, among which integrins that could be involved in the capacity of exosomes to adhere to cells but the cellular and molecular basis for their specific targeting to acceptor cells is still misunderstood. Only some examples are summarized here. Exosomes released by the human intestinal epithelial cell line T84, enriched in CD9, CD81, CD82, and A33 antigen that bind specifically human serum albumin (HSA) 64–76 peptide on HLA-DR4 molecules interact preferentially with DCs rather than with B or T lymphocytes [107]. The uptake of rat reticulocyte exosomes by macrophages is decreased by adding galectin-5 suggesting a possible function for galectin-5 in this interaction in view of the proposed role for galectins in polarity and adhesion processes [108, 109]. In the immune system, it was shown that T cells could recruit MHC class II-containing DC exosomes secreted in the extracellular milieu during cognate DC-T-cell interactions. Recruitment of these exosomes required T cell activation and was dependent on leukocyte function-associated antigen-1 (LFA-1) rather than on T cell receptor specificity. Indeed, inducing a high-affinity state of LFA-1 on resting T cells was sufficient to allow exosome binding [110]. A difficult issue to approach is, in a more general way, how exosomes present in biological fluids (e.g., in the blood) originating from very different cell types (reticulocytes, DCs, B lymphocytes, platelets) and if physiological-relevant may find their way specifically to their targets.

1.5.2 Destiny of Exosomes

Several studies have proposed that depending on the origin of exosomes and the identity of the recipient cells, exosomes may directly fuse with the plasma membrane or may be internalized by endocytosis or phagocytosis (Fig. 1.4). However, there is also evidence for the docking and/or adsorption of exosomes at the surface of their targets conferring in this way novel properties to the cells through the components they express [60, 111]. It has been proposed that exosomes from B cells at the surface of follicular DCs able them to express MHC class II molecules that may be involved in maintenance of T cell Memory. The function of DC-exosomes at the surface of T cells is not known. Direct fusion with the cell surface is difficult to probe because of the small size and dynamics of the event. A recent study has shown that low pH, which is a hallmark of malignant tumors, plays a key role in

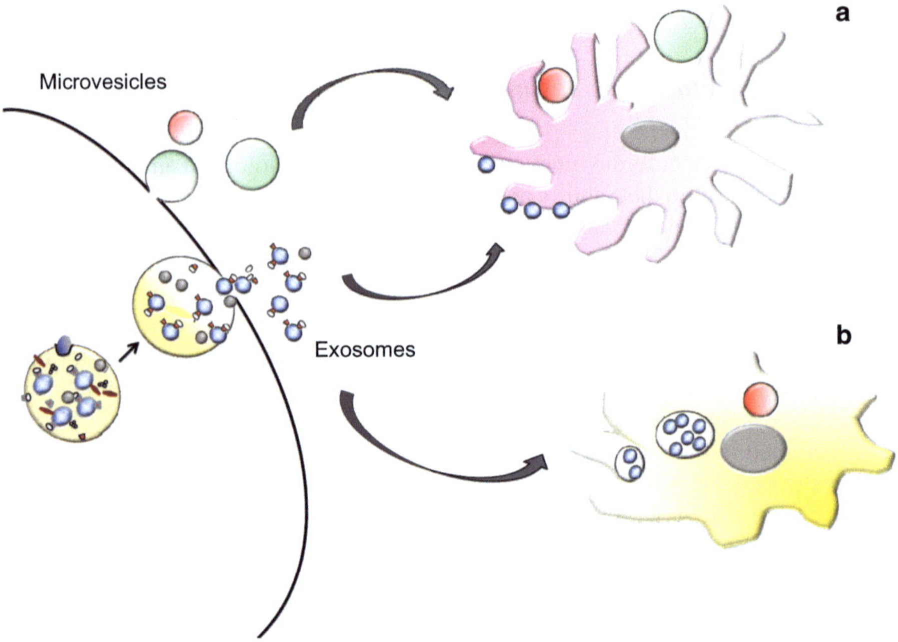

Fig. 1.4 Targeting of exosomes and microvesicles to recipient cells. Different nonmutually exclusive possibilities are represented. Exosomes and microvesicles can be adsorbed at the cell surface, fuse with the plasma membrane (**a**) or be internalized and/or undergo phagocytosis by recipient cells (**b**)

aberrantly regulating exosome traffic within the tumor mass [112]. Exosomes were purified from metastatic melanoma cell medium labeled with a lipid fluorescent probe, R18, and analyzed by spectrofluorometry and confocal microscopy. Using different pH conditions as a modifier of exosome traffic, it was shown that exosome uptake by melanoma cells appeared to occur by fusion. Membrane biophysical analysis, such as fluidity and lipid composition, indicated a high rigidity and sphingomyelin/ganglioside GM3 (N-acetylneuraminylgalactosylglucosylceramide) content in exosomes released at low pH. This was likely responsible for the increased fusion efficiency. Fusion efficiency of tumor exosomes resulted in being higher in cells of metastatic origin than in those derived from primary tumors or normal cells. Caveolin-1, for example, a protein involved in melanoma progression, appears to be delivered through exosomes released in an acidic condition [112]. Whether such fusion occurs at the plasma membrane or in internal endocytic compartments is not clear. It certainly represents a possible mechanism for transfer of genetic materials between immune cells or tumor cells via exosomes [45, 113].

Most studies performed to follow the fate of exosomes in cells used fluorescently labeled exosomes. Several of these studies provided evidence for the capture of vesicles and their accumulation in internal endocytic or phagocytic compartments, especially in phagocytic cells [114]. In these conditions, exosome internalization was dependent on the actin cytoskeleton and phosphatidylinositol 3-kinase, and could be inhibited by the knockdown of dynamin2 or overexpression of a dominant-negative form of dynamin2. Despite the dependence on dynamin, exosomes did not appear to be internalized through caveolae, macropinocytosis and clathrin-coated vesicles. Without excluding specificity of recognition through expressed molecules, the fact that exosomes can be often released as small aggregate could explain in part why these exosomes derived from K562 or MT4 cells are internalized more efficiently by phagocytes than by nonphagocytic cells. Results from other groups using PKH67 fluorescently labeled exosomes, also revealed that the vesicles are efficiently internalized by macrophages through a dynamin-dependent pathway modulated by the presence of galectin-5 on the exosome surface [108].

Once internalized exosomes could be either fated for degradation which allow liberation of their contents but they could also fuse with endosomal membranes allowing by this way mixing of their membrane components with the recipient cell and liberation of the cytosolic component, including RNAs into the cytosol.

In a recent study it was shown that exosomes isolated from the culture medium of neuronal-like PC12 cells, labeled by lipophilic dye and amino-reactive fluorophore, and further incubated with resting PC12 cells indicated that exosomes were internalized through the endocytocytic pathway, trapped in vesicles, and transported to perinuclear region [115]. Particle-tracking fluorescent vesicles suggested that the active transport of exosomes might be mediated by cytoskeleton. The proteins on exosome membrane were found to be released from exosomes and trapped in lysosome. The inverted transport of lipophilic dye from perinuclear region to cell peripheries was revealed, possibly caused by recycling of the exosome lipids. This study provides some new insights into the mechanisms of exosome uptake and intracellular fate but, as discussed previously, each cellular system and process may have their particularities. Thereby when using drugs affecting intracellular trafficking or manipulation of expression of proteins (inactivation by siRNA, for example) involved in endocytosis/phagocytosis appropriate controls are highly recommended.

1.6 Functions of Exosomes

This book is mostly dedicated to tumor exosomes, our goal in the last section of this chapter is to give the reader a general, although not exhaustive, overview of the different functions reported for exosomes over the past years in addition to the studies already highlighted in the first section.

1.6.1 Functions in the Immune System

Because of the ability of exosomes to act as antigen-presenting vesicles a great interest was given on their immunological effects that can be either stimulatory or toleregenic (reviewed in [29, 75]). Mature DCs secrete exosomes that carry antigens or MHC-peptide complexes, and induce antigen-specific immune responses by other DCs for example in the context of anticancer therapies [15]. On the other hand, exosomes secreted by DCs subjected to immunosuppressive treatments, or modified to express immunosuppressive cytokines, can instead promote tolerogenic immune responses, which could make them useful as a treatment for autoimmune diseases. Other example are the studies reporting that macrophages infected with procaryotic mycobacteria release exosomes that, depending on the mycobacteria strain, bear antigens and promote immune responses (nonpathogenic strain), or inhibit macrophage activation and cytokine secretion, thus dampening immune response (pathogenic strain) [116, 117]. Exosomes released by tumor cells bear tumor antigens that can be captured and used to present tumor antigens by DCs [83]. Nonetheless, they also bear immunosuppressive molecules, which can inactivate T lymphocytes or natural killer cells, or promote the differentiation of regulatory T lymphocytes or myeloid cells suppressing immune responses ([118] and reviewed in [119]). Thereby there are lines of evidence that exosomes secreted by tumors contribute to tumor progression by impairing antitumor immune responses, or by promoting angiogenesis or even migration to form metastases [120, 121]. This could the case for exosomes but also other types of microvesicles [120]. Other tissues or cells secrete exosomes bearing immunosuppressive molecules: placenta-derived vesicles, which bear ligands for natural killer lymphocytes, are found in pregnant women's blood circulation, and extensive analyses of components of these vesicles are underway to identify their role in the mother's tolerance to the fetus [122].

1.6.2 Exosomes in Epithelia

The capacity of epithelial cells to secrete exosomes was first described in IECs [59, 123] and as described above these polarized cells secrete apically and basolaterally different types of exosomes originating from at least two distinct MVE populations. Release of apical exosomes in the intestinal lumen suggests their involvement in a clearance process. The presence, however, of high-mobility group box 1 (HMGB1), a cytokine-like proinflammatory protein, on immunostimulated Caco-2 cell-derived apical exosomes can influence epithelial permeability [124]. Basolateral exosomes are constitutively enriched in A33 antigen and in MHC II molecules after activation by *g*IFN [35]. The presence of the porous basal membrane between epithelium and mucosal immune system suggests a role for basolateral exosomes in antigen presentation independently of direct cellular contact with effector cells. Interestingly murine epithelial cell exosomes display immunostimulatory properties against food antigens when injected intraperitoneally [35]. Earlier studies, however, reported that exosome-like structures named tolerosomes, isolated from rat IECs and serum, were shown to induce tolerance [123]. The different effect can be explained by a different site of injection of exosomes in the animal, or the activation state of the recipient cells. Tolerizing properties were further reported for exosomes extracted from the serum containing weak amounts of A33 antigens. Nevertheless, the same study also showed that "tolerosomes" induce in vivo and in vitro activation of specific T cells [123]. The presence of exosomes in serum [123] and potentially in cells entering mesenteric lymph nodes [35] may confer the ability of static epithelial cells to act at a distance. Exosomes of intestinal epithelium seem therefore to constitute antigen-carrying structures establishing a link between luminal antigens and the local immune system. They could act as sensors of the antigenic information present in the intestinal lumen, their capacity to induce immune response would be then influenced by the inflammation.

Other types of epithelia also likely secrete exosomal vesicles. In the airways, exosomes present in their bronchoalveolar fluid can bear tolerizing molecules (e.g., in mice tolerized for an allergen) or, conversely, increase proinflammatory cytokine secretion by airway epithelial cells (in asthmatic human patients) [125, 126].

1.6.3 Functions in the Nervous System

In addition to the immune system and epithelia, exosomes are likely involved in other physiological functions such as communication within the nervous system. Neurons, oligodendroglial cells, and microglia secrete exosomes, and exosomes from each source can affect other neural cells [28–30] [93, 127, 128]. Exosome release from these cells has been shown to be stimulated by different stimuli including increase in cytoplasmic Ca^{2+} [127, 128]. Lachenal and collaborators have recently shown that exosome release by neuronal cells is modulated by glutamatergic synaptic activity, suggesting that this process might be a part of normal synaptic physiology [128]. On the other hand, the release of neurotransmitters (i.e., glutamate) during neuronal activity could impact on other cells such as oligodendrocytes, stimulating exosome secretion by these cells. These exosomes could then in turn impact on neuronal function. Another proposed function for exosomes in the nervous system is myelin formation [129]. Mature oligodendrocytes enwrap axons with myelin membrane sheaths after receiving the appropriate signal(s) from the axon or its microenvironment. It was recently shown that exosomes secreted by oligodendrocytes inhibit both the morphological differentiation of oligodendrocytes and myelin formation. It was proposed in this case that neurons control the secretion of autoinhibitory oligodendroglial-derived exosomes to coordinate myelin membrane biogenesis [129]. Additional functions for exosomes in the nervous system is their involvement in neurite outgrowth and neuronal survival [130]. The data indicate that, under conditions of high neuronal activity and/or oxidative stress, synapsin I—an oligomannose-binding protein in mouse brain—can be released from glial-derived exosomes and promoted neurite outgrowth from hippocampal neurons and increased survival of cortical neurons upon hydrogen peroxide treatment or oxygen/glucose deprivation.

1.6.4 Exosomes in Disease

Despite the exciting possibility that exosomes within the nervous system certainly play an essential role in intercellular communication, several pathogenic proteins involved in CNS diseases, such as prions [131], β-amyloid peptide [132], superoxidedismutase [133] and α-synuclein [134] are released from cells in association with exosomes. These exosomes seem to be directly involved in disseminating pathogenesis. Our group has initially shown that both the cellular prion protein (PrPc) and the scrappie infectious isoform (PrPsc) in the cell culture supernatants of PrP-expressing rabbit kidney epithelial cells and Schwann cells were associated with exosomes. Furthermore, PrPsc-containing exosomes inoculated in transgenic mice induced a typical neurological disorder [131]. Further studies by other groups evidenced the association of prions with platelet-derived exosomes [135]. These findings are now supported by electron microscopical observations in situ in the gut of prion-infected mice the presence of A33 antigen-positive exosomes [136]. Exosome delivery from infected to recipient noninfected cells could be relevant to understanding also how the propagation is combined with the transconformation process. The raft-like nature of exosomal membranes [33, 42] could be a favorable environment for PrP aggregation, transconformation and amyloid fiber formation. A similar scenario could also apply to the pathology of Parkinson disease where alpha-synuclein aggregation plays a central role. Exosomes released from alpha-synuclein overexpressing SH-SY5Y cells contained alpha-synuclein and these exosomes were capable of efficiently transferring alpha-synuclein protein to normal SH-SY5Y cells. This study clearly demonstrates the importance of exosomes in both the release of alpha synuclein and its transmission between cells. These mechanisms may play an important role in the pathology of Parkinson disease and provide a suitable target for therapeutic intervention. The results together show for the first time that cell-produced alpha-synuclein is secreted via an exosomal, calcium-dependent mechanism and suggest that alpha-synuclein secretion serves to amplify and propagate Parkinson disease-related pathology [134]. Interestingly alpha-synuclein can be detected in the plasma and CSF of humans in addition to the culture media of neuronal cells. These observations extend the increase interest for exosomes as biomarkers in disease [137], which has led to the recent development of rapid assays to detect exosomes in the circulation and in other biological fluids.

1.7 Concluding Remarks

It has been extremely stimulating to follow over the past decade the increasing interest of scientists and physicians for exosomes. The field has quickly expanded since the discovery that exosomes have a wide range of regulatory functions and carry genetic materials. As pointed in the beginning of this chapter, the identification of vesicles as exosomes is based on both morphological and biochemical criteria and a particular attention should be given to the purification and the characterization methods. Also, it will be essential to finally understand the respective functions of vesicles that are endosomal (exosomes) or plasma membrane derived (Microvesicles, Ectosomes). In a more complex manner, it would be also of interest to decipher the functional relevance of the distinct exosomal subpopulations. Advances in the forthcoming years in deciphering the molecular mechanisms involved in their biogenesis, secretion, and targeting will certainly lead to a better knowledge of exosome function in physiological and pathological situations. An "exosomal-type" of communication is certainly not limited to the functions reported here as examples but also exploited by other cells to interact with each other, locally or at a distance. It is worth mentioning that not only mammalian cells exploit such membrane-based process of secretion (exosomes and microvesicles) for vehiculating components but also diverse microbial pathogens. Gram-negative bacteria, eukaryotic parasites of the kinetoplast lineage, and opportunistic fungal pathogens [20]. On the basis of recent findings with Leishmania, exosome release by an intracellular pathogen could serve as a general mechanism for effector molecule delivery from eukaryotic pathogen to host cell cytosol [138, 139]. These new findings of linking exosomes and other microvesicles to infection biology have important implications for the design of research strategies aimed at the development of novel therapeutics and vaccines in diverse areas. Another interesting perspective is the potential use of exosomal vesicles with optimized targeting properties of drugs (e.g., RNA drugs). In this perspective, it was recently reported that exosomes could deliver siRNA to the brain in mice [140] .

The scientific community interested on exosomes and other secreted membrane vesicles has quickly expanded and a society has been now launched as International Society for Extracellular Vesicles (ISEV).

Acknowledgments We thank Guillaume van Niel for reading the manuscript and for many stimulating discussions especially on the isolation of exosomes and their mechanisms of biogenesis. We are grateful to our laboratory members for their continuous support. We apologize to all colleagues whose work could not be cited. We thank Institut Curie, CNRS, ANR (Agence Nationale pour la Recherche ANR Blanc-ANR MIE), ARC (Association pour la Recherche Contre le Cancer), and Clarins for support and funding.

References

1. Johnstone RM (1992) The Jeanne Manery-Fisher Memorial Lecture 1991. Maturation of reticulocytes: formation of exosomes as a mechanism for shedding membrane proteins. Biochem Cell Biol 70:179–190
2. Pan BT, Teng K, Wu C, Adam M, Johnstone RM (1985) Electron microscopic evidence for externalization of the transferrin receptor in vesicular form in sheep reticulocytes. J Cell Biol 101:942–948
3. Pan YC, Johnstone RM (1983) Fate of the transferrin receptor during maturation of sheep reticulocytes in vitro: selective externalization of the receptor. Cell 33:967–978
4. Harding C, Heuser J, Stahl P (1983) Receptor-mediated endocytosis of transferrin and recycling of the transferrin receptor in rat reticulocytes. J Cell Biol 97:329–339

5. Harding C, Heuser J, Stahl P (1984) Endocytosis and intracellular processing of transferrin and colloidal-gold transferrin in rat reticulocytes: demonstration of a pathway for receptor shedding. Eur J Cell Biol 35:256–263
6. Johnstone RM, Adam M, Hammond JR, Orr L,Turbide C (1987) Vesicle formation during reticulocyte maturation. Association of plasma membrane activities with released vesicles (exosomes). J Biol Chem 262:9412–9420
7. Johnstone RM, Ahn J (1990) A common mechanism may be involved in the selective loss of plasma membrane functions during reticulocyte maturation. Biomed Biochim Acta 49:S70–75
8. Johnstone RM (2006) Exosomes biological significance: a concise review. Blood Cells Mol Dis 36(2):315–321
9. Raposo G, Nijman HW, Stoorvogel W, Liejendekker R, Harding CV, Melief CJ, Geuze HJ (1996) B lymphocytes secrete antigen-presenting vesicles. J Exp Med 183:1161–1172
10. Zitvogel L, Regnault A, Lozier A, Wolfers J, Flament C, Tenza D, Ricciardi-Castagnoli P, Raposo G, Amigorena S (1998) Eradication of established murine tumors using a novel cell-free vaccine: dendritic cell-derived exosomes. Nat Med 4:594–600
11. Andre F, Andersen M, Wolfers J, Lozier A, Raposo G, Serra V, Ruegg C, Flament C, Angevin E, Amigorena S et al (2001) Exosomes in cancer immunotherapy: preclinical data. Adv Exp Med Biol 495:349–354
12. Chaput N, Taieb J, Schartz NE, Andre F, Angevin E, Zitvogel L (2004) Exosome-based immunotherapy. Cancer Immunol Immunother 53(3):234–239
13. Escudier B, Dorval T, Chaput N, Andre F, Caby MP, Novault S, Flament C, Leboulaire C, Borg C, Amigorena S et al (2005) Vaccination of metastatic melanoma patients with autologous dendritic cell (DC) derived-exosomes: results of thefirst phase I clinical trial. J Transl Med 3:10
14. Morse MA, Garst J, Osada T, Khan S, Hobeika A, Clay TM, Valente N, Shreeniwas R, Sutton MA, Delcayre A et al (2005) A phase I study of dexosome immunotherapy in patients with advanced non-small cell lung cancer. J Transl Med 3:9
15. Chaput N, Thery C (2011) Exosomes: immune properties and potential clinical implementations. Semin Immunopathol 33:419–440
16. Cocucci E, Racchetti G, Meldolesi J (2009) Shedding microvesicles: artefacts no more. Trends cell Biol 19:43–51
17. Gyorgy B, Szabo TG, Pasztoi M, Pal Z, Misjak P, Aradi B, Laszlo V, Pallinger E, Pap E, Kittel A et al (2011) Membrane vesicles, current state-of-the-art: emerging role of extracellular vesicles. Cell Mol Life Sci: CMLS 68:2667–2688
18. Hess C, Sadallah S, Hefti A, Landmann R, Schifferli JA (1999) Ectosomes released by human neutrophils are specialized functional units. J Immunol 163:4564–4573
19. Sadallah S, Eken C, Schifferli JA (2011) Ectosomes as modulators of inflammation and immunity. Clin Exp Immunol 163:26–32
20. Silverman JM, Reiner N (2011) Exosomes and other microvesicles in infection biology: organelles with unanticipated phenotypes. Cell Microbiol 13:1–9
21. Trams EG, Lauter CJ, Salem NJ, Heine U (1981) Exfoliation of membrane ecto-enzymes in the form of micro-vesicles. Biochim Biophys Acta 645:63–70
22. Janowska-Wieczorek A, Wysoczynski M, Kijowski J, Marquez-Curtis L, Machalinski B, Ratajczak J, Ratajczak MZ (2005) Microvesicles derived from activated platelets induce metastasis and angiogenesis in lung cancer. Int J Cancer 113:752–760
23. Lee TH, D'Asti E, Magnus N, Al-Nedawi K, Meehan B, Rak J (2011) Microvesicles as mediators of intercellular communication in cancer—the emerging science of cellular 'debris'. Semin Immunopathol 33:455–467
24. Muralidharan-Chari V, Clancy J, Plou C, Romao M, Chavrier P, Raposo G, D'Souza-Schorey C (2009) ARF6-regulated shedding of tumor cell-derived plasma membrane microvesicles. Current Biol: CB 19:1875–1885
25. Muralidharan-Chari V, Clancy JW, Sedgwick A, D'Souza-Schorey C (2010) Microvesicles: mediators of extracellular communication during cancer progression. J Cell Sci 123: 1603–1611

26. Heijnen HF, Schiel AE, Fijnheer R, Geuze HJ, Sixma JJ (1999) Activated platelets release two types of membrane vesicles: microvesicles by surface shedding and exosomes derived from exocytosis of multivesicular bodies and alpha-granules. Blood 94:3791–3799
27. Deregibus MC, Cantaluppi V, Calogero R, Lo Iacono M, Tetta C, Biancone L, Bruno S, Bussolati B Camussi G (2007) Endothelial progenitor cell derived microvesicles activate an angiogenic program in endothelial cells by a horizontal transfer of mRNA. Blood 110:2440–2448
28. Booth AM, Fang Y, Fallon JK, Yang JM, Hildreth JE, Gould SJ (2006) Exosomes and HIV Gag bud from endosome-like domains of the T cell plasma membrane. J Cell biol 172:923–935
29. Thery C, Ostrowski M, Segura E (2009) Membrane vesicles as conveyors of immune responses. Nat Rev Immunol 9:581–593
30. Fevrier B, Raposo G. (2004) Exosomes: endosomal-derived vesicles shipping extracellular messages. Curr Opin Cell Biol 16:415–421
31. Stein JM, Luzio JP (1991) Ectocytosis caused by sublytic autologous complement attack on human neutrophils. The sorting of endogenous plasma-membrane proteins and lipids into shed vesicles. Biochem J 274:381–386
32. Thery C, Amigorena S, Raposo G, Clayton A (2006) Isolation and characterization of exosomes from cell culture supernatants and biological fluids. Curr Protoc Cell Biol Unit 3.22, pp 1–29
33. Wubbolts R, Leckie RS, Veenhuizen PT, Schwarzmann G, Mobius W, Hoernschemeyer J, Slot JW, Geuze HJ, Stoorvogel W (2003) Proteomic and biochemical analyses of human B cell-derived exosomes. Potential implications for their function and multivesicular body formation. J Biol Chem 278:10963–10972
34. Escola JM, Kleijmeer MJ, Stoorvogel W, Griffith JM, Yoshie O, Geuze HJ (1998) Selective enrichment of tetraspan proteins on the internal vesicles of multivesicular endosomes and on exosomes secreted by human B-lymphocytes. J Biol Chem 273:20121–20127
35. Van Niel G, Mallegol J, Bevilacqua C, Candalh C, Brugiere S, Tomaskovic-Crook E, Heath JK, Cerf-Bensussan N, Heyman M (2003) Intestinal epithelial exosomes carry MHC class II/peptides able to inform the immune system in mice. Gut 52:1690–1697
36. Clayton A, Court J, Navabi H, Adams M, Mason MD, Hobot JA, Newman GR, Jasani B (2001) Analysis of antigen presenting cell derived exosomes, based on immuno-magnetic isolation and flow cytometry. J Immunol Methods 247:163–174
37. Lamparski HG, Metha-Damani A, Yao JY, Patel S, Hsu DH, Ruegg C, Le Pecq JB (2002) Production and characterization of clinical grade exosomes derived from dendritic cells. J Immunol Methods 270:211–226
38. Viaud S, Ploix S, Lapierre V, Thery C, Commere PH, Tramalloni D, Gorrichon K, Virault-Rocroy P, Tursz T, Lantz O et al (2011) Updated technology to produce highly immunogenic dendritic cell-derived exosomes of clinical grade: a critical role of interferon-gamma. J Immunother 34:65–75
39. Conde-Vancells J, Rodriguez-Suarez E, Embade N, Gil D, Matthiesen R, Valle M, Elortza F, Lu SC, Mato JM, Falcon-Perez JM (2008) Characterization and comprehensive proteome profiling of exosomes secreted by hepatocytes. J Proteome Res 7:5157–5166
40. Stoorvogel W, Kleijmeer MJ, Geuze HJ, Raposo G (2002) The biogenesis and functions of exosomes. Traffic 3:321–330
41. van Niel G, Porto-Carreiro I, Simoes S, Raposo G (2006) Exosomes: a common pathway for a specialized function. J Biochem (Tokyo) 140:13–21
42. Laulagnier K, Motta C, Hamdi S, Roy S, Fauvelle F, Pageaux JF, Kobayashi T, Salles JP, Perret B, Bonnerot C et al (2004) Mast cell and dendritic cell-derived exosomes display a specific lipid composition and an unusual membrane organization. Biochem J 380:161–171
43. Subra C, Laulagnier K, Perret B, Record M (2007) Exosome lipidomics unravels lipid sorting at the level of multivesicular bodies. Biochimie 89:205–212
44. Olver C, Vidal M (2007) Proteomic analysis of secreted exosomes. Subcell Biochem 43:99–131

45. Valadi H, Ekstrom K, Bossios A, Sjostrand M, Lee JJ, Lotvall JO (2007) Exosome–mediated transfer of mRNAs and microRNAs is a novel mechanism of genetic exchange between cells. Nat Cell Biol 9:654–659
46. Zomer A, Vendrig T, Hopmans ES, van Eijndhoven M, Middeldorp JM, Pegtel DM (2010) Exosomes: fit to deliver small RNA. Commun Integr Biol 3:447–450
47. Rabinowits G, Gercel-Taylor C, Day JM, Taylor DD, Kloecker GH (2009) Exosomal microRNA: a diagnostic marker for lung cancer. Clin Lung Cancer 10:42–46
48. Michael A, Bajracharya SD, Yuen PS, Zhou H, Star RA. Illei GG, Alevizos I (2010) Exosomes from human saliva as a source of microRNA biomarkers. Oral Dis 16:34–38
49. Zhang L, Hou D, Chen X, Li D, Zhu L, Zhang Y, Li J, Bian Z, Liang X, Cai X et al (2011) Exogenous plant MIR168a specifically targets mammalian LDLRAP1: evidence of cross-kingdom regulation by microRNA. Cell Res 22:107–126
50. Mathivanan S, Fahner CJ, Reid GE, Simpson RJ (2011) ExoCarta 2012: database of exosomal proteins, RNA and lipids. Nucleic Acids Res 40:D1241–D1244
51. Gruenberg J (2001) The endocytic pathway: a mosaic of domains. Nat Rev Mol Cell Biol 2:721–730
52. Gould GW, Lippincott-Schwartz J (2009) New roles for endosomes: from vesicular carriers to multi-purpose platforms. Nat Rev Mol Cell Biol 10:287–292
53. van Ijzendoorn SC (2006) Recycling endosomes. J Cell Sci 119:1679–1681
54. Sachse M, Urbé S, Oorshot V, Strous G, Klumperman J (2002) Bilayered clathrin coats on endosomal vacuoles are involved in protein sorting toward lysosomes. Mol Biol Cell 13:1313–1328
55. Sotelo JR, Porter KR (1959) An electron microscope study of the rat ovum. J Biophys Biochem Cytol 5:327–342
56. Mobius W, Ohno-Iwashita Y, van Donselaar EG, Oorschot VM, Shimada Y, Fujimoto T, Heijnen HF, Geuze HJ, Slot JW (2002) Immunoelectron microscopic localization of cholesterol using biotinylated and non-cytolytic perfringolysin O. J Histochem Cytochem 50:43–55
57. Mobius W, van Donselaar E, Ohno-Iwashita Y, Shimada Y, Heijnen HF, Slot JW, Geuze HJ (2003) Recycling compartments and the internal vesicles of multivesicular bodies harbor most of the cholesterol found in the endocytic pathway. Traffic 4:222–231
58. White IJ, Bailey LM, Aghakhani MR, Moss SE, Futter CE (2006) EGF stimulates annexin 1-dependent inward vesiculation in a multivesicular endosome subpopulation. Embo J 25:1–12
59. van Niel G, Raposo G, Candalh C, Boussac M, Hershberg R, Cerf-Bensussan N, Heyman M (2001) Intestinal epithelial cells secrete exosome-like vesicles. Gastroenterology 121:337–349
60. Buschow SI, Nolte-'t Hoen EN, van Niel G, Pols MS, ten Broeke T, Lauwen M, Ossendorp F, Melief CJ, Raposo G, Wubbolts R et al (2009) MHC II in dendritic cells is targeted to lysosomes or T cell-induced exosomes via distinct multivesicular body pathways. Traffic 10:1528–1542
61. Hurley JH, Emr SD (2006) The ESCRT complexes: structure and mechanism of a membrane-trafficking network. Annu Rev Biophys Biomol Struct 35:277–298
62. Raiborg C, Stenmark H (2009) The ESCRT machinery in endosomal sorting of ubiquitylated membrane proteins. Nature 458:445–452
63. Babst M (2006) A close-up of the ESCRTs. Dev Cell 10:547–548
64. Williams RL, Urbe S (2007) The emerging shape of the ESCRT machinery. Nat Rev Mol Cell Biol 8:355–368
65. Martin-Serrano J, Zang T, Bieniasz PD (2003) Role of ESCRT-I in retroviral budding. J Virol 77:4794–4804
66. von Schwedler UK, Stuchell M, Muller B, Ward DM, Chung HY, Morita E, Wang HE, Davis T, He GP, Cimbora DM et al (2003) The protein network of HIV budding. Cell 114:701–713
67. Saksena S, Sun J, Chu T, Emr SD (2007) ESCRTing proteins in the endocytic pathway. Trends Biochem Sci 32:561–573

68. Roxrud I, Stenmark H, Malerod L (2010) ESCRT & Co. Biol Cell 102:293–318
69. Wollert T, Hurley JH (2010) Molecular mechanism of multivesicular body biogenesis by ESCRT complexes. Nature 464:864–869
70. Bowers K, Piper SC, Edeling MA, Gray SR, Owen DJ, Lehner PJ, Luzio JP (2006) Degradation of endocytosed epidermal growth factor and virally ubiquitinated major histocompatibility complex class I is independent of mammalian ESCRTII. J Biol Chem 281:5094–5105
71. Malerod L, Stuffers S, Brech A, Stenmark H (2007) Vps22/EAP30 in ESCRT-II mediates endosomal sorting of growth factor and chemokine receptors destined for lysosomal degradation. Traffic 8:1617–1629
72. Buschow SI, Liefhebber JM, Wubbolts R, Stoorvogel W (2005) Exosomes contain ubiquitinated proteins. Blood Cells Mol Dis 35:398–403
73. Geminard C, De Gassart A, Blanc L, Vidal M (2004) Degradation of AP2 during reticulocyte maturation enhances binding of hsc70 and Alix to a common site on TfR for sorting into exosomes. Traffic 5(3):1–13
74. Tamai K, Tanaka N, Nakano T, Kakazu E, Kondo Y, Inoue J, Shiina M, Fukushima K, Hoshino T, Sano K et al (2010) Exosome secretion of dendritic cells is regulated by Hrs, an ESCRT-0 protein. Biochem Biophys Res Commun 399:384–390
75. Bobrie A, Colombo M, Raposo G, Thery C (2011) Exosome secretion: molecular mechanisms and roles in immune responses. Traffic 12:1659–1668
76. Trajkovic K, Hsu C, Chiantia S, Rajendran L, Wenzel D, Wieland F, Schwille P, Brugger B, Simons M (2008) Ceramide triggers budding of exosome vesicles into multivesicular endosomes. Science 319:1244–1247
77. Theos AC, Truschel ST, Tenza D, Hurbain I, Harper DC, Berson JF, Thomas PC, Raposo G, Marks MS (2006) A lumenal domain-dependent pathway for sorting to intralumenal vesicles of multivesicular endosomes involved in organelle morphogenesis. Dev Cell 10:343–354
78. Stuffers S, Sem Wegner C, Stenmark H, Brech A (2009) Multivesicular endosome biogenesis in the absence of ESCRTs. Traffic 10:925–937
79. Goni FM, Alonso A (2009) Effects of ceramide and other simple sphingolipids on membrane lateral structure. Biochimica et biophysica acta 1788:169–177
80. Zöller M (2009) Tetraspanins: push and pull in suppressing and promoting metastasis. Nat Rev Cancer 9:40–55
81. van Niel G, Charrin S, Simoes S, Romao M, Rochin L, Saftig P, Marks MS, Rubinstein E, Raposo G (2011) The tetraspanin CD63 regulates ESCRT-independent and -dependent endosomal sorting during melanogenesis. Dev Cell 21:708–721
82. Raposo G, Marks MS, Cutler DF (2007) Lysosome-related organelles: driving post-Golgi compartments into specialisation. Curr Opin Cell Biol 19:394–401
83. Wolfers J, Lozier A, Raposo G, Regnault A, Thery C, Masurier C, Flament C, Pouzieux S, Faure F, Tursz T et al (2001) Tumor-derived exosomes are a source of shared tumor rejection antigens for CTL cross-priming. Nat Med 7:297–303
84. Carayon K, Chaoui K, Ronzier E, Lazar I, Bertrand-Michel J, Roques V, Balor S, Terce F, Lopez A, Salome L et al (2011) Proteolipidic composition of exosomes changes during reticulocyte maturation. J Biol Chem 286:34426–34439
85. Segura E, Amigorena S, Thery C (2005) Mature dendritic cells secrete exosomes with strong ability to induce antigen-specific effector immune responses. Blood Cells Mol Dis 35:89–93
86. Thery C, Boussac M, Veron P, Ricciardi-Castagnoli P, Raposo G, Garin J, Amigorena S (2001) Proteomic analysis of dendritic cell-derived exosomes: a secreted subcellular compartment distinct from apoptotic vesicles. J Immunol 166:7309–7318
87. de Gassart A, Geminard C, Fevrier B, Raposo G, Vidal M (2003) Lipid raft-associated protein sorting in exosomes. Blood 102:4336–4344
88. Irion U, St Johnston D (2007) Bicoid RNA localization requires specific binding of an endosomal sorting complex. Nature 445:554–558
89. Gibbings DJ, Ciaudo C, Erhardt M, Voinnet O (2009) Multivesicular bodies associate with components of miRNA effector complexes and modulate miRNA activity. Nature Cell biol 11:1143–1149

90. Stinchcombe J, Bossi G, Griffiths GM (2004) Linking albinism and immunity: the secrets of secretory lysosomes. Science 305:55–59
91. Savina A, Fader CM, Damiani MT, Colombo MI (2005) Rab11 promotes docking and fusion of multivesicular bodies in a calcium-dependent manner. Traffic 6:131–143
92. Raposo G, Tenza D, Mecheri S, Peronet R, Bonnerot C, Desaymard C (1997) Accumulation of major histocompatibility complex class II molecules in mast cell secretory granules and their release upon degranulation. Mol Biol Cell 8:2631–2645
93. Faure J, Lachenal G, Court M, Hirrlinger J, Chatellard-Causse C, Blot B, Grange J, Schoehn G, Goldberg Y, Boyer V et al (2006) Exosomes are released by cultured cortical neurones. Mol Cell Neurosci 31:642–648
94. Blanchard N, Lankar D, Faure F, Regnault A, Dumont C, Raposo G, Hivroz C (2002) TCR activation of human T cells induces the production of exosomes bearing the TCR/CD3/zeta complex. J Immunol 168:3235–3241
95. Bonifacino JS, Glick BS (2004) The mechanisms of vesicle budding and fusion. Cell 116:153–166.
96. Cai H, Reinisch K, Ferro-Novick S (2007) Coats, tethers, Rabs, and SNAREs work together to mediate the intracellular destination of a transport vesicle. Dev Cell 12:671–682
97. Stenmark H (2009) Rab GTPases as coordinators of vesicle traffic. Nat Rev Mol Cell 10:513–525.
98. Ostrowski M, Carmo NB, Krumeich S, Fanget I, Raposo G, Savina A, Moita CF, Schauer K, Hume AN, Freitas RP et al (2009) Rab27a and Rab27b control different steps of the exosome secretion pathway. Nat Cell Biol 12:19–30 (Suppl pp 11–13)
99. Hsu C, Morohashi Y, Yoshimura S, Manrique-Hoyos N, Jung S, Lauterbach MA, Bakhti M, Gronborg M, Mobius W, Rhee J et al (2010) Regulation of exosome secretion by Rab35 and its GTPase-activating proteins TBC1D10A-C. J Cell Biol 189:223–232
100. Menager M, de Saint Basile G (2007) Secretory cytotoxic granule maturation is required to their lytic contents excretion. Med Sci (Paris) 23:473–474
101. Lebrand C, Corti M, Goodson H, Cosson P, Cavalli V, Mayran N, Faure J, Gruenberg J (2002) Late endosome motility depends on lipids via the small GTPase Rab7. EMBO J 21:1289–1300
102. Jahn R, Scheller RH (2006) SNAREs–engines for membrane fusion. Nat Rev Mol Cell Biol 7:631–643
103. Rao SK, Huynh C, Proux-Gillardeaux V, Galli T, Andrews NW (2004) Identification of SNAREs involved in synaptotagmin VII-regulated lysosomal exocytosis. J Biol Chem 279:20471–20479
104. Proux-Gillardeaux V, Raposo G, Irinopoulou T, Galli T (2007) Expression of the Longin domain of TI-VAMP impairs lysosomal secretion and epithelial cell migration. Biol Cell 99:261–271
105. Liegeois S, Benedetto A, Garnier JM, Schwab Y, Labouesse M (2006) The V0-ATPase mediates apical secretion of exosomes containing Hedgehog-related proteins in Caenorhabditis elegans. J Cell Biol 173:949–961
106. Marshansky V, Futai M (2008) The V-type H^+-ATPase in vesicular trafficking: targeting, regulation and function. Curr Opin Cell Biol 20:415–426
107. Mallegol J, Van Niel G, Lebreton C, Lepelletier Y, Candalh C, Dugave C, Heath JK, Raposo G, Cerf-Bensussan N, Heyman M (2007) T84-intestinal epithelial exosomes bear MHC class II/peptide complexes potentiating antigen presentation by dendritic cells. Gastroenterology 132:1866–1876.
108. Barres C, Blanc L, Bette-Bobillo P, Andre S, Mamoun R, Gabius HJ, Vidal M (2010) Galectin-5 is bound onto the surface of rat reticulocyte exosomes and modulates vesicle uptake by macrophages. Blood 115:696–705
109. Blanc L, Vidal M (2010) Reticulocyte membrane remodeling: contribution of the exosome pathway. Curr Opin Hematol 17:177–183
110. Nolte-'t Hoen EN, Buschow SI, Anderton SM, Stoorvogel W, Wauben MH (2009) Activated T cells recruit exosomes secreted by dendritic cells via LFA-1. Blood 113:1977–1981

111. Denzer K, van Eijk M, Kleijmeer MJ, Jakobson E, de Groot C, Geuze HJ (2000) Follicular dendritic cells carry MHC class II-expressing microvesicles at their surface. J Immunol 165(3):1259–1265
112. Parolini I, Federici C, Raggi C, Lugini L, Palleschi S, De Milito A, Coscia C, Iessi E, Logozzi M, Molinari A et al (2009) Microenvironmental pH is a key factor for exosome traffic in tumor cells. J Biol Chem 284:34211–34222
113. Pegtel DM, Cosmopoulos K, Thorley-Lawson DA, van Eijndhoven MA, Hopmans ES, Lindenberg JL, de Gruijl TD, Wurdinger T, Middeldorp JM (2010) Functional delivery of viral miRNAs via exosomes. Proc Natl Acad Sci U S A 107:6328–6333
114. Morelli AE (2006) The immune regulatory effect of apoptotic cells and exosomes on dendritic cells: its impact on transplantation. Am J Transplant 6:254–261
115. Tian T, Wang Y, Wang H, Zhu Z, Xiao Z (2010) Visualizing of the cellular uptake and intracellular trafficking of exosomes by live-cell microscopy. J Cell Biochem 111:488–496
116. Giri PK, Schorey JS (2008) Exosomes derived from M. Bovis BCG infected macrophages activate antigen-specific $CD4^+$ and $CD8^+$ T cells in vitro and in vivo. PLoS ONE 3:e2461
117. Schorey JS, Bhatnagar S (2008) Exosome function: from tumor immunology to pathogen biology. Traffic 9:871–881
118. Clayton A, Mitchell JP, Court J, Mason MD, Tabi Z (2007) Human tumor-derived exosomes selectively impair lymphocyte responses to interleukin-2. Cancer Res 67:7458–7466
119. Zhang HG, Grizzle WE (2011) Exosomes and cancer: a newly described pathway of immune suppression. Clin Cancer Res 17:959–964
120. Rak J (2010) Microparticles in cancer. Semin Thromb Hemost 36:888–906
121. Hood JL, San RS, Wickline SA (2011) Exosomes released by melanoma cells prepare sentinel lymph nodes for tumor metastasis. Cancer Res 71:3792–3801
122. Mincheva-Nilsson L, Baranov V (2010) The role of placental exosomes in reproduction. Am J Reprod Immunol 63:520–533
123. Karlsson M, Lundin S, Dahlgren U, Kahu H, Pettersson I, Telemo E (2001) "Tolerosomes" are produced by intestinal epithelial cells. Eur J Immunol 31:2892–2900
124. Liu S, Stolz DB, Sappington PL, Macias CA, Killeen ME, Tenhunen JJ, Delude RL, Fink MP (2005) HMGB1 is secreted by immunostimulated enterocytes and contributes to cytomix-induced hyperpermeability of caco-2 monolayers. Am J Physiol Cell Physiol 290(4):990–999
125. Prado N, Marazuela EG, Segura E, Fernandez-Garcia H, Villalba M, Thery C, Rodriguez R, Batanero E (2008) Exosomes from bronchoalveolar fluid of tolerized mice prevent allergic reaction. J immunol 181:1519–1525
126. Qazi KR, Torregrosa Paredes P, Dahlberg B, Grunewald J, Eklund A, Gabrielsson S (2010) Proinflammatory exosomes in bronchoalveolar lavage fluid of patients with sarcoidosis. Thorax 65:1016–1024
127. Kramer-Albers EM, Bretz N, Tenzer S, Winterstein C, Mobius W, Berger H, Nave KA, Schild H, Trotter J (2007) Oligodendrocytes secrete exosomes containing major myelin and stress-protective proteins: Trophic support for axons? Proteomics Clin Appl 1:1446–1461
128. Lachenal G, Pernet-Gallay K, Chivet M, Hemming FJ, Belly A, Bodon G, Blot B, Haase G, Goldberg Y, Sadoul R (2011) Release of exosomes from differentiated neurons and its regulation by synaptic glutamatergic activity. Mol Cell Neurosci 46:409–418
129. Bakhti M, Winter C, Simons M (2011) Inhibition of myelin membrane sheath formation by oligodendrocyte-derived exosome-like vesicles. J Biol Chem 286:787–796
130. Wang S, Cesca F, Loers G, Schweizer M, Buck F, Benfenati F, Schachner M, Kleene R (2011) Synapsin I is an oligomannose-carrying glycoprotein, acts as an oligomannose-binding lectin, and promotes neurite outgrowth and neuronal survival when released via glia-derived exosomes. J Neurosci 31:275–7290
131. Fevrier B, Vilette D, Archer F, Loew D, Faigle W, Vidal M, Laude H, Raposo G (2004) Cells release prions in association with exosomes. Proc Natl Acad Sci U S A 101:9683–9688
132. Rajendran L, Honsho M, Zahn TR, Keller P, Geiger KD, Verkade P, Simons K (2006) Alzheimer's disease beta-amyloid peptides are released in association with exosomes. Proc Natl Acad Sci U S A 103:11172–11177

133. Gomes C, Keller S, Altevogt P, Costa J (2007) Evidence for secretion of Cu, Zn superoxide dismutase via exosomes from a cell model of amyotrophic lateral sclerosis. Neurosci Lett 428:43–46
134. Emmanouilidou E, Melachroinou K, Roumeliotis T, Garbis SD, Ntzouni M, Margaritis LH, Stefanis L, Vekrellis K (2010) Cell-produced alpha-synuclein is secreted in a calcium-dependent manner by exosomes and impacts neuronal survival. J Neurosci 30:6838–6851
135. Brouckova A, Holada K (2009) Cellular prion protein in blood platelets associates with both lipid rafts and the cytoskeleton. Thromb Haemost 102:966–974
136. Kujala P, Raymond CR, Romeijn M, Godsave SF, van Kasteren SI, Wille H, Prusiner SB, Mabbott NA, Peters PJ (2011) Prion uptake in the gut: identification of the first uptake and replication sites. Plos Pathogens 7(12): e1002449
137. Simpson RJ, Lim JW, Moritz RL, Mathivanan S (2009) Exosomes: proteomic insights and diagnostic potential. Expert Rev Proteomics 6:267–283
138. Silverman JM, Clos J, de'Oliveira CC, Shirvani O, Fang Y, Wang C, Foster LJ, Reiner NE (2010) An exosome-based secretion pathway is responsible for protein export from Leishmania and communication with macrophages. J Cell Sci 123:842–852
139. Clayton C, Estevez A (2010) The exosomes of trypanosomes and other protists. Adv Exp Med Biol 702:39–49
140. Alvarez-Erviti L, Seow Y, Yin H, Betts C, Lakhal S, Wood MJ (2011) Delivery of siRNA to the mouse brain by systemic injection of targeted exosomes. Nature Biotechnol 29:341–345

Chapter 2
The Role of Exosomal Shuttle RNA (esRNA) in Cell-to-Cell Communication

Cecilia Lässer, Maria Eldh and Jan Lötvall

Abstract Cell-to-cell communication can occur in several ways, with or without cell contact. Exosomes play a role in one of the most recently discovered and versatile cell-to-cell communications, which do not require cell contact and that can act over long distances. The RNA content, mRNA and microRNA, is protected by the exosomes rigid membranes, which makes it possible for cells to communicate long-distance RNA messages via the circulation system. Their mRNA content differs substantially from their mother cell mRNA content, whereas their microRNA content seems to reflect their cellular origin more. This chapter reviews the role of exosomes in cell-to-cell communication and in particular the role of exosomal shuttle RNA (esRNA). This is a new and rapidly expanding field of research that has given cell-to-cell communication an increased complexity and that has great potential within both diagnostic and therapeutic applications.

2.1 Introduction

Exosomes were discovered in the early 1980s and were then mainly studied for their role in discarding unwanted proteins during the maturation of reticulocytes [1, 2]. In the late 1990s, exosomes were shown to have immune regulatory effects [3, 4], which introduced exosomes as an important factor in cell-to-cell communication. In 2007, mast cell-derived exosomes were shown to not only contain proteins, but also to contain functional microRNA (miRNA) and messenger RNA (mRNA) that could be transferred to recipient cells [5]. This RNA was named exosomal shuttle RNA (esRNA) [5].

All cells communicate with their surrounding environment through many different pathways, including growth factors, cytokines, hormones, chemokines, and surface-to-surface communication via membrane-bound proteins and lipids. The discovery of transferrable RNA in exosomes further increases the complexity of cell-to-cell communication, especially in relation to the presence of miRNA in exosomes, which can induce RNA interference (RNAi). RNAi is a natural process within living cells that participates in the control of gene activity. Two types of small RNA molecules

J. Lötvall (✉) · C. Lässer · M. Eldh
Krefting Research Centre, University of Gothenburg, Gothenburg, Sweden
e-mail: jan.lotvall@gu.se

H.-G. Zhang (ed.), *Emerging Concepts of Tumor Exosome–Mediated Cell–Cell Communication,* DOI 10.1007/978-1-4614-3697-3_2,

are important for RNAi, including miRNA and small interfering RNA (siRNA). These small RNAs (20–25 nucleotides) can bind to specific mRNAs and decrease their activity by preventing the mRNA from producing its corresponding protein.

Not only exosomes, but also other larger extracellular vesicles such as microvesicles [6] and apoptotic bodies [7] also contain RNA. Furthermore, both high-density lipoproteins [8] and Argonaute 2 [9], an RNA-binding protein complex, have been suggested to carry and protect extracellular RNA. However, most of the reports that show functional effects of extracellular RNA have studied exosomes.

2.2 Presence of Exosomal Shuttle RNA (esRNA)

The field of esRNA is seeing an explosive growth over the last few years, after the initial report in 2007. In the first year after the discovery of esRNA, only three studies were published supporting the finding that exosomes contained RNA [10–12]. Two of these studies were investigating the RNA content of exosomes circulating in the plasma of cancer patient, suggesting a role for esRNA to function as biomarkers for ovarian and glioblastoma cancer [10, 12]. The third study reported the presence of esRNA in a prostate cancer cell line. Together, these studies highlight the importance of esRNA in the field of malignant disease, which has further boosted the field. In 2009, an additional five studies on esRNA were published [13–17]. Two of these reports supported the previous findings that esRNA could be used as a biomarker [14, 15]. Interestingly, the possible importance of esRNA in the fields' innate immunity [13] and reproduction [16] was also suggested, supporting the importance of esRNA in a wider range of cell-to-cell communication, beyond malignant disease.

Over the last 5 years, approximately 40 different publications have documented the presence of RNA in exosomes and cover diverse fields, including cell biology, immunology, cancer, neurology, stem cell function, and beyond. There is, however, a confounding in the literature, as the nomenclature in the field of exosomes and microvesicles is not homogenous and sometimes outright confusing. Exosomes are sometimes referred to as microvesicles and vice versa, and as many different isolation methods are used. Some investigators will have a mixture of both exosomes (40–100 nm) and larger microvesicles in their studies of extracellular RNA. The authors of this chapter suggest that the name "microvesicles" should only be used for vesicles larger than 100 nm that are of nonendosomal origin, and protocols that lead to mixtures of exosomes and microvesicles should be referred to as "extracellular vesicles." Depending on how stringent the classification of "exosomes" in each publication is, different number of publications can be identified. All extracellular vesicles seem to contain RNA, but the differences in their biogenesis, loading of RNA, and purification methods will most probably influence the specific RNA profiles identified. In this chapter, we are focusing on RNA in exosomes, with exosomes defined as vesicles smaller than 100 nm, and with evidence of having been produced in the endosomal pathway.

Table 2.1 microRNA (miRNA)

Exosomal source	Species	Number of miRNA	Method	Reference
The mast cell line, MC/9	Mouse	121	miRCURY LNA array	[5]
Plasma from patients with ovarian cancer	Human	218	Microarray	[10]
The Jurkat-derived J77 T cell line	Human	236	Agilent micro-RNA microarray	[18]
The Raji B cell line	Human	67	Agilent micro-RNA microarray	[18]
Peripheral blood monocyte-derived dendritic cells	Human	148	Agilent micro-RNA microarray	[18]
Bone marrow-derived immature dendritic cells	Mouse	144	Illumina miRNA expression array	[19]
Bone marrow-derived mature dendritic cells	Mouse	197	Illumina miRNA expression array	[19]
The hepatocellular carcinoma cell line, Hep3B	Human	134	qRT-PCR	[20]
The hepatocellular carcinoma cell line, PLC/PRF/5	Human	144	qRT-PCR	[20]

The Internet-based database, ExoCarta (http://www.exocarta.org/), lists at this time 764 miRNA as found in exosomes. This database is mainly based on the findings in exosomes derived from mouse mast cells (2007) [5], plasma of ovarian cancer patients (2008) [10], human dendritic cells, and B and T lymphocytes (2011) [18]. However, many miRNA have been identified in several of these studies and approximately 450 unique miRNA have been identified in exosomes. Recently, Montecalvo et al. [19] published miRNA data identifying 202 miRNA in exosomes from immature and mature dendritic cells and Kogure et al. [20] have identified 134 and 144 miRNA, respectively, in two different hepatocellular carcinoma cell lines, although these two datasets are not yet in ExoCarta. All studies reporting compressive lists of miRNA in exosomes till date are summarized in Table 2.1, but, of course, other studies have also reported the presence of individual miRNAs.

The presence of different mRNAs in exosomes has been clearly documented in many studies and is also reported in ExoCarta. The presence of mRNA in exosomes was earlier shown to be fundamentally different form the exosome-producing cell, and is not a random sample of the cell's cytoplasmatic mRNA, arguing for specific packaging mechanisms of RNA into exosomes. The total number of reported mRNAs in exosomes exceeds 2,300, however, these data are based on two studies only, Valadi et al. 2007 [5] and Nazarenko et al. 2010 [21]. Table 2.2 shows these 2 studies, but furthermore, it highlights that other studies have also detected large amounts of mRNAs in exosomes, which are not presented in ExoCarta.

Table 2.2 Messenger RNA (mRNA)

Exosomal source	Species	Number of mRNA	Method	Reference
The mast cell line, MC/9	Mouse	1,300	Affymetrix microarray	[5]
Culture of primary cells from glioblastoma tumor	Human	27,000	Agilent 44 K whole genome microarray	[12]
The colorectal adenocarcinoma cells, SW480	Human	11,327	Illumina Human-6 v2 expression	[28]
Saliva	Human	509	Affymetrix microarray	[24]
BDX-derived pancreatic adenocarcinoma	Rat	1,500	Illumina RatRef-12 chip	[21]
The mast cell line, MC/9	Mouse	?	Affymetrix microarray	[32]

2.3 Exosomal Shuttle RNA (esRNA)

2.3.1 Profile of the esRNA

When RNA was first discovered in exosomes, it was clearly shown that the exosomal mRNA content differs substantially from the RNA profile of the donor cells [5]. Also, the donor cells and exosomes' total RNA profile differed, where the exosomal RNA profile is lacking the 18S and 28S ribosomal RNA (rRNA) peaks that are highly prominent in the donor cells. Exosomes are instead enriched with smaller RNA such as mRNA and miRNA [5]. Since then, most studies have reported the same RNA profile with no or little rRNA in exosomes from plasma [10, 12, 22], amniotic fluid [23], saliva [22, 24], hepatocellular carcinoma cell line [20], embryonic kidney cell line [25], breast milk [22], and nasal lavage fluid [26]. However, there are studies reporting RNA profiles with more prominent rRNA peaks in samples collected from urine [23, 27] and colorectal cancer cell lines [28]. Current research seems to suggest that the RNA profiles may differ in exosomes and larger extracellular vesicles, where vesicles with a diameter >100 nm seem to contain a greater fraction of rRNA. However, the RNA profile for other nonexosomes vesicles have to be further investigated before this can be conclusively stated.

2.3.2 Horizontal Transfer of esRNA

In the first study showing the presence of RNA in exosomes, Valadi et al. [5] also showed that exosomes are able to transfer their RNA to recipient cells. This was shown by culturing mouse mast cells with radioactive uridine, prior to exosome isolation. The exosomes isolated from these cells, containing radiaoactive RNA, were then cocultured with recipient cells. The radioactive RNA was then quantified in the recipient cells, which showed that mouse mast cell-derived exosomes can transfer their RNA to other mast cells.

Subsequently, several studies have demonstrated that exosomes from many different cellular sources can be taken up by a variety of recipient cells [12, 20, 22, 29], commonly by labeling them with a fluorescent membrane dye, for example, PHK67. For the RNA to be functional in the recipient cell, exosomes cannot only be internalized by the recipient cell, but the RNA also needs to be transported to the cytosol. Montecalvo et al. [19] showed that the membrane of the exosomes can fuse with the plasma membrane and that the content of the exosomes can be released into the cytosol of the recipient cell. The second step, showing that the exosomes are able to transfer their luminal cargo to a recipient cell, was important to prove as adherence or hemifusion would not result in delivery of the RNA into the recipient cell's cytoplasm. To prove that the fusion was complete, luciferin, which is not able to cross lipid membranes, was captured inside the exosomes. Activity could be measured in the recipient transgenic luciferase cells already after 8 min, which indicates the delivery of the luminal cargo of the exosomes to the recipient cell, including miRNA and mRNA [19].

2.3.3 Functionality of Horizontally Transferred esRNA

It has also been important to show that the transfer of RNA from one cell to another via exosomes leads to functional RNA-related biological effects in the recipient cell. When exosomal RNA was first identified, the functionality of the exosomal mRNA was proven by using an in vitro translation assay, as exosomes do not themselves have the complete machinery to produce proteins. Thus, the exosomal mRNAs isolated from mouse mast cells were used as templates in an in vitro translation assay, where it was shown that it could be translated into proteins. Furthermore, both the mRNA and the protein content of the mouse mast cell-derived exosomes were analyzed before these exosomes were added to human mast cells. Newly produced mouse proteins could subsequently be identified in the human cells that were only present as mRNA in the donor mouse exosomes, and not as proteins. The combination of these data thus shows that the exosomal mRNA is functional as it can be translated into proteins in the recipient cells [5].

More recently, it has been further proven that mRNA in exosomes are functional as two independent studies have shown that an mRNA coding for a luciferase reporter gene can be transferred via exosomes and lead to luciferase activity in a recipient cell. Interestingly, the luciferase activity in the recipient cell was dependent on the amount of exosomes that were added [20] and increased over time [12].

It has also been shown that not only mRNA can be transferred to recipient cells via exosomes, but also that miRNAs can be transferred to and be functional in recipient cells. Thus, Epstein-Barr virus (EBV) transformed B lymphocytes release exosomes containing mature EBV-miRNAs that can be delivered and internalized by recipient cells [30]. Interestingly, when incubating the EVB-miRNA containing B lymphocyte exosomes with cells transfected with a luciferase vector carrying an EBV-miRNA-regulated sequence they showed an 80 % reduction in luciferase activity in the transfected recipient cells. Cells expressing disrupted EBV-miRNA-binding sites were shown to have significantly less luciferase activity reduction, indicating that exosomes can regulate miRNA-mediated functional gene repression of specific target mRNAs in recipient cells [30].

Subsequent to this Montecalvo et al. [19] have further confirmed the delivery of functional miRNA via exosomes. This was also shown by transfecting the recipient cells with a luciferase reporter gene, with copies of the complementary target sequence for a specific miRNA, which resulted in an inhibited luciferase activity when incubated with exosomes containing the specific miRNA. The functionality of exosomal miRNA has also been shown by transfecting cells with lentiviral vector short hairpin RNA (LV-sh). Exosomes isolated from cells transfecting with LV-shCD81 were shown to induce a downregulation of the surface expression of CD81 in the recipient cells [31].

In conclusion, these studies demonstrate that the miRNA encapsulated and transferred via exosomes is functional in the recipient cells and can exhibit its regulatory effect on a targeted mRNA, which leads to a changed cellular phenotype.

2.4 Specific Loading of RNA into Exosomes

Several papers have shown that the levels of mRNA [5, 12, 32] and miRNA [5, 18] in exosomes and their donor cells correlate poorly, suggesting that the cell sorts specific RNA species into the lumen of exosomes. Thus, the RNA in exosomes is not related to just an engulfment of arbitrary RNA from the cytoplasm, which, however, could be true for larger vesicles with different production process. The mechanism behind this sorting is not yet known, but complementary papers in the same issue of *Nature Cell Biology* in 2009 possibly shed some light into these questions [17, 33]. Gibbings et al. [17] showed that miRNA were mainly detected together with RNA-inducing silencing complex (RISC) proteins, GW182 and Argonaute 2 (AGO2), in compartments assembled with endosomes and multivesicular bodies (MVB). Interestingly, GW182 and AGO2 are two major components of RISC and are required for effective gene silencing [34, 35]. In addition, these were not associated with lysosomes, endoplasmic reticulum (ER), or P-bodies. Furthermore, these investigators showed that the miRNA-silencing capacity and the sorting of GW182 into the MVB, was dependent on MVB-associated endosomal sorting complex required for transport (ESCRT) proteins, such as Alix. The authors therefore suggest that the MVB is the location for assembling and loading of RISC. These observations may be important findings for the understanding of the loading of RNA into exosomes as the exosomes are formed in the MVB and also contain high levels of GW182 [17].

In addition, another study also showed that inhibition of Alix led to loss of the suppressive effect of miRNA [25], which supports the findings above. However, the inhibition of Alix did not affect the secretion of miRNA. Instead, the authors demonstrated that inhibition of neutral sphingomylinases (nSMase) that did not have an effect on the activity of miRNA, but rather had an effect on the secretion of miRNA [25].

nSMase is an enzyme involved in the hydrolysis of sphingomyelin to form ceramide and has previously been shown that nSMase is also important for the budding of intracellular vesicles into the MVB [36]. Inhibition of nSMase left the intracellular levels of miRNA unchanged, but reduced secretion of not only miRNA, but also exosomal proteins. Furthermore, overexpression of nSMase induced secretion of miRNA, without affecting the intracellular miRNA levels. It was not exclusively shown that the secreted miRNA was inside exosomes, but as another study has shown that inhibition of nSMase led to unchanged miRNA levels intracellular, but a significant reduction of miRNA in secreted exosomes [20], this is also likely an important factor in the loading of RNA into exosomes.

Together, these data emphasize the importance of the multivesicular bodies and their cooperative proteins as well as sphingomyelin for correct loading and function of miRNA, and this machinery may play an important role in the selection and loading of specific miRNA into exosomes, which can fundamentally influence their biological function.

2.5 Biological Role of Exosomal Shuttle RNA (esRNA)

The biological function of exosomes is fundamentally dependent on the cell that produces these exosomes, as well as the current state of that cell, as this will affect the loading of proteins, mRNA, and miRNA into the exosomes. Exosomes have been shown to have many functions, which primarily have been ascribed to the exosomal proteins, for example, Fas ligand on the surface of tumor exosomes can induce apoptosis in T lymphocytes [37]. Fewer studies have so far determined the biological function of the exosomal RNA. However, as it is predicted that one miRNA can interfere with 100–200 mRNA [38, 39] and also that mRNAs can affect the protein production of the recipient cells, the potential biological role of esRNA is vast.

2.5.1 Protective Effect Against Oxidative Stress

In 2010, Eldh et al. [32] showed, by microarray analysis, that the exosomal mRNA content does not only differ from their donor cells but also depends on which condition they are released under. Furthermore, exosomes released under different conditions have different effects on the recipient cells. Thus, exosomes released from cells grown under the condition of oxidative stress have been shown to provide recipient cells with a tolerance to further oxidative stress, showing that the exosomes condition the recipient cells in some ways. This was shown by a greater viability in cells receiving exosomes released by other cells grown under oxidative stress, compared with cells receiving exosomes released from cells grown under normal condition. The protective effect of the exosomes produced under oxidative stress was removed by exposing the exosomes to UV light, which destroys the RNA, arguing that the esRNA is involved in the protective effect.

2.6 Induce Invasiveness of Tumor Cells

The microenvironment surrounding a tumor plays an important role in the tumor progression. Large amount of macrophages are often found in breast cancer and an increased macrophage infiltration is commonly associated with a poor prognosis. Macrophages induce proliferation and angiogenesis and enhance the invasiveness and metastasis of tumors [40, 41]. Previously, it has been demonstrated that soluble factors such as cytokines and growth factors released by macrophages can induce these effects. In addition, it was recently shown that exosomal miRNA from macrophages could upregulate the invasiveness of breast cancer cell in vitro [42]. The study by Yang et al. [42] showed that exosomes from IL-4-activated macrophages contained miR-223 and that these exosomes promoted invasion of breast cancer cells (SKBR3) in a transwell invasion assay. When antisense oligonucleotides for miR-223 was added, the invasion induced by the exosomes was decreased. Furthermore, miR-223 could target and reduce *Mef2c*, which led to an increase of β-catenin in the nucleus. The authors suggested that miR-223 was transferred from macrophages to breast cancer cells, via exosomes, where it affected the *Mef2c*-β-catenin pathway, which led to an increased invasiveness of the breast cancer cells. Disruption of the exosomal communication between macrophages and breast cancer cells may play an important role in the prevention of metastasis and could be a potential target in breast cancer therapy.

2.6.1 Increased LDL Levels in Mouse Plasma

The fact that RNA can be transferred between cells via exosomes has taken the cell-to-cell communication concept to another level of complexity. In 2011, this communication mechanism was shown to be even more complex when Zhang et al. [43] found plant microRNA in human serum of healthy Chinese subjects. MIR156a and MIR168a were the two exogenous plant microRNAs that were found in the highest concentrations and have shown to be enriched in rice. They showed that these plant microRNA were upregulated in mice feed with a rice diet, both in the serum and in the liver. Furthermore, more than half of the exogenous plant microRNA found in mouse serum was found to be contained within microvesicles. By in vitro studies they showed that colon cells transfected with MIR168a release microvesicles containing MIR168a. Also, that these microvesicles can be taken up by liver cells and upregulate their MIR168a concentration. Bioinformatics analysis identified low-density lipoprotein reporter adaptor protein 1 (LDLRAP1) as the most conserved putative-binding site for MIR186a. Interestingly, they showed that exogenous plant microRNA can regulate mammalian gene expression. This was shown by an enrichment of MIR168a in mouse serum and liver, in mice feed with rice, and an inhibition of MIR168a target gene LDLRAP1 in the liver of these mice. Most significantly, this resulted in a decrease in the removal of LDL from the mouse plasma. The authors, therefore, hypothesize that when food is processed in the gastrointestinal tract, the intestinal epithelial cells have the capacity to take up the plant miRNA and pack it into vesicles, which are released into the circulation to reach other organs such as the liver, suggesting that regulatory plant miRNA could travel to and affect different cells [43].

2.6.2 Allow Tumor Cells to Maintain Their Oncogenesis

Let-7 is commonly viewed as a tumor suppressor miRNA and loss of the family members of let-7 indicates poor survival [44]. It has been shown that the metastatic gastric cancer cell line, AZ-P7a, expresses high levels of let-7 miRNAs, both intracellular and in their exosomes. In contrast, in other cancer cell lines analyzed, the level of let-7 miRNAs was high in the cells but low in the exosomes. The authors hypothesized that AZ-P7a cells selectively and actively secreted let-7 miRNA family into the extracellular environment via exosomes, which would decrease the antitumor effect on the inside of the cells and lead them to maintain their oncogenesis and invasiveness, as AZ-P7 is highly metastatic [45].

These findings are interesting as it has been shown that presence of let-7 miRNA in plasma vesicles is related to survival in nonsmall-cell lung cancer [46]. When the patients were divided into let-7f low and high groups, the low-let-7f group had an overall survival rate of 30 % at 46 months, while the high-level group had a survival rate of 8 % (p-value 0.038). Together, these two studies suggest that tumors use exosomes to release let-7f, which may allow the cell to become more tumorigenic, thus affecting survival. Furthermore, the second study is the first to show that the miRNA of extracellular vesicles can be used as a prognostic tool and to predict survival in cancer patients. Importantly, it was not clearly established that the vesicles studied were indeed exosomes and not other larger extracellular vesicles [46].

2.7 Diagnostic and Therapeutic Potential

2.7.1 Biomarkers

Exosomes have shown to have many qualities that make them excellent candidates as biomarkers. Qualities such as being easily sampled from patients by relatively noninvasive means as they can be extracted from several different body fluids such as plasma[47], urine [48], breast milk [49], BAL [50], saliva [24], and NAL [26]. Another feature that makes exosomes a prime candidate for biomarkers is that they have been shown to be upregulated in serum of patients with ovarian and lung cancer [10, 15]. Furthermore, these tumor exosomes were shown to contain miRNA earlier associated with ovarian and lung tumor, respectively. Thus, the miRNA isolated from the tumor exosomes were shown to reflect the miRNA profile in the tumor tissue. In addition, for the ovarian cancer patients, the miRNA correlating with the tumor miRNA was also upregulated in comparison with both patients with benign ovarian tumors and healthy subjects [10].

Plasma exosomes from patients with glioblastoma have been demonstrated to contain the mRNA for the mutated protein EGFRvIII. The mutated mRNA was found in about 50 % of the tumors, in approximately 25 % of the plasma exosomes of the glioblastoma patients and in 0 % of the healthy controls. Furthermore, the mRNA could not be detected in the plasma exosomes 2 weeks after removal of the tumor, indicating that the tumor was the source of the exosomes [12] and again suggesting a potential role of the RNA content to be used as biomarkers.

In addition, miRNA from exosomes isolated from urine and saliva has also been investigated for the potential use as biomarkers for Sjögren syndrome and prostate cancer, respectively [14, 51], and RNA isolated from exosomes found in amniotic fluid can be used for fetal sex determination [23].

Taken together, multiple studies show that exosomes can be used for profiling of many different diseases such as cancers, and that they can even be used in baby gender determination.

2.7.2 Gene Delivery Vehicles

The fact that exosomes function as natural vectors for the transfer of genetic material between cells has made them interesting candidates as gene therapy vectors. Exosomes have the advantage of being endogenous and to be able to escape the immune system. Vesicles such as viruses, nanoparticles, and liposomes all function as vectors to a certain extent, but will sooner or later be discovered by the immune system and trigger an immune response, which will lead to a rejection or inflammation of affected recipient cells. Another challenge in gene therapy is the delivery, which to have the greatest effect and least side effects, has to be able to be specific. A group in Oxford showed for the first time that RNA-loaded exosomes have therapeutic potential by a knock down of the levels of BACE1 in the brain, which is a gene implicated in Alzheimer's disease [52].By harvesting bone marrow from mice and culturing the immature dendritic cells, they transfected these cells with a peptide-targeting neuron cells, called RVG, fused to a membrane protein, Lamp2b, known to be enriched in exosomal membranes. These transfected cells produced exosomes with the RVG peptide on their surface. The modified exosomes were then loaded with therapeutic siRNA, designed to knock down the BACE1 gene, and injected back into the mice. Mice receiving the exosomes displayed a reduced level of BACE1 in the brain. This study showed both that exosomes can be modified to be able to be derived to a specific cell/organ and that they can be loaded with a cargo, which can be successfully delivered to the target. Further studies are, of course, required to show any equivalent function in humans, which, of course, raise several methodological and safety concerns and require extensive preparative studies.

In addition, it has also been found that $CD34^+$ cells release exosomes containing proangiogenic miRNA, as $CD34^+$ exosomes induced angiogenic activity in endothelial cells in vitro and vessel growth in vivo in mice. Although it was not established that it was the miRNA that transferred the effect. This natural therapeutic potential of using $CD34^+$ exosomes to induce angiogenesis could improve the outcome of ischemic injuries and improve recovery [53] and also further enhance the potential use of exosomes in therapy.

2.8 Summary and Conclusions

Exosomes have been proven to present an important route of cell-to-cell communication by the delivery of functional RNA species, which fundamentally affect the biological function of a recipient cell. Exosomal shuttle RNA is likely to have vast regulatory functions in the human body under healthy circumstances, but may also be used as biomarkers for many diseases, and may explain progression of severe diseases such as cancers. Last, as exosomes are natural vectors for delivering RNA to cells, they are likely to be useful for the delivery of therapeutic RNAi, such as siRNAs or microRNAs in many diseases. The field of understanding the function and possibilities of esRNA is just beginning its journey and will keep researchers in many fields of biology and medicine occupied for years to come.

References

1. Harding C, Heuser J, Stahl P (1983) Receptor-mediated endocytosis of transferrin and recycling of the transferrin receptor in rat reticulocytes. J Cell Biol 97(2):329–339
2. Pan BT, Teng K, Wu C et al (1985) Electron microscopic evidence for externalization of the transferrin receptor in vesicular form in sheep reticulocytes. J Cell Biol 101(3):942–948
3. Zitvogel L, Regnault A, Lozier A et al (1998) Eradication of established murine tumors using a novel cell-free vaccine: dendritic cell-derived exosomes. Nat Med 4(5):594–600
4. Raposo G, Nijman HW, Stoorvogel W et al (1996) B lymphocytes secrete antigen-presenting vesicles. J Exp Med 183(3):1161–1172
5. Valadi H, Ekström K, Bossios A et al (2007) Exosome-mediated transfer of mRNAs and microRNAs is a novel mechanism of genetic exchange between cells. Nat Cell Biol 9(6):654–659
6. Grange C, Tapparo M, Collino F et al (2011) Microvesicles released from human renal cancer stem cells stimulate angiogenesis and formation of lung premetastatic niche. Cancer Res 71(15):5346–5356
7. Zernecke A, Bidzhekov K, Noels H et al (2009) Delivery of microRNA-126 by apoptotic bodies induces CXCL12-dependent vascular protection. Sci Signal 2(100):ra81
8. Vickers KC, Palmisano BT, Shoucri BM et al (2011) MicroRNAs are transported in plasma and delivered to recipient cells by high-density lipoproteins. Nat Cell Biol 13(4):423–433

9. Arroyo JD, Chevillet JR, Kroh EM et al (2011) Argonaute2 complexes carry a population of circulating microRNAs independent of vesicles in human plasma. Proc Natl Acad Sci U S A 108(12):5003–5008
10. Taylor DD, Gercel-Taylor C (2008) MicroRNA signatures of tumor-derived exosomes as diagnostic biomarkers of ovarian cancer. Gynecol Oncol 110(1):13–21
11. Lehmann BD, Paine MS, Brooks AM et al (2008) Senescence-associated exosome release from human prostate cancer cells. Cancer Res 68(19):7864–7871
12. Skog J, Wurdinger T, van Rijn S et al (2008) Glioblastoma microvesicles transport RNA and proteins that promote tumour growth and provide diagnostic biomarkers. Nat Cell Biol 10(12):1470–1476
13. Kesimer M, Scull M, Brighton B et al (2009) Characterization of exosome-like vesicles released from human tracheobronchial ciliated epithelium: a possible role in innate defense. FASEB J 23(6):1858–1868
14. Nilsson J, Skog J, Nordstrand A et al (2009) Prostate cancer-derived urine exosomes: a novel approach to biomarkers for prostate cancer. Br J Cancer 100(10):1603–1607
15. Rabinowits G, Gercel-Taylor C, Day JM et al (2009) Exosomal microRNA: a diagnostic marker for lung cancer. Clin Lung Cancer 10(1):42–46
16. Luo SS, Ishibashi O, Ishikawa G et al (2009) Human villous trophoblasts express and secrete placenta-specific microRNAs into maternal circulation via exosomes. Biol Reprod 81(4):717–729
17. Gibbings DJ, Ciaudo C, Erhardt M et al (2009) Multivesicular bodies associate with components of miRNA effector complexes and modulate miRNA activity. Nat Cell Biol 11(9):1143–1149
18. Mittelbrunn M, Gutierrez-Vazquez C, Villarroya-Beltri C et al (2011) Unidirectional transfer of microRNA-loaded exosomes from T cells to antigen-presenting cells. Nat Commun 2:282
19. Montecalvo A, Larregina AT, Shufesky WJ et al (2012) Mechanism of transfer of functional microRNAs between mouse dendritic cells via exosomes. Blood 119(3):756–766
20. Kogure T, Lin WL, Yan IK et al (2011) Intercellular nanovesicle-mediated microRNA transfer: a mechanism of environmental modulation of hepatocellular cancer cell growth. Hepatology 54(4):1237–1248
21. Nazarenko I, Rana S, Baumann A et al (2010) Cell surface tetraspanin Tspan8 contributes to molecular pathways of exosome-induced endothelial cell activation. Cancer Res 70(4):1668–1678
22. Lässer C, Alikhani VS, Ekström K et al (2011) Human saliva, plasma and breast milk exosomes contain RNA: uptake by macrophages. J Transl Med 9:9
23. Keller S, Ridinger J, Rupp AK et al (2011) Body fluid derived exosomes as a novel template for clinical diagnostics. J Transl Med 9:86
24. Palanisamy V, Sharma S, Deshpande A et al (2010) Nanostructural and transcriptomic analyses of human saliva derived exosomes. PLoS One 5(1):e8577
25. Kosaka N, Iguchi H, Yoshioka Y et al (2010) Secretory mechanisms and intercellular transfer of microRNAs in living cells. J Biol Chem 285(23):17442–17452
26. Lässer C, O'Neil SE, Ekerljung L et al (2011) RNA-containing exosomes in human nasal secretions. Am J Rhinol Allergy 25(2):89–93
27. Miranda KC, Bond DT, McKee M et al (2010) Nucleic acids within urinary exosomes/microvesicles are potential biomarkers for renal disease. Kidney Int 78(2):191–199
28. Hong BS, Cho JH, Kim H et al (2009) Colorectal cancer cell-derived microvesicles are enriched in cell cycle-related mRNAs that promote proliferation of endothelial cells. BMC Genomics 10:556
29. Vallhov H, Gutzeit C, Johansson SM et al (2011) Exosomes containing glycoprotein 350 released by EBV-transformed B cells selectively target B cells through CD21 and block EBV infection in vitro. J Immunol 186(1):73–82
30. Pegtel DM, Cosmopoulos K, Thorley-Lawson DA et al (2010) Functional delivery of viral miRNAs via exosomes. Proc Natl Acad Sci U S A 107(14):6328–6333
31. Pan Q, Ramakrishnaiah V, Henry S et al (2011) Hepatic cell-to-cell transmission of small silencing RNA can extend the therapeutic reach of RNA interference (RNAi). Gut. 2011 Dec 23 (Epub ahead of print)

32. Eldh M, Ekström K, Valadi H et al (2010) Exosomes communicate protective messages during oxidative stress; possible role of exosomal shuttle RNA. PLoS One 5(12):e15353
33. Lee YS, Pressman S, Andress AP et al (2009) Silencing by small RNAs is linked to endosomal trafficking. Nat Cell Biol 11(9):1150–1156
34. Ding L, Han M (2007) GW182 family proteins are crucial for microRNA-mediated gene silencing. Trends Cell Biol 17(8):411–416
35. Kawamata T, Tomari Y (2010) Making RISC. Trends Biochem Sci 35(7):368–76
36. Trajkovic K, Hsu C, Chiantia S et al (2008) Ceramide triggers budding of exosome vesicles into multivesicular endosomes. Science 319(5867):1244–1247
37. Abusamra AJ, Zhong Z, Zheng X et al (2005) Tumor exosomes expressing Fas ligand mediate $CD8^{+}$ T-cell apoptosis. Blood Cells Mol Dis 35(2):169–173
38. Lim LP, Lau NC, Garrett-Engele P et al (2005) Microarray analysis shows that some microRNAs downregulate large numbers of target mRNAs. Nature 433(7027):769–773
39. Krek A, Grun D, Poy MN et al (2005) Combinatorial microRNA target predictions. Nat Genet 37(5):495–500
40. Pollard JW (2008) Macrophages define the invasive microenvironment in breast cancer. J Leukoc Biol 84(3):623–630
41. Leek RD, Harris AL (2002) Tumor-associated macrophages in breast cancer. J Mammary Gland Biol Neoplasia 7(2):177–189
42. Yang M, Chen J, Su F et al (2011) Microvesicles secreted by macrophages shuttle invasion-potentiating microRNAs into breast cancer cells. Mol Cancer 10:117
43. Zhang L, Hou D, Chen X et al (2012) Exogenous plant MIR168a specifically targets mammalian LDLRAP1: evidence of cross-kingdom regulation by microRNA. Cell Res 22(1):541–552
44. Boyerinas B, Park SM, Hau A et al (2010) The role of let-7 in cell differentiation and cancer. Endocr Relat Cancer 17(1):F19–F36
45. Ohshima K, Inoue K, Fujiwara A et al (2010) Let-7 microRNA family is selectively secreted into the extracellular environment via exosomes in a metastatic gastric cancer cell line. PLoS One 5(10):e13247
46. Silva J, Garcia V, Zaballos A et al (2011) Vesicle-related microRNAs in plasma of nonsmall cell lung cancer patients and correlation with survival. Eur Respir J 37(3):617–623
47. Caby MP, Lankar D, Vincendeau-Scherrer C et al (2005) Exosomal-like vesicles are present in human blood plasma. Int Immunol 17(7):879–887
48. Pisitkun T, Shen RF, Knepper MA (2004) Identification and proteomic profiling of exosomes in human urine. Proc Natl Acad Sci U S A 101(36):13368–13373
49. Admyre C, Johansson SM, Qazi KR et al (2007) Exosomes with immune modulatory features are present in human breast milk. J Immunol 179(3):1969–1978
50. Admyre C, Grunewald J, Thyberg J et al (2003) Exosomes with major histocompatibility complex class II and co-stimulatory molecules are present in human BAL fluid. Eur Respir J 22(4):578–583
51. Michael A, Bajracharya SD, Yuen PS et al (2010) Exosomes from human saliva as a source of microRNA biomarkers. Oral Dis 16(1):34–38
52. Alvarez-Erviti L, Seow Y, Yin H et al (2011) Delivery of siRNA to the mouse brain by systemic injection of targeted exosomes. Nat Biotechnol 29(4):341–345
53. Sahoo S, Klychko E, Thorne T et al (2011) Exosomes from human CD34(+) stem cells mediate their proangiogenic paracrine activity. Circ Res 109(7):724–728

Chapter 3
Exosomal Lipids in Cell–Cell Communication

Michel Record

Abstract Exosomes feature a high protein-to-lipid ratio as reported by Subra et al. [1]. However, the lipid phase and lipid mediators participate in biological activity of these vesicles. The lipid composition, membrane organization, metabolic pathways operating within exosomes, and the effects of exosomal bioactive lipids in target cells are described.

3.1 Exosome-Mediated Intercellular Communication Involves Lipidic Molecules

Exosomes released by cells can be internalized by other cells and participate in intercellular signaling by means of the subset of the numerous proteins and lipids they contain [2]. Peripheral proteins such as LFA1 and ICAM1 or heat shock proteins (HSP) such as Hsp 70 and 90, which interact with receptors such as CD91 participate in recognition of exosomes by target cells [3]. CD91, which belongs to a low-density lipoprotein receptor-related protein (LRP) receptor family can mediate further endocytosis of the exosomes. Engagement of this type of receptor also regulates some pathways of cell lipid metabolism such as phospholipase A2 activity and cholesterol efflux via ABCA1 transporters [4].

Among lipid molecules, exosomes contain phosphatidylserine (PS) on their outer surface due to the lack of phospholipid transmembrane asymmetry [5]. PS can interact with a receptor, TIM-4 [6], allowing internalization of exosomes with accumulation of cholesterol in the recipient cell [7]. Exosomes are also enriched in lysophosphatidylcholine (LPC), which can bind IgM-type immunoglobulins on peripheral apoptotic cells, allowing exosomes uptake by these cells and their elimination during the apoptotic cell degradation [8]. This LPC-mediated exosomes uptake accounts for the clearance of exosomes from the circulating network. On the reverse, immunocompetent exosomes can stay for a couple of weeks in lymph nodes [9], and their

M. Record (✉)
Team "Sterol Metabolism and Therapeutic Innovation in Oncology,"
Cancer Research Center of Toulouse, UMR1037 INSERM-Université Toulouse,
3-CNRS ERL 5294, BP3028, CHU Purpan, Toulouse 31300, France
e-mail: michel.record@inserm.fr

H.-G. Zhang (ed.), *Emerging Concepts of Tumor Exosome–Mediated Cell–Cell Communication,* DOI 10.1007/978-1-4614-3697-3_3,

content in LPC might favor dendritic cell maturation and antigen presentation. Thus, lipidic molecules participate in exosomes cell-to-cell communication.

3.2 Exosomal Membrane Lipids and Interaction with Target Cells

Exosome-mediated cell-to-cell communication appears to involve three different processes on target cells, that is: (1) remaining on the cell surface, (2) fusing with the plasma membrane, and (3) be internalized by endocytosis processes involving either receptor-mediated endocytosis, phagocytosis, or macropinocytosis [3].

To remain on the target cell periphery and cover the cell surface, exosomes must not fuse with the cell membrane, which is consistent with the observation that the lipid phase of exosomes is more rigid than a typical plasma membrane, at least at neutral pH [5]. The high content of exosomal membrane in sphingomyelin and cholesterol (Table 3.1, Fig. 3.1) and in saturated fatty acids, either in a free form or incorporated into phospholipids, accounts for the rigidity of the exosomal membrane. About 20 % of the major phospholipids from mast cell exosomes contain only palmitic acid (16:0) and stearic acid (18:0) [5]. Those fatty acids termed "saturated" because of the lack of double bonds in their structure make a gel phase at 37 °C when included into phospholipids.

The fluidity of the lipid phase is the key event of membrane fusion. At the fusion-initiation point, proteins deviate from each other allowing the formation of large areas of only lipidic molecules, which will undergo interdigitation to trigger membrane lipid mixing. For membrane fusion, each membrane must exhibit a similar fluidity, which was observed only when exosomes are in an acidic pH environment [5], similar to that of late endosomes (pH 5.5). Consistent with this observation, fusion between exosomes and a target cell membrane was reported at the acidic pH of a tumor environment [13].

Exosomal internalization operates by various processes, and among the receptors involved in exosomal endocytosis, receptors of the TIM family, which binds the phospholipid PS, have been involved in T cell exosomes uptake by monocytes. This uptake of exosomes leads to the accumulation of cholesterol in monocytes with phenotypic alterations of the cells [7].

Exosomal fate in target cells might depend upon whether exosomes circulate as a single entity or as aggregates. In many situations, exosomes appear to accumulate outside the producing cells as aggregates [14], in a way that recall the organization of some retroviruses, which bud out the cells arranged in a "biofilm" [15]. Exosomes and, for instance, HIV feature similitudes in size and share a common step in their biogenesis via the MVB compartment. One can conceive that, similarly to retroviruses, exosomes aggregation favors their spreading. Formation of exosomes aggregates might be subsequent to calcium present in physiological media and which can create bridges between anionic phospholipids such as PS or phosphatidic acid (PA) from several single exosomes.

Table 3.1 Exosomal lipids. Lipid composition of exosomes from various sources. Data from the literature of exosomal lipid composition from reticulocytes at various maturation steps [10], mast cells [5], immunocompetent cells (B cell [11], dendritic cells [5]), and from central nervous system (oligodendrocytes [12]) are reported

Lipid		B cells	Mast cells	Dendritic cells	Reticulocytes		Oligodendrocytes
Family	Class				Day 2[a]	Day 7[a]	
		Total percentage (fold increase/parental cells)					
Neutral lipids							
Sterols	Free cholesterol	42.1 (2.14)	14.1 (1)		41.4 (0.92)	45.9 (1)	2.2
	Esterified cholesterol				8.5 (1.7)	3.5 (0.7)	
Diglycerides	Diglycerides		5.6 (0.5)				
Ceramides	Ceramides				0.3	0.8	3.2
	Hexosylceramide						2.2
Phospholipids							
Sphingolipides	Sphingomyelin	23 (1.8)	12 (2.8)	20 (2.2)	10	6.2	15 (1.5)
Glycerophospholipids							
	Phosphatidylcholine	20.3 (0.6)	21 + 5.2[b] (0.66)	26 (0.6)	25.2	25.3	40 (0.66)
	Lysophosphatidylcholine		5.6 (1.4)	9 (0.7)			
	Phosphatidylethanolamine	14.6 (0.53)	18 + 3.5[b] (1.08)	26 (1.13)	9.6	16	20 (1)
	Phosphatidylserine + phosphatidylinositol			19 (1.6)	5.0	2.3	
	Phosphatidylserine only		7.4 (1.6)				25 (3)
	Phosphatidylinositol only		7 (1.5)				nd
Polyglycerophospholipid	Bis(monoacylglycero)phosphate (BMP, also called LBPA)		0.6 (1)				
Total		100	100	100	100	100	100

Total lipid content was normalized to 100 %. The ratio between exosomal lipids and parental cell lipids is plotted between parenthesis. Lipids are listed as neutral lipids and phosphate-containing lipids (phospholipids). Backbone of phospholipids is either sphingosine (sphingomyelins) or one glycerol molecule (glycerophospholipids), or two glycerol molecules (polyglycerophospholipids)

[a]Refers to the period of cell maturation

[b]Refers to the amount of disaturated species in PC (5.2 %) and PE (3.5 %) classes (% of each total classes)

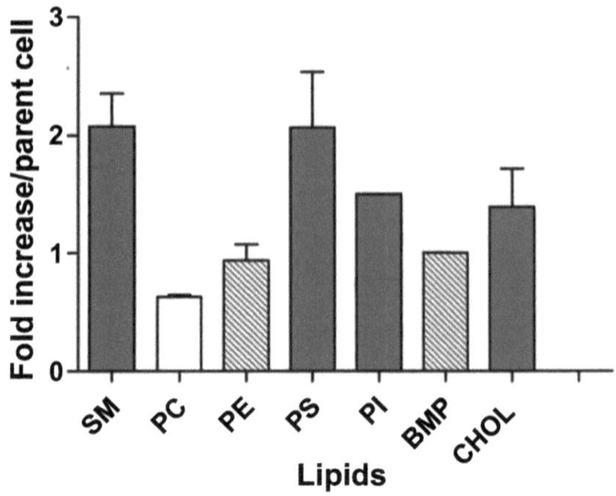

Fig. 3.1 Typical exosome lipid composition ratios between exosomes and parent cell phospholipid contents (data taken from Table 3.1). *Black histograms* represent ratios *above* 1, *dashed histograms* ratios of 1, and *clear histograms* ratio below 1. *SM* sphingomyelin, *PC* phosphatidylcholine, *PE* phosphatidylethanolamine, *PS* phosphatidylserine, *PI* phosphatidylinositol, *BMP* Bis(Monoacylglycero)phosphate, *CHOL* free cholesterol

3.3 Lipids in Exosomal Membrane Organization

3.3.1 Lipid Composition of Exosomes

As compared with parental cells, exosomal content increases in sphingomyelin (SM), PS, phosphatidylinositol (PI), and free cholesterol (CHOL), whereas phosphatidylethanolamine (PE) and bis(monoacylglycero)phosphate (BMP) stay constant, and phosphatidylcholine (PC) is decreased by two (Fig. 3.1). On average, there is four times more sphingomyelin than phosphatidylcholine in exosomes, which is the opposite of parental cells. In mammalian cells, there is about three times more phosphatidylcholine than sphingomyelin. Another reverse balance could occur between diglycerides (DG) and ceramides, but more analysis is required to establish that observation. This reverse balance between choline-containing phospholipids (SM and PC) is typical of exosomes and monitors an active phospholipid sorting during exosomes biogenesis from the MVB-limiting membrane.

The increase, both in sphingomyelin and cholesterol, suggests the formation of rafts, which have been in fact isolated from exosomal membrane [16]. However, to assimilate the whole exosomes as a raft, as reported in some works, is abusive since other lipids account for the membrane composition.

3.3.2 Role of Membrane Lipid Organization in Exosomes Interaction with Target Cells

Nuclear magnetic resonance (NMR) analysis has shown that exosomal membrane is bilayer, which displays an apparent asymmetry in the transmembrane distribution of phospholipids [5]. The difference in the nature of phospholipids between the two

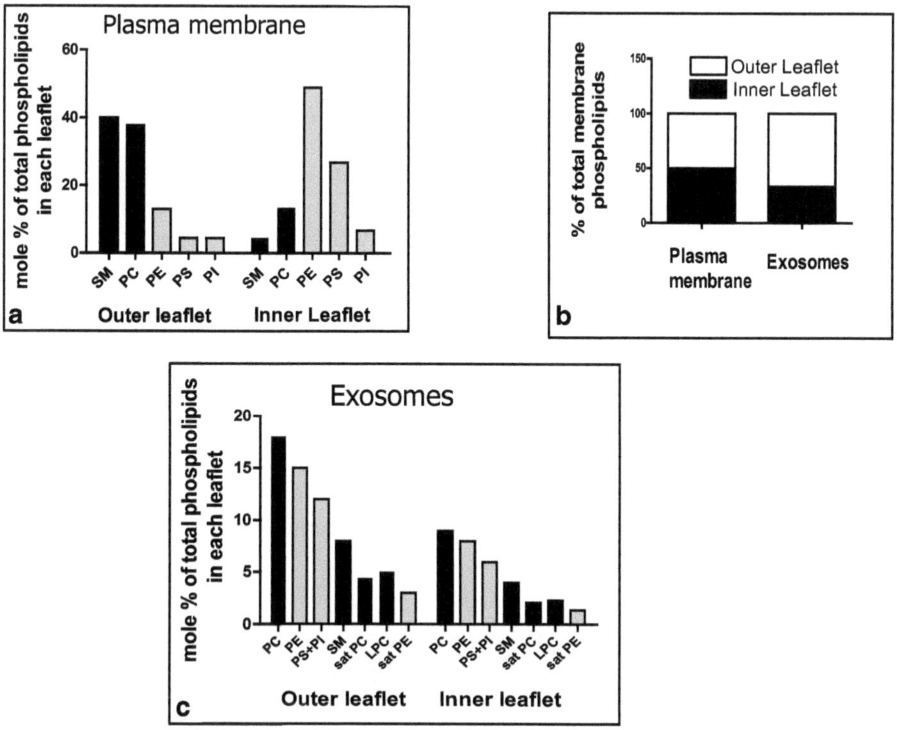

Fig. 3.2 Comparative phospholipid composition of each leaflet of a typical plasma membrane and the exosomal membrane. **a** Typical phospholipid composition of each leaflet of the erythrocyte plasma membrane. The strong difference in composition between leaflets is termed "transmembrane" asymmetry of phospholipids. *In black*, choline-containing phospholipids; *in gray*, aminophospholipids and phosphatidylinositols. **b** Relative amount of phospholipids between the two leaflets of an erythrocyte plasma membrane and the two leaflets of the exosomal membrane. **c** Phospholipid composition of each leaflet of the exosomal membrane. *In black*, choline-containing phospholipids; *in gray*, aminophospholipids and phosphatidylinositols. *SatPC*, *SatPE* saturated fatty acid-containing phosphatidylcholines and phosphatidylethanolamines, respectively

leaflets of a cellular membrane is physiologically relevant since, for instance, the amount of PS on the outer leaflet during platelet activation account for their aggregation capability [17]. On the contrary, the presence of phosphoinositides in the inner leaflet of a cell membrane favors their involvement in multiple intracellular signaling pathways [17]. The canonical transmembrane distribution of phospholipids in mammalian plasma membrane was established in human erythrocyte [18] and is reported in Fig. 3.2a. The outer leaflet is enriched in zwitterionic choline-containing phospholipid whereas the inner leaflet contains mainly charged phospholipids, namely aminophospholipids (phosphatidylethanolamine, phosphatidylserine) and phosphoinositides. Aminophospholipids are maintained in the inner leaflet by aminotranslocases (flipases) and are anchored on cytoskeleton proteins. Upon cell activation, a phospholipid scramblase will allow phospholipid mixing between the two membrane leaflets, disrupting the transmembrane asymmetry and allowing PS exposure on the outer leaflet of the cell membrane [17].

In plasma membrane, the relative amount of phospholipids in each leaflet is 50 % [17]. Instead in exosomes, because of the strong membrane curvature, the outer leaflet contains more phospholipids than the inner leaflet. Considering an exosomes of 60 nm as diameter, it was calculated that two-third of the total phospholipids were in the outer leaflet (Fig. 3.2b) [5]. This leads to an apparent transmembrane asymmetry of PE based on their chemical reactivity to an impermeant reagent [5]. In addition, it was demonstrated that exosomes contain a phospholipid scramblase [19], a lipid transporter involved in the lipid mixing between the two leaflets. Although the presence of actin in exosomes has been reported, there is no organized cytoskeleton in these vesicles. Altogether, these observations suggest the lack of a qualitative transmembrane asymmetry of phospholipids in the exosomal membrane, but instead a quantitative asymmetry subsequent to a higher amount of phospholipids in the outer leaflet, as shown in Fig. 3.2c. Subsequently, PS is present in the outer layer of the exosomes, although not in a higher amount as compared with the typical plasma membrane of erythrocytes (see Figs. 3.2a–c). Because of the membrane curvature, PS in the exosomal outer leaflet (Fig. 3.2c) are in enough quantity to become accessible to annexin V labeling and interaction with the PS receptor on target cells.

3.4 Exosomes are Both Intercellular Carriers of Bioactive Lipids and an Autonomous Lipid Metabolism Entity

Exosomes contain bioactive lipids such as prostaglandins and leukotrienes, which can originate from the parent cells, but they also contain many active enzymes of lipid metabolism able to generate those bioactive lipids . Phospholipids and sterols reported in Table 3.1 as constituent of the exosomal membrane can be converted into bioactive lipids via specific enzymes. Exosomes thus appear as carriers of lipidic molecules but also as a functional unit of lipid metabolism.

3.4.1 The Phospholipase A2 and the Transcellular Metabolism of Eicosanoids via Cyclooxygenase and Lipoxygenase Pathways

Glycerophospholipids hydrolyzed by a phospholipase A2 (PLA2) release free fatty acids such as arachidonic acid (AA) and a lysophospholipid named LPC. AA can be processed into eicosanoids, which divide into leukotrienes and prostaglandins (see Fig. 3.3).

The 5-lipoxygenase (5-LOX) transforms arachidonic acid into another set of oxidized derivatives called leukotrienes. LTB4 is involved in inflammation and promotes cancer progression. LTC4 and LTD4 promote angiogenesis [20].

Exosomes contain some of the enzymes involved in the eicosanoid pathway (Fig. 3.3) [19, 21]. The three classes of phospholipases A2 (calcium-dependent

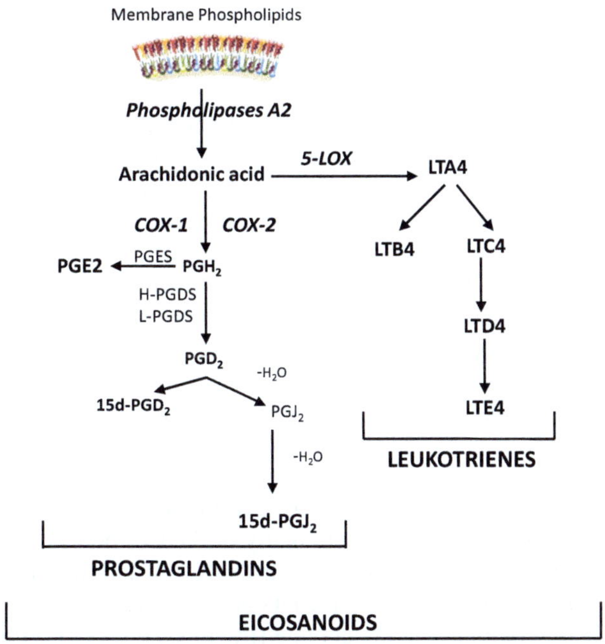

Fig. 3.3 Eicosanoid bioactive lipids derived from arachidonic acid. Phospholipases A2 hydrolyze mainly phosphatidylcholine to release free arachidonic acid, further transformed by the constitutive (Cox1) or inducible (Cox2) cyclooxygenases. Prostaglandin E2 (PGE2) synthesized by PGE synthase (PGES) is proinflammatory and immunosuppressive, whereas the 15deoxy-prostaglandin J2 (15d-PGJ2) is antiinflammatory and provokes tumor regression. *PGDS* PGD synthases. (For review see [20])

cPLA2, calcium-independent iPLA2, secreted sPLA2) are present as well as the cyclooxygenases COX, involved in the first step of transformation of arachidonic acid into the prostaglandin PGH2. Also, the presence of PGE synthase has been reported [22]. The two prostaglandins PGE2 and 15d-PGJ2 were detected in similar amounts in rat basophil leukemia-derived exosomes [19]. Enzymes for the other pathway of AA processing, i.e., for leukotriene biosynthesis, are also embedded in exosomes since incubation with the leukotriene intermediate LTA4 leads to the formation of LTC4 as the major product [21]. The enzyme LTC4 synthase is indeed present in exosomes, as well as the LTA4 hydrolase, which generates the LTB4 compound [21]. However, the 5-lipoxygenase (Fig. 3.3) appears not to be present in exosomes since leukotrienes are not detected from arachidonic acid supplied to leukocyte-derived exosomes, but detected when arachidonic acid is supplied to leukocytic cells. This indicates that exosomes participate in the transcellular metabolism of leukotrienes [23], in which some biosynthesis steps are located in leukocytes and the others in endothelial cells, to prevent uncontrolled formation of inflammation mediators. Exosomes might also participate in the transcellular metabolism of prostaglandins via exosomes subpopulations, which could carry various subsets of prostaglandin-related enzymes between cells [24].

The expression level of enzymes involved in eicosanoid metabolism appear to vary between exosome-producing cells, but this might also depend on which stages of cell proliferation or differentiation the exosomes are recovered, since in the case of reticulocyte-derived exosomes their protein and lipid content vary according to the differentiation stages of parent cells [10].

3.4.2 The Phospholipase D/Diglyceride Kinase/Phosphatidate Phosphatase Pathway in Exosomes

Some of the phospholipases A2 present in exosomes can hydrolyze PA into lysophosphatidic acid (LPA). Both are mitogenic compounds, PA activating the mTOR machinery [25], while LPA triggers cell proliferation via a G protein-coupled receptor [26]. PA is enriched in exosomes (Laulagnier K, Record M unpublished data) and this could result from the activation of exosomal phospholipase D by monomeric G proteins. Exosomes are highly enriched in active GTPases [19]. Subsequently, PA can be hydrolyzed into diglycerides by a phosphatidate phosphatase (PAP), also called lysophospholipid phosphatase (LPP1, see Fig. 3.4) [26]. PAP is commonly associated to PLD activity since both PA and DG compounds are produced upon PLD activation. In addition, PLD2 itself triggers exosomes release [27]. PAP (LPP1) is present in exosomes and PAP has also been characterized as a coactivator of the nuclear receptor PPARα [28].

However, PA can also derive from the conversion of DG by a diglyceride kinase (Fig. 3.4) [29]. The functionality of the two pools of PA, which derived from PLD activity and that generated via DGK, appears different. They differ in their fatty acid composition. Also, the DG pools formed by the PAP/LPP1 or phospholipase C acting on phosphatidylcholine (PC-PLC; Fig. 3.4) differ in the nature of their fatty acid content, and differ from the fatty acid composition of DG generated by the canonical phospholipase C hydrolyzing phosphatidylinositols.

3.4.3 The Bis(Monoacylglycero)Phosphate/Cholesterol Pathway

The combination of phospholipase D, phospholipase C (PC-PLC), and phospholipases A2 activities in exosomes can be involved in the biosynthesis of LBPA, whose correct nomenclature name is Bis(Monoacylglycero)Phosphate (BMP; Fig. 3.4). It has been shown that both phospholipase D and phospholipases A2 are activated by GTPases of the Ras superfamily in exosomes [19], and such activation could occur inside recipient late endosomes in which exogenous exosomes are internalized. Late endosomes including multivesicular bodies feature an acid pH and the Ras-GDP-Mg complex starts to unfold at acidic pH [30, 31], allowing exchange between GDP and GTP with subsequent activation of the GTPase. Phospholipases D, C, and A2 also operate at acid pH and their combined activity for generating BMP precursors is documented [32]. BMP is fusogenic in acidic conditions [33], and the exosomal phospholipase-mediated BMP biosynthesis would allow exogenous exosomes to fuse with the limiting membrane of a recipient endosome. This fusion process will then trigger the release of exosome-vehiculated material from a "cell A" into the cytosol of a "cell B," therefore regulating a key step in the cell-to-cell communication. This fusion will be facilitated by the presence of diglycerides in exosomes since this lipid lowers the lateral membrane pressure of phospholipids and favors intermembrane lipid mixing [34].

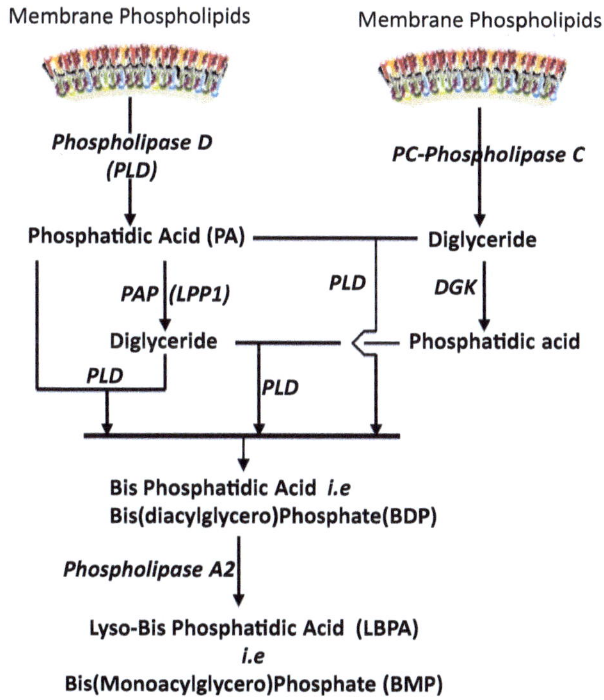

Fig. 3.4 Bioactive lipids derived from phopholipases D and C activities. The two isoforms of phospholipases D (PLD1 and PLD2) hydrolyze phosphatidylcholine (PC) into the mitogenic compound phosphatidic acid (PA). PA is a second messenger interacting with raf and mTOR. PA is partly converted into diglyceride by a PA phosphatase. Diglycerides (DG) can also originate from the activity of a phospholipase C acting on phosphatidylinositols but also on phosphatidylcholines (PC-Phospholipase C), and converted into PA by means of a DG kinase (DGK). Enzymes involved in these pathways have been identified in exosomes. PLDs display a second enzymatic activity, termed transphosphatidylation, catalyzing the nucleophilic attack of the phosphate moiety of PA by a primary alcohol such as a diglyceride. Transphosphatidylation allows the formation of a polyglycerophospholipid termed BisDiacylglyceroPhosphate, a precursor of BMP (LBPA) involved in MVB and intralumenal vesicle(ILV) formation. This part of the metabolic scheme occurs in the limiting membrane of the MVB, and BMP allows the ILV formation. ILVs will be named "exosomes" once released out of the cell

It is worthy to note that there is a functional link between BMP and cholesterol [35]. The metabolism of these two compounds is completely unrelated, but accumulation of cholesterol into MVB leads to an enrichment of BMP in the same cell compartment [36]. The supply of the BMP precursor phosphatidylglycerol (PG) by internalization of PG-containing liposomes augment BMP content in MVB [37], suggesting that the final steps of biosynthesis of this phospholipid occur in the MVB. Using the same approach, the supply of liposome-containing BMP into MVBs regulate cholesterol endosomal level, under the control of Alix, a BMP-interacting protein involved in the formation of exosomes from the MVB membrane [38]. Therefore, it is conceivable that BMP triggers the enrichment of cholesterol reported in exosomes (Fig. 3.1). The balance between cholesterol and BMP appears critical in lipid pathologies related to cholesterol accumulation into MVBs, such as the Niemann-Pick type C (NPC) disease [38]. This pathology can be mimicked by incubating cells with U18666 A, an amphiphilic aminosterol. In NPC cells, addition of BMP-enriched liposomes enhances exosome biogenesis and their release allows cholesterol elimination from MVBs [38].

Exosomes as a unit of lipid metabolism is now documented regarding phospholipid/fatty acid-derived bioactive lipids, but whether exosomes exhibit some sterol metabolism requires further studies. Exosomes are enriched in cholesterol content as compared with the parental cells (Fig. 3.1) and also contain some cholesterol esters [10]. Cholesterol esters are involved in cell growth and invasion [39]. A summary of bioactive activity of lipids borne by exosomes is reported in Table 3.2.

Although exosomes can potentially produce many of the bioactive lipids mentioned, the amount of bioactive compounds they transport appears higher than exosomal biosynthesis capacity, which is limited by the amount of metabolic precursors they carry, such as arachidonic acid. Besides their lipid biosynthetic properties, exosomes have to be considered as transporters of parental cell lipid mediators to target cells.

Altogether, the role of lipids in exosome-mediated cell-to-cell communication is exemplified, for instance, in the observation that exosomes from T lymphocytes trigger cholesterol accumulation into monocytes via the phosphatidylserine receptor [7].

3.5 Functional Effects of Exosomal Lipid Transfer Between Cells

3.5.1 Exosomes as a Concentrate of Molecules Potentially Involved in Cell-to-Cell Communication

From about 200 studies performed so far on exosomes, proteomic analysis have identified 4,049 proteins, and transcriptomic studies unveiled the presence of 1,639 mRNA and 764 miRNA [40]. Extensive lipidomic analysis have not been performed so far. An average of 700 nmol phospholipids/mg protein was measured in exosomes [5]. The diversity of phospholipid molecules is in the range of 250 molecular species, from mouse liver or brain, identified within six different classes of phospholipids [41]. Molecular species of free fatty acids [19] and sterols metabolites account for an additional diversity. Thus, exosomes contain a high number of potentially signaling molecules, and can trigger a large diversity of signals as compared with single molecule agonists. Exosomes behave as a multisignaling device.

Table 3.2 Primary and secondary bioactive lipids in exosomes

Parental lipids	Primary bioactive lipids	Secondary bioactive lipids	Functional properties
Cholesterol (a)	Cholesterol esters (b)		Raft formation (a) Proliferative (b)
Sphingomyelin (a)	Ceramide (b)	Sphingosine-phosphate (c)	Raft formation (a) Trigger apoptosis (b) Proliferative (c)
Phosphatidylcholine (a)	Arachidonic acid (b)	Eicosanoids (prostaglandins, leukotrienes) (c)	Inflammation (c) Cancer (c)
Phosphatidyethanolamine (a)	Docosahexaenoic acid (b)		Antioxidant (b) Many positive beneficial effects in health (b)
Phosphatidylinositol (a)	Diglyceride (b)		Activates protein kinases C (b)
Phosphatidylcholine (a)	Diglyceride (b)	Phosphatidic acid (PA) (c)	Favor membrane fusion (c) Proliferative (c)
Phosphatidylcholine (a)	Lysophosphatidylcholine (b)		Allows exosome recognition (b)
	Phosphatidic acid (b)	Lysophosphophatidic acid (LPA) (c) Diglyceride (DG) (d)	Mitogenic compounds (c) Sustained Protein kinases C activation (d)
Phosphatidylglycerol (a)	Bis(monoacylglycero)phosphate (BMP) (b)		Allows membrane fusion (b) Exosome biogenesis (b)

Primary bioactive lipids (b) derive from parental lipids (a) and secondary bioactive lipids (c) from primary ones (b). Phosphatidic acid (PA) can generate either LPA (c) or diglycerides (d) through activities of phospholipases A2 or PA phosphatase respectively

Because of their high number, the functional role of most of the molecules embedded in exosomes has not been experimentally established, but one can reasonably propose that they trigger the effect(s) they are known for, when taken individually. The requirement of numerous researches to be conducted to answer that point makes exosomes as a fascinating field.

Functional effects of exosomal lipids in cell-to-cell communication will also depend upon which part of the target cells they interact with. It has been shown that exosomes can first interact with the plasma membrane for some period of time and then be internalized via various internalization processes [19, 42]. Thus, exosomes feature the particularity to potentially trigger functional effects both at the periphery and inside of a target cell.

3.5.2 Interaction of Exosomal Lipids with Target Cell Periphery

As already mentioned, exosomal PS binds to specific peripheral receptors [6] and LPC is the agonist of the G2A receptor coupled to G proteins [43]. Among the secondary bioactive lipids carried by exosomes (Table 3.2), eicosanoids can potentially interact with their respective GPCR [20], except the prostaglandin 15d-PGJ2, which binds the nuclear receptor PPARγ.

3.5.2.1 PS and TIM Receptors

Exosomal PS can bind TIM-1 and TIM-4 receptors [6]. TIM stands for T cell/transmembrane, immunoglobulin, and mucin, indicative of the various structure domains of these receptors. TIM receptors are a family of three proteins in humans (TIM-1, TIM-3, TIM-4), which play a critical role in regulating the immune response [44]. TIM-1 functions as costimulator for T cell activation whereas TIM-3 triggers apoptosis of Th1 cells. TIM-4 is exclusively expressed in antigen-presenting cells (APC). It has been proposed that PS on exosomes can bind separately TIM-1 and TIM-4 receptors, forming a bridge, which could make closer APC and T cells [44]. PS-dependent endocytosis allows immunocompetent exosomes to be internalized and to enhance the presenting capacity of APCs.

3.5.2.2 LPC and its G Protein-Coupled Receptor G2A

Exosomes feature a substantial content in LPC [5]. Whether exosomal LPC binds the G2A receptor on target cell remains to be established but if so, this interaction could trigger some of the positive effects of exosomes in the immune system. LPC is a chemoattractant for lymphocytes [45] and induces the maturation of dendritic cells, and one can consider that immunocompetent exosomes, which accumulate in lymph nodes [9], could enhance antigen presentation via LPC-mediated processes.

Consistently, the secreted phospholipase A2 IIa, which generates LPC, has been shown to induce dendritic cell maturation [46], and precisely the sPLA2 IIa is borne by exosomes [19].

3.5.2.3 Exosomal PS and LPC as an "Eat-Me" Signal

The steady state of efficient exosomes in either their positive effect (i.e., enhancing immune response, reticulocyte maturation) or their negative effects (i.e., promoting tumor growth) depends upon their amount in the intercellular space as well as in biological fluids. When considering quantitatively the daily reticulocyte maturation in the human body, it was calculated that the amount of exosomes released would be in the range of several kilograms per day [2], suggesting the presence of an efficient elimination process in the organism. Besides being involved in exosomal functional effects, the two phospholipids PS and LPC also participate in their elimination by triggering an "eat-me" signal to phagocytes.

Indeed, TIM-4 is also involved in the engulfment of apoptotic cells and can mediate that way the phagocytosis of exosomes by macrophages [6]. In the case of reticulocyte exosomes, LPC binds immunoglobulins of IgM type, which subsequently activate the complement factors such as C1q, which then allows C3b/bi deposition on the exosomal surface [8]. *In fine,* exosomal membrane becomes recognizable by complement receptors CR3 and CR4 receptors expressed on macrophages and exosomes are further eliminated by the spleen. LPC can originate from the activity of an exosomal calcium-independent phospholipase A2 (iPLA2), which is activated by reactive oxygen species and caspase 3 [8]. An iPLA2 processed by caspase 3 has also been detected in rat basophil leukemia cells (RBL-2H3)-derived exosomes [19].

3.5.2.4 Eicosanoids Receptors

GPCR Leukotrienes (Fig. 3.3) LTC4, LTD4, and LTE4 interact with Gi-coupled receptors termed CysLT1 and CystLT2 receptors, the latter working through p38MAPK-dependent pathway. LTB4 interacts with a specific receptor, BLT1, also coupled to Gi.

The prostaglandin PGE2 (Fig. 3.3) initiates diverse array of signaling pathways [47] related to the existence of four GPCRs, EP1, EP2, EP3, and EP4 receptors. The EP1 and EP3 receptors stimulate calcium signaling through Gq- and Gi-mediated pathways, respectively, whereas the EP2 and EP4 receptors use Gs proteins to stimulate adenylate cyclase and induce an increase in cAMP.

PPARγ The downstream product of the prostaglandin pathway (Fig. 3.3), 15d-PGJ2 has been established as the ligand of the nuclear receptor PPARγ, and is the most potent endogenous ligand [48]. Other natural ligands of PPARγ include specific fatty acids (arachidonic acid complexed to FABP), components of oxidized low-density lipoproteins (9-HODE, 13-HODE), and the lysophosphatidic acid (LPA, *see*

Table 3.2). PPARγ activation induces apoptosis and inhibits cancer cell proliferation and angiogenesis. The 15d-PGJ2 prostaglandin displays antiinflammatory properties.

3.5.3 *Internalization of Exosomal Lipids into Intracellular Target Cell Compartments: Exosomes as a New Type of Lipid Carrier*

It has been shown that exogenous exosomes can be internalized by cells via various receptors such as the LRP1, i.e., CD91, which binds the Heat Shock Proteins Hsp90/Hsp 70, a chaperon protein enriched in exosomes [49]. This process allows exosomes to transport the bioactive lipids they contain inside cells. So far the transport of these lipids has been assigned to soluble carrier proteins such as serum albumin, or a plasma membrane transporter, but not all bioactive lipids have been attributed to specific transporters. In the case of exosome endocytosis, many bioactive lipid molecules, such as fatty acids or eicosanoids inserted into the exosome membrane, or bound to proteins in the lumen of the vesicle, can enter the cell at the same time. Regarding 15d-PGJ2, exosomes are the first transport system described, which can drive this prostaglandin through the plasma membrane, and then channel the compound to the cytosol to interact with its receptor, PPARγ. Then, the complex 15d-PGJ2/PPARγ can move to the nucleus to regulate transcription.

It has to be borne in mind that uptake of exosomes also brings inside the cell prostaglandins or leukotrienes whose signaling normally occurs on the cell periphery via GPCRs. A functional role of intracellular PGE2 and leukotrienes remains hypothetical. Only one report emphasizes a possible activity of intracellular PGE2 to induce a Bax-dependent apoptosis in a Bax-expressing subset of glioblastoma tumors [50]. In these experiments, PGE2 was microinjected. Uptake of PGE2-containing exosomes by cells provides a natural way to supply intracellular PGE2.

However, the normal fate of PGE2 is to be secreted via a plasma membrane transporter [51] and PGE2 triggers a coordinated immunosuppression allowing tumor development [20]. Exosomes derived from murine mammary adenocarcinoma cell lines (TS/A; 4T-1) and the respective tumors contain PGE2 whose amount is correlated with the induction of myeloid-derived suppressor cells (MDSC)[52]. These cells downregulate T cell functions in vivo. Inactivating exosomal PGE2 by a specific antibody partly reverses MDSC proliferation [52]. Recently, a cell death-induced tumor repopulation pathway including caspase 3/calcium-independent phospholipase A2 (iPLA2)/PGE2 has been proposed as a mechanism for tumor recurrence following radiotherapy [53]. Precisely, all the components of this pathway (caspase3-activated iPLA2, arachidonic acid, PGE2) have been found in exosomes [19]. Thus, the balance between the immunosuppressive PGE2 and the antitumoral 15d-PGJ2 in tumor exosomes is a key point for their outcome in tumor development.

Precursor of eicosanoids is the fatty acid arachidonic acid. Fatty acids are normally transported in the circulation bound to carrier proteins such as serum albumin; they next become inserted into the cell plasma membrane from where they are taken up by intracellular fatty acid-binding proteins (FABP) and dispatched inside the cell. Exosomes appear as another mechanism to make fatty acids crossing the plasma membrane by means of an endocytosis process. Exosomes are a reservoir of a large panel of free fatty acids accumulated in their membrane or bound to protein carrier present in the lumen of the vesicles [19]. Exosomes transport mostly saturated fatty acids, but also monounsaturated and polyunsaturated ones. Among those unsaturated species, arachidonic acid is one of the most represented, but the docosahexaenoic acid (DHA) is also present. Exosomal fatty acid concentrations are elevated and overcome the capacity of the exosomes itself to generate fatty acid from its own phospholids, indicating that part of these fatty acids originate from the parent cell [19].

3.5.4 Exosomes Allow Concentration of Bioactive Lipids into Recipient Cells

Noteworthy, the concentration of eicosanoids in exosomes is much higher than that of parental cells, and since exosomes accumulate into a cell compartment of recipient cells smaller than the cells themselves, exosomes allow to concentrate bioactive lipids in target cells.

Indeed, on a basis of cell-equivalent, exosomes from a rat basophil leukemia cell carry around 15 times more prostaglandins (PGE2, 15d-PGJ2) than parental cells [19]. Similarly, exosomes from dendritic cells biosynthesize five times more leukotriene LTC4 than the cells, and macrophages (MDM) release exosomes with a very strong LTC4 synthase activity, between 50 and 100 times higher than the parental cells [21].

As compared with a single agonist lipidic molecule, which would be released out of a cell, dispersed in the intercellular space and then be "dissolved" in the whole cell volume of a target cell, exosomes vectorize numerous molecules from a donor cell to a small compartment of a recipient cell, namely the endosome compartment. Internalization of exosomes into endosomes of target cell indeed concentrates the bioactive lipids they carry because of the difference in volume between the total endosome compartment and the whole cell volume [19]. At an average, there are 30 late endosomes (MVB) of 600 nm diameter in a rat basophil leukemia cell of 15 μm diameter average. The total volume of those 30 MVBs is about 500 times inferior to the whole cell volume, and subsequently the concentration of the bioactive lipids inside the MVB compartment is 500 times higher than if they had been dispersed in the whole cell volume. This concentration effect of intercellular exosomes transport is illustrated in Fig. 3.5.

A final 0.1 μM concentration of 15d-PGJ2 is fairly weak (Fig. 3.5), but a concentration of 57 μM is in the range of biological effects of the prostaglandin such as PPARγ activation related to antitumoral properties toward colorectal or breast cancer tumors [54]. Similar calculation applied to other prostaglandins and fatty acids indicate that the endosomal compartment can contain 30 μM of PGE2, 1 mM DHA and 4 mM arachidonic acid.

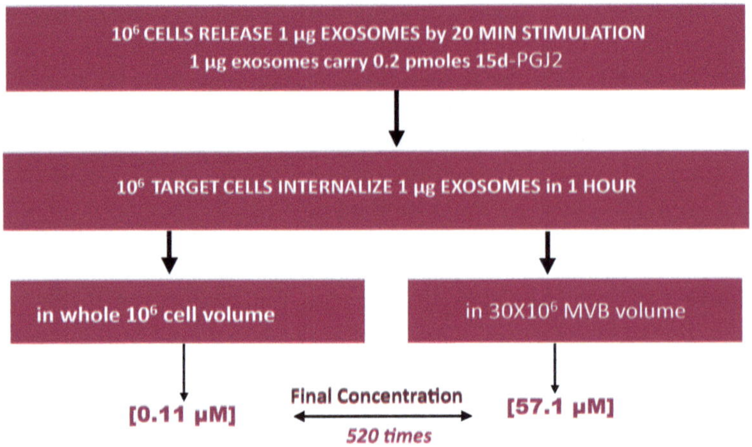

Fig. 3.5 Exosomal cell-to-cell communication concentrate bioactive lipids into endosomes of target cells. Exosome transfer 0.2 pmole of the prostaglandin 15d-PGJ2 between 10^6 cells, leading to a final concentration of 0.1 μM in the whole cell volume of recipient cells, but to a concentration of 57 μM in the volume of the whole endosomes contained in these 10^6 cells. (Data taken from [19])

3.5.5 *Endosomes-Accumulating Exosomes as a Hub to Dispatch Exosomal Lipids into Several Compartments*

Uptake of exosomes into the endocytic track of target cells has been reported in many works. Accumulation of exosomes into recipient endosomes highlights a possible role of those endosomes as a hub to dispatch bioactive molecules into several cell compartments (Fig. 3.6). In that case, exosomes have to fuse with the membrane of the endosome in which they are trapped, to release the exosomal material into the cytosol. This process appears involved for antigen cross-presentation when extra-cellular tumor antigens are vehiculated by exosomes [55]. Similarly, miRNA have to be released out of exosomes trapped into endosomes to target the transduction machinery of recipient cells. It has been reported (Table 3.2) that exosomes contain fusogenic lipids such as BMP and phosphatidic acid, together with the facilitating role of diglycerides. The pH inside endosome is 5.5 and BMP is fusogenic at that pH, while PA displays fusogenic properties in the presence of calcium. As already mentioned, the combination between exosomal phospholipases A2 and transphos-phatidylation activity of phospholipases D in an acidic environment can lead to the biosynthesis of BMP (Fig. 3.4), whereas the hydrolytic activity of PLD will generate PA (Fig. 3.4). Therefore, exosomes appear to contain the machinery allowing them to fuse with the endosome membrane and to release their material into the cytosol. Although the fusion between exosomes and the plasma membrane of a target cell occurring in neutral pH conditions cannot be rule out, this event appears as minor because of the biophysical characteristics of the exosomal membrane comparatively to the plasma membrane (Figs. 3.1 and 3.2; Chap. 2). It must be borne in mind that natural-occurring cell–cell fusion is quantitatively a rare event.

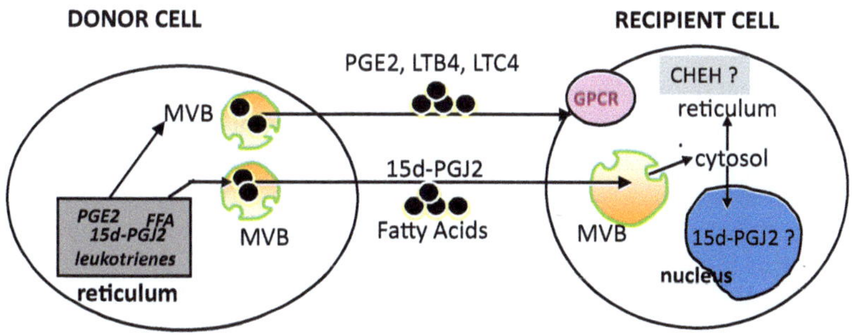

Fig. 3.6 Exosomes as cell-to-cell "signalosomes" involving bioactive lipids. Leukotrienes and prostaglandins are synthesized by enzymes in the endoplasmic reticulum of the donor cell. Those enzymes and their respective products are also channeled into exosomes, which make them as an autonomous entity. Exosomal cell-to-cell communication will then operate between MVB of donor cell toward peripheral G protein-coupled receptor of recipient cells, but also from MVB to MVB. Indirect evidences suggest that internalized exosomes in recipient MVB can fuse with the MVB membrane and deliver its content in the cytosol, with possible targeting of the released molecules to receptors or enzymes present in the endoplasmic reticulum or the nucleus. *CHEH* cholesterol epoxyde hydrolase

3.5.5.1 Adressing 15d-PGJ2 and Arachidonic Acid to the Recipient Cell Nucleus

Because of the high concentration of arachidonic acid (AA) in the exosomes, part of it should be bound to the FABP transporters also present inside the exosomes. Following exosomes fusion with the MVB membrane, this complex is released in the recipient cell cytosol where it has the possibility to interact with the receptor PPARγ present in a ligand-free state. Also, the cytosolic PPARγ receptor can directly bind to the 15d-PGJ2 present in high concentration in exosomes (Fig. 3.5). Then, both PPARγ/15d-PGJ2 and the complex FABP-AA/PPARγ can move to the nucleus where they will regulate transcription.

As shown in Fig. 3.2b, exosomes from rat basophil leukemia and from oligodendroglial cell contain high amounts of disaturated phosphatidylcholine [12, 19]. Among those species, exosomes contain substantial amount of palmitoyl-stearoyl–phosphocholine (16:0–18:0-GPC), a phospholipid, which is the agonist of another nuclear receptor, PPARα [56], whose one coactivator is the PA-phosphatase PAP (LPP1; Fig. 3.4) [28] previously mentioned to be present in exosomes. It is known that cytosols also contain a phospholipid transfer protein (PLTP), able to "extract" a specific phospholipid from a membrane and to make it available to another target. Thus, PLTP could "extract" the 16:0–18:0-GPC from the exosomal membrane and bring it to the PPARα receptor. Therefore, exosomes exhibit many possibilities to regulate nuclear receptors in target cells by means of lipidic molecules.

3.5.5.2 Addressing Fatty Acids to Enzymes in the Endoplasmic Reticulum

Although present in exosomes, the cyclooxygenases and the downstream machinery for prostaglandin synthesis are also located in the endoplamic reticulum of recipient cells. By delivering arachidonic acid in the cytosol of target cells where it can be addressed to endoplasmic reticulum, exosomes can provide additional substrate to alleviate prostaglandin synthesis in these target cells. The resulting amount of prostaglandins in these cells will be enhanced by this process. Therefore, exosomal trafficking in a closed area like tumor microenvironment might enrich that environment in the immunosuppressive prostaglandin PGE2. Indeed, exosomes isolated from a mammary tumor contain higher amounts of PGE2 than exosomes recovered from the same cells in culture [52].

Exosomes allow to accumulate millimolar concentrations of another fatty acid, DHA (22:6) into recipient endosomes. DHA is a fatty acid of the ω-3 series reported to trigger many beneficial effects in health, among which to help breast cancer recovery [57]. DHA supplied by exosomes could be vehiculated by FABP to the antiestrogen-binding site (AEBS) located in the endoplasmic reticulum, whose ligands such as DHA induce active cell death and autophagy of human cancer cells, through modulation of cholesterol metabolism [58]. The AEBS has been characterized recently as a complex of two cholesterogenic enzymes, the D8D7 isomerase and the DHCR7 reductase, whose assembly creates a third enzymatic activity, namely the cholesterol epoxyde hydrolase (CHEH; Fig. 3.6) [59], able to hydrolyze cholesterol epoxyde, an oxysterol [60]. Thus, fatty acid vehiculated between cells by exosomes could regulate some pathways of cholesterol metabolism.

3.6 The Lipid Phase of the Exosomal Membrane as a Vector of Bioactive Lipids

Whereas proteins of interest involved in exosomes functions have to be addressed to exosomes during their biogenesis in the cell, some amphiphilic molecules can be inserted into the lipid phase of the exosomal membrane during *in vitro* manipulations, and this opens the possibility to supplement exosomes with appropriated lipidic molecules. The LPC, which provokes lymphocyte chemotaxis and dendritic cells differentiation, penetrates easily into membranes. Also, prostaglandins like the 15d-PGJ2, which exhibits antitumor and antiinflammatory properties, should easily "dissolve" into the lipid phase of exosomes membrane *in vitro*, with subsequent potential therapeutic effect *in vivo*, providing that exosomal biodistribution be adequate when injected in the organism. The biodistribution of injected exosomes has not been investigated in depth so far. Exosomes are, however, able to cross the blood–brain barrier and therapeutic exosomes have been tentatively elaborated by loading curcumin in their membrane [61]. Curcumin is a lipidic molecule with polyphenolic structure and which displays antioxydant and antiinflammatory properties, and also inhibits tumor growth. Intranasal administration of curcumin-enriched exosomes allow these vesicles to reach rapidly the brain in mice and to protect them from LPS-induced brain inflammation, or to delay brain tumor growth [61].

Therefore, ex vivo manipulation of exosomes with membrane-penetrating lipids should allow to intervene in cell-to-cell communication by means of exosomal lipids.

3.7 Conclusion

Cell-to-cell lipid transfer by exosomes involves a large variety of molecules with potential functional effects, which have been reviewed in this chapter. Further studies are required to establish which of the bioactive molecules concentrated into exosomes participate in their function in vivo. Exosomes as carriers of eicosanoids and fatty acids is the most established role of exosomal lipids in cell–cell communication. This is a new pathway in the transport of these mediators, which opens new concepts to understand their role in various pathophysiologies. Exosomes provide a way to intervene in diseases. Undertaking pharmacological studies to control the relative amounts of "negative (tumor promoter)" versus "positive (immunocompetent)" exosomes is the challenge to come. In this challenge, bioactive lipids will be present since beside the eicosanoid pathways, the sterol metabolism will emerge as a source of new antitumor molecules [62]. Indeed, oxysterol derivatives are able to generate antitumor exosomes [63], and two of the seven oxysterol-binding proteins, Osh6p and Osh7p, are associated with the protein-sorting machinery for exosomal biogenesis [64].

References

1. Subra C, Laulagnier K, Perret B, Record M (2007) Exosome lipidomics unravels lipid sorting at the level of multivesicular bodies. Biochimie 89:205–212
2. Record M, Subra C, Silvente-Poirot S, Poirot M (2011) Exosomes as intercellular signalosomes and pharmacological effectors. Biochem Pharmacol 81:1171–1182
3. Thery C, Ostrowski M, Segura E (2009) Membrane vesicles as conveyors of immune responses. Nat Rev Immunol 9:581–593
4. Zhou L, Choi HY, Li WP, Xu F, Herz J (2009) LRP1 controls cPLA2 phosphorylation, ABCA1 expression and cellular cholesterol export. PLoS ONE 4:e6853
5. Laulagnier K, Motta C, Hamdi S, Roy S, Fauvelle F, Pageaux JF et al (2004) Mast cell- and dendritic cell-derived exosomes display a specific lipid composition and an unusual membrane organization. Biochem J 380:161–171
6. Miyanishi M, Tada K, Koike M, Uchiyama Y, Kitamura T, Nagata S (2007) Identification of Tim4 as a phosphatidylserine receptor. Nature 450:435–439
7. Zakharova L, Svetlova M, Fomina AF (2007) T cell exosomes induce cholesterol accumulation in human monocytes via phosphatidylserine receptor. J Cell Physiol 212:174–181
8. Blanc L, Barres C, Bette-Bobillo P, Vidal M (2007) Reticulocyte-secreted exosomes bind natural IgM antibodies: involvement of a ROS-activatable endosomal phospholipase iPLA2. Blood 110:3407–3416
9. Luketic L, Delanghe J, Sobol PT, Yang P, Frotten E, Mossman KL et al (2007) Antigen presentation by exosomes released from peptide-pulsed dendritic cells is not suppressed by the presence of active CTL. J Immunol 179:5024–5032

10. Carayon K, Chaoui K, Ronzier E, Lazar I, Bertrand-Michel J, Roques V et al (2011) The proteo-lipidic composition of exosomes changes during reticulocyte maturation. J Biol Chem 286:34426–34439
11. Wubbolts R, Leckie RS, Veenhuizen PT, Schwarzmann G, Mobius W, Hoernschemeyer J et al (2003) Proteomic and biochemical analyses of human B cell-derived exosomes. Potential implications for their function and multivesicular body formation. J Biol Chem 278:10963–10972
12. Trajkovic K, Hsu C, Chiantia S, Rajendran L, Wenzel D, Wieland F et al (2008) Ceramide triggers budding of exosome vesicles into multivesicular endosomes. Science 319:1244–1247
13. Parolini I, Federici C, Raggi C, Lugini L, Palleschi S, De Milito A et al (2009) Microenvironmental pH is a key factor for exosome traffic in tumor cells. J Biol Chem 284:34211–34222
14. Fomina AF, Deerinck TJ, Ellisman MH, Cahalan MD (2003) Regulation of membrane trafficking and subcellular organization of endocytic compartments revealed with FM1-43 in resting and activated human T cells. Exp Cell Res 291:150–166
15. Pais-Correia AM, Sachse M, Guadagnini S, Robbiati V, Lasserre R, Gessain A et al (2010) Biofilm-like extracellular viral assemblies mediate HTLV-1 cell-to-cell transmission at virological synapses. Nat Med 16:83–89
16. de Gassart A, Geminard C, Fevrier B, Raposo G, Vidal M (2003) Lipid raft-associated protein sorting in exosomes. Blood 102:4336–4344
17. Lhermusier T, Chap H, Payrastre B (2011) Platelet membrane phospholipid asymmetry: from the characterization of a scramblase activity to the identification of an essential protein mutated in Scott syndrome. J Thromb Haemost 9:1883–1891
18. Zachowski A (1993) Phospholipids in animal eukaryotic membranes: transverse asymmetry and movement. Biochem J 294 (Pt 1):1–14
19. Subra C, Grand D, Laulagnier K, Stella A, Lambeau G, Paillasse M et al (2010) Exosomes account for vesicle-mediated transcellular transport of activatable phospholipases and prostaglandins. J Lipid Res 51:2105–2120
20. Wang D, Dubois RN (2010) Eicosanoids and cancer. Nat Rev Cancer 10:181–193
21. Esser J, Gehrmann U, D'Alexandri FL, Hidalgo-Estevez AM, Wheelock CE, Scheynius A et al (2010) Exosomes from human macrophages and dendritic cells contain enzymes for leukotriene biosynthesis and promote granulocyte migration. J Allergy Clin Immunol 126:1032–1040 (40 e1–4)
22. Valadi H, Ekstrom K, Bossios A, Sjostrand M, Lee JJ, Lotvall JO (2007) Exosome-mediated transfer of mRNAs and microRNAs is a novel mechanism of genetic exchange between cells. Nat Cell Biol 9:654–659
23. Folco G, Murphy RC (2006) Eicosanoid transcellular biosynthesis: from cell-cell interactions to in vivo tissue responses. Pharmacol Rev 58:375–388
24. Laulagnier K, Vincent-Schneider H, Hamdi S, Subra C, Lankar D, Record M (2005) Characterization of exosome subpopulations from RBL-2H3 cells using fluorescent lipids. Blood Cells Mol Dis 35:116–121
25. Foster DA (2007) Regulation of mTOR by phosphatidic acid? Cancer Res 67:1–4
26. Samadi N, Bekele R, Capatos D, Venkatraman G, Sariahmetoglu M, Brindley DN (2011) Regulation of lysophosphatidate signaling by autotaxin and lipid phosphate phosphatases with respect to tumor progression, angiogenesis, metastasis and chemo-resistance. Biochimie 93:61–70
27. Laulagnier K, Grand D, Dujardin A, Hamdi S, Vincent-Schneider H, Lankar D et al (2004) PLD2 is enriched on exosomes and its activity is correlated to the release of exosomes. FEBS Lett 572:11–14
28. Reue K, Brindley DN (2008) Thematic review series: glycerolipids. Multiple roles for lipins/phosphatidate phosphatase enzymes in lipid metabolism. J Lipid Res 49:2493–2503
29. Alonso R, Mazzeo C, Rodriguez MC, Marsh M, Fraile-Ramos A, Calvo V et al (2011) Diacylglycerol kinase alpha regulates the formation and polarisation of mature multivesicular bodies

involved in the secretion of Fas ligand-containing exosomes in T lymphocytes. Cell Death Differ 18:1161–1173
30. Cheng H, Sukal S, Callender R, Leyh TS (2001) gamma-phosphate protonation and pH-dependent unfolding of the Ras.GTP.Mg2+ complex: a vibrational spectroscopy study. J Biol Chem 276:9931–9935
31. Heo J, Gao G, Campbell SL (2004) pH-Dependent perturbation of Ras-guanine nucleotide interactions and Ras guanine nucleotide exchange. Biochemistry 43:10102–10111
32. van Blitterswijk WJ, Hilkmann H (1993) Rapid attenuation of receptor-induced diacylglycerol and phosphatidic acid by phospholipase D-mediated transphosphatidylation: formation of bisphosphatidic acid. Embo J 12:2655–2662
33. Kobayashi T, Beuchat MH, Chevallier J, Makino A, Mayran N, Escola JM et al (2002) Separation and characterization of late endosomal membrane domains. J Biol Chem 277:32157–32164
34. van Rossum DB, Oberdick D, Rbaibi Y, Bhardwaj G, Barrow RK, Nikolaidis N et al (2008) TRP_2, a lipid/trafficking domain that mediates diacylglycerol-induced vesicle fusion. J Biol Chem 283:34384–34392
35. Kobayashi T, Beuchat MH, Lindsay M, Frias S, Palmiter RD, Sakuraba H et al (1999) Late endosomal membranes rich in lysobisphosphatidic acid regulate cholesterol transport. Nat Cell Biol 1:113–118
36. Delton-Vandenbroucke I, Bouvier J, Makino A, Besson N, Pageaux JF, Lagarde M et al (2007) Anti-bis(monoacylglycero)phosphate antibody accumulates acetylated LDL-derived cholesterol in cultured macrophages. J Lipid Res 48:543–552
37. Bouvier J, Zemski Berry KA, Hullin-Matsuda F, Makino A, Michaud S, Geloen A et al (2009) Selective decrease of bis(monoacylglycero)phosphate content in macrophages by high supplementation with docosahexaenoic acid. J Lipid Res 50:243–255
38. Chevallier J, Chamoun Z, Jiang G, Prestwich G, Sakai N, Matile S et al (2008) Lysobisphosphatidic acid controls endosomal cholesterol levels. J Biol Chem 283:27871–27880
39. Paillasse MR, de Medina P, Amouroux G, Mhamdi L, Poirot M, Silvente-Poirot S (2009) Signaling through cholesterol esterification: a new pathway for the cholecystokinin 2 receptor involved in cell growth and invasion. J Lipid Res 50:2203–2211
40. Mathivanan S, Fahner CJ, Reid GE, Simpson RJ (2011) ExoCarta 2012: database of exosomal proteins, RNA and lipids. Nucleic Acids Res:1–4
41. Taguchi R, Ishikawa M (2010) Precise and global identification of phospholipid molecular species by an orbitrap mass spectrometer and automated search engine lipid search. J Chromatogr A 1217:4229–4239
42. Feng D, Zhao WL, Ye YY, Bai XC, Liu RQ, Chang LF et al (2010) Cellular internalization of exosomes occurs through phagocytosis. Traffic 11:675–687
43. Meyer zu Heringdorf D, Jakobs KH (2007) Lysophospholipid receptors: signalling, pharmacology and regulation by lysophospholipid metabolism. Biochim Biophys Acta 1768:923–940
44. Freeman GJ, Casasnovas JM, Umetsu DT, DeKruyff RH (2010) TIM genes: a family of cell surface phosphatidylserine receptors that regulate innate and adaptive immunity. Immunol Rev 235:172–89
45. Radu CG, Yang LV, Riedinger M, Au M, Witte ON (2004) T cell chemotaxis to lysophosphatidylcholine through the G2A receptor. Proc Natl Acad Sci U S A 101:245–250
46. Perrin-Cocon L, Agaugue S, Coutant F, Masurel A, Bezzine S, Lambeau G et al (2004) Secretory phospholipase A2 induces dendritic cell maturation. Eur J Immunol 34:2293–2302
47. Wang D, Dubois RN (2010) The role of COX-2 in intestinal inflammation and colorectal cancer. Oncogene 29:781–788
48. Mansure JJ, Nassim R, Kassouf W (2009) Peroxisome proliferator-activated receptor gamma in bladder cancer: a promising therapeutic target. Cancer Biol Ther 8:6–15
49. Skokos D, Botros HG, Demeure C, Morin J, Peronet R, Birkenmeier G et al (2003) Mast cell-derived exosomes induce phenotypic and functional maturation of dendritic cells and elicit specific immune responses in vivo. J Immunol 170:3037–3045
50. Lalier L, Cartron PF, Pedelaborde F, Olivier C, Loussouarn D, Martin SA et al (2007) Increase in PGE2 biosynthesis induces a Bax dependent apoptosis correlated to patients' survival in glioblastoma multiforme. Oncogene 26:4999–5009

51. Chan BS, Satriano JA, Pucci M, Schuster VL (1998) Mechanism of prostaglandin E2 transport across the plasma membrane of HeLa cells and Xenopus oocytes expressing the prostaglandin transporter "PGT." J Biol Chem 273:6689–6697
52. Xiang X, Poliakov A, Liu C, Liu Y, Deng ZB, Wang J et al (2009) Induction of myeloid-derived suppressor cells by tumor exosomes. Int J Cancer 124:2621–2633
53. Huang Q, Li F, Liu X, Li W, Shi W, Liu FF et al (2011) Caspase 3-mediated stimulation of tumor cell repopulation during cancer radiotherapy. Nat Med 17:860–866
54. Sertznig P, Seifert M, Tilgen W, Reichrath J (2007) Present concepts and future outlook: function of peroxisome proliferator-activated receptors (PPARs) for pathogenesis, progression, and therapy of cancer. J Cell Physiol 212:1–12
55. Amigorena S, Savina A (2010) Intracellular mechanisms of antigen cross presentation in dendritic cells. Curr Opin Immunol 22:109–117
56. Chakravarthy MV, Lodhi IJ, Yin L, Malapaka RR, Xu HE, Turk J et al (2009) Identification of a physiologically relevant endogenous ligand for PPARalpha in liver. Cell 138:476–488
57. Bougnoux P, Hajjaji N, Maheo K, Couet C, Chevalier S (2010) Fatty acids and breast cancer: sensitization to treatments and prevention of metastatic re-growth. Prog Lipid Res 49:76–86
58. de Medina P, Payre B, Boubekeur N, Bertrand-Michel J, Terce F, Silvente-Poirot S et al (2009) Ligands of the antiestrogen-binding site induce active cell death and autophagy in human breast cancer cells through the modulation of cholesterol metabolism. Cell Death Differ 16:1372–1384
59. de Medina P, Paillasse MR, Segala G, Poirot M, Silvente-Poirot S (2010) Identification and pharmacological characterization of cholesterol-5,6-epoxide hydrolase as a target for tamoxifen and AEBS ligands. Proc Natl Acad Sci U S A 107:13520–13525
60. de Medina P, Paillasse MR, Segala G, Khallouki F, Brillouet S, Dalenc F et al (2011) Importance of cholesterol and oxysterols metabolism in the pharmacology of tamoxifen and other AEBS ligands. Chem Phys Lipids 164:432–437
61. Zhuang X, Xiang X, Grizzle W, Sun D, Zhang S, Axtell RC et al (2011) Treatment of brain inflammatory diseases by delivering exosome encapsulated anti-inflammatory drugs from the nasal region to the brain. Mol Ther 19:1769–1779
62. de Medina P, Paillasse MR, Payre B, Silvente-Poirot S, Poirot M (2009) Synthesis of new alkylaminooxysterols with potent cell differentiating activities: identification of leads for the treatment of cancer and neurodegenerative diseases. J Med Chem 52:7765–7777
63. Silvente-Poirot S, Record M, de Medina P, Subra C, Payré B, Paillasse MR et al (2011) Dendrogenin A modifies agressive melanoma cells to secrete exosomes with antitumor activity. International Workshop on Exosomes, January 19–22, Paris, France
64. Wang P, Zhang Y, Li H, Chieu HK, Munn AL, Yang H (2005) AAA ATPases regulate membrane association of yeast oxysterol binding proteins and sterol metabolism. EMBO J 24:2989–2999

Chapter 4
The Functional Importance of Tetraspanins in Exosomes

Sanyukta Rana and Margot Zöller

Abstract Exosomes are small vesicles that derive from the release of limiting membranes from multivesicular bodies (MVB). They carry and transfer selected membrane and cytoplasmic proteins, mRNA, and microRNA into target cells. Due to this shipping of information, exosomes are considered to be a most promising therapeutic tool for multiple diseases.

Whereas knowledge on the composition of exosomes is rapidly increasing, the mode of selective recruitment into exosomes as well as target cell selection is poorly understood. We suggest that at least part of this task is taken over by tetraspanins. Tetraspanins are a highly conserved family of proteins that are involved in a multitude of biological processes. This is a sequel of their special arrangement in glycolipid-enriched membrane microdomains, where tetraspanins form a net between themselves and neighboring molecules. These membrane domains, called tetraspanin-enriched microdomains (TEM), are prone for internalization. The early endosomes become integrated into MVBs, the primary source of exosomes. Thereby tetraspanin and tetraspanin-associated molecules become enriched in exosomes.

We will discuss here how the protein assembly of exosomes and possibly the recruitment of microRNA can be regulated by tetraspanins and how tetraspanins account for target cell selection and uptake of exosomes.

Exosomes herald an unanticipated powerful path of cell–cell communication. An answer to how exosomes collect and transfer information will allow the use of nature's concept to cope with malfunctions.

Abbreviations

ADAM	A disintegrin and metalloproteinase
AP	Adapter protein
APC	Antigen-presenting cells
DC	Dendritic cells
ECL	Extracellular loop

M. Zöller (✉) · S. Rana
Department of Tumor Cell Biology, University Hospital of Surgery,
Im Neuenheimer Feld 365, 69120 Heidelberg, Germany
e-mail: margot.zoeller@uni-heidelberg.de

Department of Tumor Progression and Immune Defense, German Cancer Research Center, Heidelberg, Germany

H.-G. Zhang (ed.), *Emerging Concepts of Tumor Exosome–Mediated Cell–Cell Communication,* DOI 10.1007/978-1-4614-3697-3_4,

EMT	Epithelial-mesenchymal transition
ERM	Ezrin-radixin-moesin protein family
FAK	Focal adhesion kinase
HCV	Hepatitis C virus
HIV	Human immunodeficiency virus
HSP	Heat shock protein
Mϕ	Macrophage
MFG	Milk fat globulin
MHC	Major histocompatibility genes
MMP	Matrix metalloproteinase
MT1-MMP	Membrane-type 1 MMP
MVB	Multivesicular body
NK	Natural killer cell
PI3K	Phosphatidylinositol-3-kinase
PI4KII	Type II phosphoinositol 4 kinase
PKC	Protein kinase C
PLCγ	Phospholipase C gamma
Src	Cytoplasmic phosphotyrosine kinase
RISC	RNA-induced silencing complex
TCR	T cell receptor
TEM	Tetraspanin-enriched membrane microdomain
TH	T helper cell
TM	Transmembrane
uPAR	Urokinase receptor

4.1 Introduction

Tetraspanins are small membrane proteins expressed in many species ranging from sponges to mammals, each organism disposing on a large number of tetraspanin members [68]. Tetraspanins are involved in a multitude of biological processes such as fertilization, parasite and viral infection, synaptic contacts at neuromuscular junctions, platelet aggregation, maintenance of skin integrity, immune response induction, metastasis suppression, and tumor progression [91, 92, 153, 273, 274]. This multitude of functions ascribed to tetraspanins is due to their mode of activity [91, 143]. Tetraspanins act as molecular facilitators by associating with themselves and additionally with a multitude of transmembrane (TM) and intracellular proteins in glycolipid-enriched membrane microdomains, which are prone for internalization [91]. It is this latter feature, which accounts for the enrichment of tetraspanins in exosomes.

However, whether tetraspanins only are enriched in exosomes due to their engagement in vesicle formation or whether tetraspanins are functionally important components of exosomes has not yet received much attention. We will discuss this question after introducing the structure of tetraspanins, the tetraspanin web, the structure—function relationship between tetraspanins and associated molecules and the sorting of tetraspanins into endosomes. We will proceed to introduce our point of view that the tetraspanin web in exosomes is essential for exosome—target cell selection. Finally, we will discuss whether exosomal tetraspanins are functionally engaged in exosome activities, which will be exemplified for exosomal activity in immune response induction, cell fusion, angiogenesis, and metastasis.

4.2 Structure and Organization of Tetraspanins

4.2.1 The Structure of Tetraspanins

Tetraspanins are small molecules that cross the membrane four times, such that the short N- and C-terminal tails are located within the cytoplasm. They have a small extracellular loop (ECL1) between the TM regions TM1 and TM2, a small intracellular loop between TM2 and TM3, and a large extracellular loop (ECL2) between TM3 and TM4. The essential signature of tetraspanins, which differentiates them from other 4-TM molecules, are the highly conserved cysteines in this large extracellular loop. Furthermore, the ECL2 is subdivided into a more constant and a variable region. The more constant region likely accounts for tetraspanin dimerization. The variable region is important for interactions with nontetraspanin partner molecules. The tertiary structure of tetraspanins is stabilized by polar residues in the TM regions [236, 218]. Palmitoylation (S-acylation of proteins with palmitate) of intracellular, juxtamembrane cysteines likely is required for initiating tetraspanin–tetraspanin web formation, which can also be supported by palmitoylation of associating integrins. Protein palmitoylation in general regulates subcellular traffic and/or protein degradation [189] or can protect the molecules from lysosomal degradation and link them to cholesterol and gangliosides [19, 31, 219, 248, 249].

4.2.2 The Tetraspanin Web

Tetraspanins form complexes by interacting between themselves as well as a large variety of TM and cytosolic proteins [92, 143, 6], which are essential for the functional activity of tetraspanins. Integrins, particularly $\alpha3\beta1$, $\alpha4\beta1$, and $\alpha6\beta1$, are the most prominent nontetraspanin partners. The association with other integrins such as $\alpha6\beta4$ is more restricted [16]. Tetraspanins also associate with growth factor receptors [231, 175], G protein-coupled receptors (GPCR), 7-TM receptors that signal through small heterotrimeric GTP-binding proteins, and activate multiple signaling pathways through second messengers such as cAMP, Ca^{2+}, and inositol trisphosphate (IP3) [146], claudins [136], a family of 4-TM proteins originally described as components of tight junctions [252], nonintegrin TM molecules associated with cell-matrix or cell–cell adhesion, such as CD44 and EpCAM [135], and Ig superfamily members including two members of a new class that contains a conserved Glu-Phe-Ile (EWI) motif [47], EWI-F/FPRP/CD9-P1, and EWI-2 [30, 234, 36, 6]. Of special importance, also with respect to exosomal activities of tetraspanins, are the associations with several proteases and peptidases [135], where particularly the association with matrix metalloproteinases (MMP) coordinates pericellular proteolysis and invasion [283]. Prominent cytosolic signal transduction molecules coimmunoprecipitating with tetraspanins are protein kinase C (PKC), a type II phosphatidylinositol 4 kinase (PI4KII) and phospholipase C gamma (PLCγ) [18, 288, 35, 92, 6]. Several isoforms of PKC-mediate serine and threonine phosphorylation in many different protein substrates, thereby propagating various signal transduction pathways, which lead to transcription factor activation [264]. PI4KII generates phosphoinositol-4-phosphates, precursors of important regulatory phosphoinosites. Besides regulating signaling events, PI4KII is also important in vesicular trafficking and lipid transport [80]. PLCγ catalyzes hydrolysis of phospholipid PIP_2 to generate the signaling molecules inosito-triphosphate (IP3) and diaglycerol (DAG) [267].

The interactions of tetraspanins with themselves and this large array of different nontetraspanin molecules are grouped in three categories according to the strength of the detergent required to disrupt the association. Direct protein–protein interactions, the so-called type I interactions, resist stringent detergents. They are rare and comprise some tetraspanin-homodimers (tri- and tetramers) and few heterointeractions as the interactions of CD151 with some integrins and claudin-1 [287, 233, 128] and the interactions of CD9, CD81, and Tspan8 with EWI proteins [234, 235, 36]. These direct interactions may proceed through the ECL2 and/or the TM regions [127]. The majority of tetraspanin–integrin and tetraspanin–tetraspanin interactions are type II interactions, maintained under milder lysis conditions. Palmitoylation of tetraspanins and/or the associating proteins are essential for this type of interaction as well as for weak type III interactions, e.g., with several kinases [19, 31, 296, 92].

Besides these primary interactions, tetraspanins also associate with cholesterol and gangliosides [32, 171, 182], anionic glycosphingolipids characterized by sialic acid residues [286]. Thereby higher order tetraspanin complexes are formed in microdomains, termed tetraspanin-enriched membrane microdomains (TEM) [91]. TEM share several features with lipid rafts, but are independent membrane microdomains that, different from lipid rafts, are disrupted by Triton X-100 at 4 °C [92, 135, 248].

4.3 Vesicular Traffic of Tetraspanins and Exosomes

Exosomes are small vesicles delivered by many cells in the organism [44, 58, 112, 185]. Exosomes derive from multivesicular bodies (MVB) from where the individual vesicles are released as exosomes in the extracellular space [44, 132, 163, 257]. MVBs are assembled from vesicles sorted from the trans-Golgi network or from internalized membranes. MVB fuse with lysosomes for protein degradation [232] or their intraluminal vesicles fuse with the plasma membrane and are released as exosomes [44, 237, 163, 226]. The latter process, first described for the release of the transferrin receptor during reticulocyte maturation, was considered as a mode to eliminate obsolete proteins [184]. Exosome research became highly stimulated, when it was noted that antigen-presenting cells (APC) release exosomes derived from MVB of the major histocompatibility genes (MHC) class II compartment, which function similar to APCs and stimulate T cells in vitro and in vivo [44, 5, 3, 3, 216]. Exosome research received an additional boost with the discovery that exosomes contain mRNA and microRNA. This so-called shuttle RNA can become transferred into target cells, where the horizontal transfer of exosomal genetic material between cells can induce exogenous gene expression and mediate RNA silencing [199, 45, 255, 25, 132].

4.3.1 Recruitment of Tetraspanins into Exosomes

According to their origin, the exosomal protein profile is rich in molecules located in membrane domains prone for internalization such as rafts and tetraspanin-enriched microdomains as well as molecules engaged in fission, scission, and vesicular transport [163, 195, 196, 299]. In fact, tetraspanins such as CD9, CD37, CD53, CD63, CD81, CD82, CD151, and Tspan8 are abundantly recovered in intracellular vesicles [91, 17, 216, 163] and are constitutive components of exosomes, where CD63 is frequently used as a marker to differentiate exosomes from other extracellular vesicles [51, 192, 299]. Tetraspanin enrichment in exosomes is well in line with the preferential internalization of oligomeric proteins and/or proteins localized in cholesterol-rich membrane domains. However, the pathway, whereby tetraspanins become recruited into exosomes has not yet been fully unraveled and there is evidence that targeting into exosomes may differ for individual tetraspanins and their associated molecules [1, 197].

One pathway of tetraspanin internalization relies on a well-defined internalization motif. Some tetraspanins possess a tyrosine-based sorting motif, a sequence of Tyr-Xaa-Xaa-ϕ, where ϕ stands for an AA with a bulky hydrophobic side chain, in the C-terminal cytoplasmic domain [158]. By this sorting, motif tetraspanins become prone for delivery to intracellular compartments [158, 17]. However, some tetraspanins enriched in exosomes do not possess a sorting motif (CD9) or an inappropriately located sorting motif (Tspan8) [17]. This was a first hint toward individual tetraspanins possibly following different routes of internalization. For CD63, it was shown that the C-terminal domain interacts with μ4, μ2, μ3A, and μ3B subunits of adaptor protein (AP) complexes, linking the traffic of CD63 to the clathrin-dependent pathway [204]. Alternatively, CD63 may be sorted via its association with syntenin-1 [111]. Trafficking of tetraspanins can also rely on AP-independent pathways. To give an example: CD151 is involved in the internalization of $\alpha3\beta1$. CD151 has a sorting motif (YRSL) and may promote internalization through the clathrin-dependent pathway [227, 270]. Alternatively, CD151 may link the complex to PKCα, which regulates integrin traffic [178, 292] or CD151 may connect $\alpha3\beta1$ to other TM proteins in TEM with an internalization sequence [178, 292]. Other examples, where internalization does not depend on a sorting motif in the C-terminal tail, are CD81 and Tspan8. Thus, it has been described that membrane-associated RING-CH (MARCH), TM ubiquitin ligases, modulate intracellular trafficking of TM proteins and sequester CD44 and CD81 in endolysosomal vesicles [12]. Tspan8 associates via its N-terminal domain with intersectin-2, a multimodular complex involved in clathrin-coated pit internalization [225] and dynamin [198]. Intersectins contain Eps15 homology domains, where binding of Eps15 domains promotes endocytosis [123]. Intersectins also interact with dynamin and synaptojanin [278], which regulates the fission process [54, 124]. Tspan8 internalization is promoted by PMA and is accompanied by a rearrangement of the Tspan8 web, which includes the recruitment of α4β1. Thus, under stress condition, Tspan8 along with α4β1 becomes recruited into distinct membrane microdomains [198] that resemble classical rafts [142] rather than cholesterol- and glycolipid-enriched membrane microdomains mostly formed by tetraspanins [92, 282]. The molecular pathways that account for this membrane microdomain reorganization remain to be explored.

Taken together, by their location in cholesterol-enriched membrane microdomains and their oligomerization and multimolecular complex formation tetraspanins are prone for internalization and recruitment into MVBs and exosomes. According to the present state of knowledge, the C-terminal sorting motif of some tetraspanins is not essential, although it may contribute to the recruitment into endosomes. Furthermore, there is evidence that the tetraspanin web becomes recruited as an entity into endosomes, which raises the question, whether tetraspanins have bearing on the exosome composition?

4.3.2 The Impact of Tetraspanins on the Composition of Exosomes

It is well known that the relative abundance of proteins, mRNA, and miRNAs differs between exosomes and donor cells. This implies active sorting into MVB [132], which is suggested to be directed by monoubiquitinylation [79, 104, 229], localization in cholesterol-rich membrane microdomains, or higher order oligomerization via endosomal sorting complex required for transport (ESCRT) proteins [56, 229] or a zip code in the 3′-UTR (mRNA) [155] or by physical and functional coupling of RNA-induced silencing complexes (RISCs) to the components of the sorting complex, where GW182-containing GW bodies are sorted into MVB, which promotes continuous assembly and disassembly of membrane-associated miRISCs (miRNA-loaded RISC) [73, 138, 300]. As exosomes are detected in all body fluids, the selective enrichment of "marker" proteins as well as of miRNA makes exosomes a very attractive means for noninvasive diagnosis [169, 203, 272].

There are hints that the recruitment of tetraspanins into endosomes, MVB, and exosomes is accompanied by additional selection processes of associating proteins, but also mRNA and possibly miRNA. This hypothesis is supported by indirect, but quite strong, evidences. Thus, proteins and mRNA are differentially recruited in two tumor lines that differ only in transfection with an empty versus a Tspan8 cDNA-containing vector [177]. Several proteins with equal expression in both cells have been selectively enriched in Tspan8-containing exosomes, such as galactoside-binding protein-3 and VCAM-1 among others [177]. Differences at the mRNA level have been even more striking with 322 mRNA differing by a factor of >3 in the two lines [177]. As outlined above, stress-induced internalization is accompanied by a reorganization of TEM such that Tspan8 associates with $\alpha4\beta1$, whereas CD9 remains $\alpha3\beta1$ associated [198]. The stress-induced membrane microdomain reorganization with $\alpha4\beta1$ recruitment into the proximity of Tspan8 allows for rapid recycling of $\alpha4\beta1$ [95, 114], which could well support the pronounced release of $\alpha4$-associated Tspan8 into exosomes. Indeed, this internalization-induced tetraspanin web rearrangement is maintained during the intracellular transport and is recovered in exosomes [95, 177]. In another rat tumor line and additionally in human tumors, Tspan8 associates with $\alpha6\beta4$ upon PMA treatment, which is accompanied by a strong reduction in the association with $\alpha3\beta1$. Notably, the Tspan8 web as rearranged during internalization is maintained during the intracellular transport and is recovered in exosomes [95, 177]. Finally, it should be mentioned that release of miRNA is actively controlled through a ceramide-dependent machinery associated with exosome secretion [125]. Taking into account the known interaction between gangliosides and tetraspanins [81], it is tempting to speculate that tetraspanins may also contribute to miRNA recruitment.

In brief, there is evidence that tetraspanins are not only a constitutive component of exosomes, but contribute—due to the tetraspanin web—to the selectivity of protein uptake in exosomes. Whether tetraspanins are also engaged in the selective uptake of mRNA and microRNA remains to be explored.

4.3.3 Tetraspanins and Exosome Target Cell Selection

Selective binding and uptake of exosomes has repeatedly been suggested [162, 256, 257, 299] and has been demonstrated, besides others, for exosome uptake by endothelial cells [162, 14, 177] and for tumor exosomes uptake in premetastatic lymph nodes [101, 115].

Exosomes are characterized by phosphatidyl serine exposure on the outer membrane of exosomes, where phosphatidyl serine is involved in exosome budding in the late endosomes [63]. Phosphatidyl serine can also trigger exosome uptake by binding to scavenger receptors, integrins, and complement receptors [290]. In line with this suggestion, we noted that exosomes bind very rapidly to macrophages (Mϕ)and that binding and uptake are efficiently blocked by anti-CD11b (Zech et al. submitted). However, it has also been noted that AnnexinV may not consistently block exosome uptake or induce only a partial blockade [238, 208]. Exosome uptake in vivo also was not dictated by scavenger receptors (Zech et al. submitted; Rana et al. submitted). These findings and the selectivity of exosome binding argue for phosphatidyl serine likely facilitating (unspecific) binding that may not be followed by exosome uptake.

Already in 2004, it was described that allogeneic exosomes are taken up by dendritic cells (DC), Kupffer cells, and some Mϕ in the spleen. The authors pointed out that exosome targeting is mediated by milk fat globulin (MFG)-E8, CD11a, CD54, phosphatidyl serine, CD9, and CD81 on exosomes and requires αvβ3, CD11a, and CD54 as ligands on DC [174]. These findings suggested that exosome binding and uptake by target cells likely varies depending on the pathway of endocytosis as well as by the protein pattern on exosomes and target cells [10, 52]. Indirect evidences pointed, in addition, toward a contribution of tetraspanins [192, 299].

We recently explored this question in detail by pulldown assays and antibody blocking studies and found that exosomes bind selectively via TEM-localized and tetraspanin-associated adhesion molecules, where the adhesion molecules and the associated tetraspanins vary depending on the exosome source. The most surprising finding considering the target cell ligands has been the large overlap with exosomal binding structures, which included annexins, chaperons, molecules involved in vesicular transport and consistently tetraspanins and adhesion molecules. In stromal cells, mostly CD151 and Tspan8, in hematopoietic cells mostly CD81 and CD9 are engaged in exosome binding. Exosomal adhesion molecules frequently involved in target cell binding and uptake are β1 integrins, where the selectivity depends on the associated α chain. At the target cell αM, αL, αX, several CAMs, but also CD44 are involved (Zech et al. submitted; Rana et al. submitted). These findings strongly support the suggestion that the internalization process in the exosome donor cell and the exosome uptake by target cells use similar fusion/fission machineries [132, 246].

Considering exosome binding, we finally want to mention an additional aspect that has been rather neglected so far. Exosomes bind with high avidity several matrix proteins. Matrix protein binding is selective and requires defined tetraspanin—adhesion molecule complexes, where besides integrins, CD44 plays a major role (unpublished finding). As outlined below, we consider it important to further explore the cross-talk between exosomes and the matrix, as the matrix is not only a structural element, but a storage of bioactive compounds and an essential component in tissue repair as well as in the cross-talk between tumor cells and the stroma [262, 211, 247]. In all these processes, exosomal proteases may play an important role.

Thus, exosomes display exquisite target cell selectivity in vitro and in vivo, which is based on target cell ligand interactions with exosomal tetraspanin-associated receptors. Notably, the tetraspanin complexes used for targeting correspond to those arranged during internalization. Maintenance of internalization complexes and reuse of these complexes for exosome uptake appear to be a common theme [14, 163]. Importantly, the engagement of GEM-located and tetraspanin-associated adhesion molecules also provides an explanation for the target cell selectivity that is difficult to imagine to rely exclusively on single adhesion molecules, which frequently are expressed on many cells.

4.4 Major Functional Activities of Cellular and Exosomal Tetraspanins

One of the first reviews on tetraspanins from Shoshana Levy's group "The tetraspanin superfamily: molecular facilitators," [152] still hits the nail's head. Tetraspanins act via modulating, stabilizing, or preventing activities of their associated molecules, where cellular penetration, invasion, and fusion are the main functional activities [91].

The engagement of tetraspanins in motility relies to a great extent on their association with integrins, where tetraspanins promote cell migration by integrin compartmentalization, internalization and recycling, or by modulating integrin signaling [16, 236, 92, 143], which supports cell spreading and migration including cable formation on gelatinous matrices [16, 90, 92, 140]. Tetraspanin-induced activation of integrin-mediated signaling pathways includes phosphatidylinositol-3-kinase (PI3K), ERK1/2, and Cdc34 activation [214, 289, 273, 274, 92, 143]. By associating with EWI proteins, tetraspanins exert opposing effects on cell polarity and migration. This is due to EWI proteins binding to ezrin-radixin-moesin (ERM) proteins, which negatively regulates ERM protein phosphorylation [277] that is required for a conformational change allowing the ERM proteins to link TM molecules with the actin cytoskeleton [251]. Accordingly, EWI-2 silencing augments ERM phosphorylation, cell polarity, and migration [210].

Regulation of invasiveness via tetraspanins relies on their association with peptidases [193, 135], A disintegrin and metalloproteinase (ADAMs), particularly ADAM10 and ADAM17 [8], MMP [281], and uPAR [13] or could proceed through tetraspanin-dependent modulating MMP transcription and secretion [84]. Taking into account the association between tetraspanins and integrins, major ligands for extracellular matrix proteins, particularly the association of tetraspanins with MT1-MMP, also known as MMP14, has been demonstrated to play an important role in pericellular lysis [281, 131].

Possibly, the best known feature of tetraspanins is the infertility of CD9-knockout mice, where sperm cannot fuse with the CD9-deficient egg [116]. A CD81 deficiency is also accompanied by reduced egg–sperm fusion [206]. Mostly, CD9, CD81, and CD82 are also involved in virus-induced syncytia formation and cell entry.

Morphogenic features of tetraspanins [293, 91, 149, 279] have been frequently associated with angiogenesis induction, where an involvement of CD82, CD151, and CO-029 via their association with $\alpha3\beta1$, $\alpha4\beta1$, and $\alpha6\beta1$ has been described [71, 72, 85, 88, 239, 293].

Exosomal proteins maintain their functional activity, such as antigen presentation; peptide and protein cleavage [216]. Thus, it becomes likely that exosomal tetraspanins maintain their activity in migration, invasion, membrane fusion, and morphogenesis. Taking into account their enrichment in exosomes, one can even speculate on pronounced activity. Nonetheless, in advance of discussing potential exosomal tetraspanin activities, we want to stress two points. First, distinct from the engagement of tetraspanins in exosome assembly and target cell binding/uptake, only few studies have directly evaluated a functional contribution of tetraspanins to exosome activities. Second, and most importantly, exosomes contain functionally active mRNA and miRNA [125, 151, 163], such that exosome uptake can severely alter the target cell that may culminate in conversion to a malignant phenotype [105, 196, 243, 268]. According to our experience, it is, indeed, the transfer of microRNA that most severely affects the exosome target cell. This implies that evaluating functional activities of exosomal proteins, including tetraspanins, in target cells may only hit the tip of an iceberg. This in mind, we will discuss the potential contribution of tetraspanins to selected exosomal activities in immune response [5, 144], cell-to-cell spread of infectious agents [216, 110], angiogenesis, and metastasis [188, 299].

4.4.1 Exosomal Tetraspanins and Immune Response

One of the first reports on tetraspanin described the involvement of the tetraspanin CD81 in B cell immune responses [23]. Meanwhile, there is abundant evidence that tetraspanins are important in nonadaptive immune responses mediated by natural killer cells (NK), neutrophils and Mϕ, as well as T and B cell responses [113, 261]. Last not least, tetraspanins play a major role in antigen presentation by dendritic cells [41]. The engagement of tetraspanins in immune responses has been supported by defects in humoral or cellular immune responses in CD37, CD81, CD151, Tspan32, and CD9/CD81 as well as CD37/Tspan32-deficient mice [273, 274, 242, 152, 122].

As mentioned, one of the first observations on the engagement of tetraspanins in immune response induction described the association of CD81 with CD19 as part of the B cell signal transduction complex, which lowers the threshold for B cell activation [33, 102]. Though resting B cells do not deliver measurable amounts of exosomes, B cells, particularly when stimulated by CD40 and the IL4 receptor, secrete exosomes [213]. B cell-derived exosomes are characterized by abundant expression of major histocompatibility class I and class II molecules (MHC I, MHC II), heat shock protein (HSP)70 and HSP90, α4, CD45, moesin, tubulin, actin, G1α2, B220, surface Ig, CD19, and CD81 [275, 213]. It is important to note that the distribution of MHC I and II, but also of α4 in the cell membrane differs from the distribution in exosomes, such that these molecules are mostly recovered in the soluble fraction of the cell membrane, but are recovered in cholesterol-enriched TEM in exosomes. The authors suggested this finding to be indicative for a selective recruitment toward tetraspanins during internalization and vesicle formation [275]. B cell-derived exosomes are trapped besides others by follicular dendritic cells [213]. Thereby follicular dendritic cells, which do not express MHC II become competent in activating T helper cells (TH) [275]. Furthermore, B cell-derived exosomes have been demonstrated to present allergen peptides and activate allergen-specific T cell to proliferate and produce the TH2 cytokines IL5 and IL13. The exosomes had the typical phenotype of activated B cell exosomes with MHCI and II, costimulatory molecules and CD81-associated CD19. Thus, B cell-derived exosomes can play a critical role in allergic immune responses [2]. Besides CD81, CD63 has been described to be expressed on exosomes of an Epstein-Barr virus- transformed B cell line. Although a CD63 deficiency did not affect the capacity of these exosomes to stimulate EBV antigen-specific TH, TH proliferation was increased in the absence of CD63, which was accompanied by a significant reduction in exosome delivery [190]. Though reduced exosome secretion can be explained by a deficit in CD63, the mechanism underlying the increased proliferative response awaits to be explored. Transfer of microRNA that interferes with components of the T cell activation signaling cascade would be one possibility. In concern of the impact of exosomes on B cells, we observed that exosomes derived from a pancreatic adenocarcinoma had no impact on B cell activation by T-dependent or T-independent antigens. Nonetheless, exosomes promoted a slight downregulation of syk and fyn. We interpret the finding in the sense that despite an unimpaired primary response exosomal microRNA may possibly interfere with B cell response regulation.

Various tetraspanins (CD9, CD37, CD53, CD81, CD82, CD151, Tspan32) are involved in T cell activation. CD81 and CD82 accumulate at the site of T cell receptor (TCR) engagement [170, 40], where in tetraspanin-deficient mice mostly a hyperproliferative response has been observed. This points toward a regulatory role of tetraspanins in T cell activation [273, 274, 242, 259]. CD81 and CD37 are also involved in the differentiation of TH in TH1 and TH2 [260, 43]. For both, the regulatory role of tetraspanins in T cell activation and the tetraspanin promoted differentiation in TH1 or TH2, the underlying mechanisms remain to be explored.

In macrophages, tetraspanins have been described to promote or inhibit activation [161, 37]. Inhibition appears to be mediated by the interaction of CD37 and CD63 with the C-type lectin dectin-1, which recognizes fungal wall components [157, 168]. The interaction of CD37 and CD63 with dectin-1 leads to inhibition of IL6 and TNFα secretion, probably by regulating syk downstream of dectin-1 [168]. Instead, CD9 regulates lipopolysaccharide-TLR4 signaling, which is accompanied by upregulated TNFα production [37]. CD9 also increases Mϕ activity by associating with Fcγ receptors [117].

In an elegant study, the group of Gabrielson explored the impact of exosomes in bronchoaveloar lavage fluid on sarcoidosis, a TH1-driven granulomatous, likely autoimmune disease, which preferentially affects the lung [64]. They noted that exosomes from patients with sarcoidosis differ significantly from exosomes of healthy donors expressing upregulated levels of MHC I and II, HSP70, neuregulin, CD9, CD63, and CD81. Exosomes stimulated IFNγ and IL13 production by leukocytes and IL8 production by epithelial cells [194]. As MHC I and II are associated with tetraspanins, and both MHC molecules and tetraspanins are concomitantly upregulated, the authors argue that the exosomal tetraspanin-MHC complex contributes to the activation of TH1. They also suggest that IL8 upregulation on epithelial cells may follow a different pathway, where TNFα-expressing exosomes have been reported to induce IL8 production [194, 181] and CD9 has been described to promote TNFα production via TLR4 signaling [37]. The authors conclude that exosomes contribute to the pathogenesis of this autoimmune disease, which should be taken into account in treatment strategies [194].

DC are the main APCs. They take up and digest antigens to present the peptides in MHC II to CD4$^+$ TH. MHC II molecules are not randomly present in the plasma membrane, but organized in lipid rafts and TEM [263], where several tetraspanins associate with MHC II and costimulatory molecules [130, 50], which leads to clustering of the antigen-presenting complex. This, in turn, as described for CD9, facilitates TCR clustering and activation of signal transduction [254]. But, other tetraspanins, particularly CD37, CD151, and Tspan32, have been described to negatively affect antigen presentation [69, 220]. This could possibly be due to tetraspanins promoting antigen endocytosis and trafficking of MHC II, e.g., upon pathogen binding CD63 becomes internalized and translocated to the endosomal-lysosomal compartment [157, 17, 192].

Notably, DC-derived exosomes can replace DC in immune response induction and the therapeutic option of exosomes has first been explored in the context of DC-derived exosomes as a vaccine in cancer therapy [41]. As this important aspect of exosomes will be elaborated in detail in another chapter of this book, here we will just mention that DC-derived exosomes are particularly rich in CD9 and CD81, CD82, MHC I and II and that MHC molecules colocalize and associate with tetraspanins during vesicle formation [9]. Finally, as already mentioned, DC also take up exosomes secreted by other cells, including tumor cells, which they internalize and process for presentation, where exosomal CD9 and CD81 are important components for uptake by DC [174]. Thus, DC use exosomes as a source of antigen and produce exosomes that suffice for T cell activation, both features expanding the operational range of DC [253].

The particular contribution of exosomal tetraspanins in DC relies on enriching and clustering MHC and costimulatory molecules in TEM, which lowers the threshold for T cell activation. It is likely, though not yet proven, that the association of tetraspanins with adhesion molecules will support exosome targeting.

Similar to dendritic cells in lymphoid organs, the microglia, brain-resident cells derived from the bone marrow secrete exosomes that express CD9 and CD81 [193]. In an organ such as the brain, where cell motility is very limited, exosomes may become especially important for intercellular communication, i.e., the transport of information between neurons and microglia [201]. The authors verify this aspect for the exchange of nutritients, where exosomes display an intact metabolic pathway necessary for anaerobic glycolysis and lactate production. Particularly notable in view the engagement of tetraspanins in exosome activities, microglial exosomes, distinct from DC, express the dipeptidase CD13, which is associated with CD9 [198]. Exosomal CD13 is functionally active and cleaves encephalin. The cleaved neuropeptides cannot bind neuronal opioid receptor. Neurons express CD13 on the presynaptic membrane, which could control neuropeptides at the synaptic level. With microglial exosomes also expressing CD13, the catabolism of neuropeptides can also become controlled at distant sites [193].

In brief, though the aspect of functional activity of tetraspanins in exosomes in immune response regulation has not yet received much attention, the work done so far, supports an active engagement of exosomal tetraspanins. Supporting immune response induction or regulation relies—in dependence on the leukocyte subpopulation—on the association of tetraspanins with accessory or response inhibitory molecules (T cells, B cells) or MHC molecules (APC) in TEM. The TEM location implies besides membrane molecule clustering also apposition of signaling molecule. The tetraspanin association with adhesion molecules will account for target cell selection and will strengthen the contact between exosomes and target cells further facilitating the modulation of the target cell's activation state. Last not least, exosome uptake will have a severe impact on gene expression in the target cell via exosomal microRNA. This aspect has, to our knowledge, not yet been explored, but exosomal microRNA most likely will be a dominating factor in exosomal immune response regulation.

4.4.2 Tetraspanins, Exosomes, and Fusion

Tetraspanins are involved in cell fusion [91, 143, 197, 55]. The most convincing example has been the incapability of CD9- and CD81-deficient eggs to fuse with sperm [206, 139, 176]. Besides egg–sperm fusion, tetraspanins are also involved in monocyte fusion leading to giant cells or osteoclast [108, 187] and under pathophysiological conditions in viral infections [86, 258]. Tetraspanins are also involved in bacterial infections as well as parasite and fungal entry [143, 258, 205, 245, 59].

With respect to the engagement of tetraspanins in nonviral infections, tetraspanins are engaged in internalization and recruitment into intracellular vesicles, but it is not yet known, whether there is an engagement of exosomal tetraspanins. To give a few examples: CD9 enhances diphtheria toxin binding and endocytosis, where CD9 interacts with the diphtheria toxin receptor proHB-EGF [109, 87]. Type1 fimbriated *E. coli* bind to the tetraspanins uroplakin Ia and Ib, which allows them to enter bladder epithelia and cause urinary tract infection [276, 265]. CD81 is involved in *Plasmodium falciparum* and *Plasmodium yoelii* infection of hepatocytes, such that CD81-deficient mice are resistant to malaria sporozoite infection. CD81 is required for entry and formation of a vacuole allowing parasite differentiation [224, 227].

Several viruses use tetraspanins for their assembly as well as for transfer. Let us start with the hepatitis C virus (HCV), which—as an exception—uses a tetraspanin as ligand. Tetraspanins mostly act via lateral interactions [53]. On the contrary, CD81 is a ligand for the hepatitis C RNA virus, where the major envelope protein (HCV-E2) interacts with the large ECL of CD81 [191, 277, 48, 89, 202]. This allows HCV to invade hepatocyte, but also B cells, T cells, NK, and DC [98], where the infection of immune cells explains the poor immune response to HCV. Other tetraspanin-associated molecules can also be involved. Thus, EWI-2 blocks HCV entry. As EWI-2 is not expressed on hepatocytes, HCV entry becomes facilitated [172]. Notably, HCV binding to CD81 promotes uptake and recruitment into MVB and exosomes [162], which facilitates the transfer of HCV [271]. Instead, in the absence of CD81, HCV envelop proteins are retained in the endoplasmic reticulum. Instead, in the presence of CD81, HCV envelope proteins pass through the GOLGI, mature, and are found complexed to CD81 in MVB and extracellular in exosomes. Exosomes containing

CD81-HCV complexes are also recovered from patients' sera. Exosomal CD81-HCV complexes fuse with target cells allowing for the spread of HCV. Thus, HCV essentially depend on CD81 for maturation and spread [57].

The retrovirus human immunodeficiency virus (HIV)-1 has a tropism for $CD4^+$ T cells and APCs. It kills $CD4^+$ T cells and uses APC for storage and transmission. In T cells, CD9 and CD81 support HIV-1-induced membrane fusion [74, 7]. It has been shown that clustering of CD9 correlates with multimerization of the major viral structural component Gag. Interestingly, though the raft-associated lipid GM1 and the raft-associated protein CD55 were recruited toward the viral assembly/budding site, these associations were transient, whereas the confinement toward TEM was consistent. The authors conclude that viral components direct the formation of microdomains for specific functions and that tetraspanin webs are particularly suited to fulfill the pathogen's requirement [129]. Furthermore, HIV-1 envelope glycoprotein and core protein Gag are enriched in MVB . However, whether by this means tetraspanins become engaged in HIV-1 transmission is controversial [78, 129]. The tetraspanins CD53, CD63, CD81, CD82, and CD151 are also engaged in HIV-1 uptake by Mϕ and DC [98, 42], where HIV-1 is delivered to the tetraspanin-enriched so-called viral endosome and recruited to the infectious synapse [159] . Distinct from T cells, in Mϕ and DC, CD9 and CD81 are supposed to function as an HIV-1 exit gateway. In fact, DC are crucial in early stages of HIV infection. They capture and internalize HIV at mucosal surfaces and transfer the virus through the infectious synapse to $CD4^+$ T cells. After uptake, HIV-1 colocalizes in early endosomes with CD81, CD82, and CD9. Replication of HIV-1 in cis-infection (within DC) was also observed, envelope-independent, in tetraspanin- and AP3-enriched vesicles, where downregulation of AP3 reduced the viral particle release [67]. Finally, there is evidence for an engagement of tetraspanins in HIV-1 envelope protein-driven fusion at the infectious synapse [180, 228].

The engagement of CD9, CD81, and CD82 in virus-induced syncytia formation and membrane fusion events has also been demonstrated by the susceptibility to feline immunodeficiency virus and canine distemper virus, where CD9 is involved in the postinfectious release [269, 160, 39]. CD82 is engaged in human T leukemia virus (HTLV) transmission [165]. Human papilloma virus enters epithelial cells via TEM, where anti-CD63 and anti-CD151 block virus entry, likely via their association with α6, which has been proposed as a candidate receptor for HPV16 [230].

Taken together, several pathogens make use of tetraspanins for cell entry and virus-induced syncytia formation frequently depends on tetraspanins, mostly CD9, CD81, and CD82. Furthermore, though distinct pathogens may use different mechanisms of entry, even when the same tetraspanin(s) is engaged, as has been demonstrated for Plasmodium and hepatitis C virus [224, 89, 202], at least for some viral pathogens, it has been proven that their invasion essentially requires TEM and is independent of caveolae [230]. Finally, viruses use tetraspanin-enriched vesicles for assembly and exosomal tetraspanins for transfection.

4.4.3 Tetraspanins, Exosomes, Angiogenesis, and Hemostasis

4.4.3.1 Tetraspanins and Angiogenesis

Angiogenesis defines the process of new capillary formation from a preexisting vasculature, which is crucial to supply a growing organism with oxygen. Accordingly, tissue remodeling/wound repair and the rapid growth of tumor cells essentially require blood supply [62, 96]. Angiogenesis is initiated by angiogenic factors, angiogenin, epidermal growth factor, IL8, TNFα, TGFα, TGFβ, and VEGF [97], that bind to endothelial cell (EC) receptors including VEGFR-1, -2, -3, and neuropilins. Stimulated EC grow and secrete matrix-degrading enzymes that digest the basement membrane surrounding the vessel. The junctions between EC become altered and EC migrate toward the source of the angiogenic stimulus. At this stage, sprouting EC are reorganized to form tubes and assemble a new basement membrane. The formation of a lumen is driven by interactions between EC and the extracellular matrix. Molecules involved in this process are, among others, galectin-2, CD31 (PECAM-1), and VE-cadherin [99]. An increasing body of evidence suggests that tetraspanins may directly regulate the development and functions of the vascular system and the pathogenesis of vascular diseases [291].

There is ample evidence that CD151 is important in angiogenesis induction [49, 239, 293, 291]. Though patients with mutations in the CD151 gene and CD151-knockout mice showed no obvious defects in vasculogenesis [120, 273, 209], defects are seen in angiogenesis. CD151 supports EC invasiveness, migration, cable formation, matrigel contraction, tube formation, and sprouting, activities which are all impaired in CD151-knockout mice [239]. CD151 regulates eNOS, Akt, and Rac activity needed for EC adhesion on laminin and angiogenesis [294] and promotes collagenolytic activity via the TEM association of MT1-MMP [281, 239, 295]. Overexpression of CD151 also promotes revascularization and improves blood perfusion in ischemia [133]. The activity of CD151 in angiogenesis is mostly related to associated integrins, particularly laminin-binding integrins [148, 291] and tetraspanin-associated proteases, such as MMP9 [222], MMP7 [84], and MT1-MMP [281].

Besides CD151, CD9 also plays a role in angiogenesis. In EC, CD9 modulates migration [121, 280] and leukocyte transendothelial migration [11, 150]. CD9 is also expressed in smooth muscle cells, the major constituent of muscular arteries. Phenotypic changes upon damage lead to neointima formation, where neointimal hyperplasia contributes to restenosis and vessel occlusion by thrombi [46]. Upon injury, upregulated CD9 expression promotes smooth muscle cell proliferation and migration [126, 141] and by associating with the fibronectin receptor α5β1 tissue remodeling [215, 126].

So far, it has not been explored for CD151 and CD9, whether their contribution to angiogenesis is strictly cell membrane-dependent or can be mediated by exosomes. This question has been explored for Tspan8. Tspan8 is a strong angiogenesis inducer [26] that suffices to induce the angiogenic switch in the absence of angiogenic factors [34, 71]. Stimulated by Tspan8-induced angiogenesis in tumor-distant organs, we searched for induction of angiogenic factors by Tspan8. Though this search did not provide any convincing result, we noted that Tspan8-expressing exosomes are sufficient for angiogenesis induction as well as for endothelial cell progenitor maturation [71, 177]. Essential for cellular and exosomal Tspan8 angiogenesis induction is the association of Tspan8 with $\alpha4\beta1$ in exosomes, where exosomal Tspan8-$\alpha4\beta1$ complexes preferentially target EC and EC progenitors [198]. Tspan8-$\alpha4\beta1$ complex-containing exosomes are internalized by EC and induce VEGF-independent regulation of several angiogenesis-related genes, which is accompanied by enhanced proliferation, migration and sprouting of EC, and by maturation of EC progenitors [177, 198, Rana et al. unpublished].

4.4.3.2 Tetraspanins and Hemostasis

Tumor vessels frequently have thin walls, an incomplete basement membrane, and decreased numbers of pericytes. As a consequence, tumor vessels are leaky, which allows for the extravasation of plasma proteins that form a scaffold for newly migrating EC. In turn, the leakiness of the EC layer facilitates initiation of thrombus formation. Spontaneously occurring focal hemorrhages are a common feature of tumor vessels [65] and a prothrombotic state that can culminate in disseminated intravascular coagulation is frequent in cancer patients, where angiogenesis and the leakiness of the new vessels are considered to be important [38]. The coagulation cascade is initiated when tissue factor, the principal initiator of coagulation, which is provided by many tumor cells, becomes exposed to plasma components. The cascade ends with platelet-bound prothrombin becoming converted to thrombin that initiates clot formation by catalyzing fibrinogen cleavage and fibrin polymerization. Activated platelets, in turn, support angiogenesis by releasing proangiogenic factors such as VEGF and angiopoietin-1. Thrombin also supports angiogenesis by cleaving PAR-1 on EC thereby inducing activation and secretion of proteases including MMPs and uPA [250]. Taken together, overshooting angiogenesis support thrombus formation and the coagulation cascade provides a feedback for angiogenesis [207].

CD63, CD9, and CD151 are abundantly expressed in platelets [77, 60, 217]. Whereas CD151 is required for efficient platelet activation/aggregation [134, 183], $CD9^{-/-}$ mice show alteration in blood coagulation, where CD9 appears to prevent excessive thrombus growth, but does not appear to play a critical role in primary hemostasis [156]. CD9 induces platelet activation via G protein activation, phosphoinosite hydrolysis, and inhibition of adenylate cyclase [77]. It promotes platelet aggregate stability and enhances fibrinogen binding [107]. CD9 is enriched in platelet–platelet contact sites and localizes within alpha granules and in pseudopodia [24]. A direct role of CD63 in platelet aggregation, though most abundantly expressed in platelets, remains to be defined [83]. It possibly could be involved in focal adhesion kinase (FAK) activation and spreading [107]. CD151-knockout mice-provided evidence that CD151, which associates in platelets with $\alpha IIb/\beta3$, accounts for platelet spreading on fibrinogen and for delayed clot retraction [134]. Tspan32-knockout mice also show a prolonged bleeding time, reduced platelet aggregation, and delayed clot retraction. The mechanism has not yet been defined [75]. Finally, Tspan8, though not expressed on platelets, can induce thrombus formation and disseminated intravascular coagulation [34]. The underlying mechanism remains to be elaborated. However, there is evidence that exosomal Tspan8 promotes platelet activation.

Platelet-derived exosomes constitute about 70–90 % of circulating exosomes in the plasma [15] with a life span of about 30 min [61]. The procoagulant activity of platelet-derived exosomes is well known. Specifically, it has been suggested that exosomes provide negatively charged phospholipids, which are required for factor IXa and Xa activation [221]. Though still controversial, the therapeutic efficacy of antiglycoprotein αIIb/β3 could be a consequence of altered platelet exosome formation [173, 200]. In addition, it has been shown in a model of lipopolysaccharide-induced septic shock that platelet-derived exosomes express NO synthase II, a NADPH oxidase subunit, protein disulfide isomerase, and Nox. EC exposed to these platelet-derived exosomes respond with caspase3 activation and become apoptotic [66]. Thus, platelet-derived exosomes contribute to physiological coagulation. Under pathological conditions, they additionally damage the surrounding tissue. Whether platelet-derived exosomal CD151, CD9, Tspan32, and CD63 contribute to the procoagulant activity and/or whether tetraspanins of tumor-derived exosomes initiate undue platelet activation, remains to be explored.

Taken together, the engagement of tetraspanins in angiogenesis and thrombosis has only recently received attention. Nonetheless, platelet accounting for the majority of exosomes in the serum and tetraspanins being constitutive components of exosomes holds promise for a wealth of information in the near future.

4.4.4 Tetraspanins, Exosomes, and Metastasis

Metastasis formation is the final result of a cascade of events that primary tumor cells pass through by changing their phenotype and by their cross-talk with the tumor environment. In epithelial tumors, the metastatic cascade may be initiated through a process called epithelial-mesenchymal transition (EMT) of cancer stem cells/cancer-initiating cells [22, 284], followed by migration of isolated tumor cells, intravasation, extravasation, settlement, and growth in distant organs [70]. Molecules involved in tumor progression are cell–cell and cell-matrix adhesion molecules, matrix-degrading enzymes and their inhibitors. In addition, chemotactic factors released from the degraded matrix and chemokine receptors expressed by the metastasizing tumor cell play an important role [4]. Finally, two tetraspanins, CD151 and Tspan8, are known to promote metastasis.

Tetraspanins are proposed to contribute to the metastatic cascade by their involvement in cell motility due to their association with integrins [16, 236, 92, 143], regulation of invasiveness due to their association with peptidases [135, 198] and MMP [131, 281], or through modulating MMP transcription and secretion [84].

The first evidence for CD151 as a metastasis-promoting molecule derived from a study in which an anti-CD151 antibody inhibited metastasis of a human epidermoid carcinoma line in a chick embryo model. The antibody inhibited cell migration, but not cell adhesion or growth [244]. Meanwhile, an association between high CD151 expression and a poor prognosis has been described for many cancers [298, 299, 266].

CD151-promoted motility of metastasizing tumor cell migration is mostly due to its association with laminin-binding integrins α3β1, α6β1, and α6β4, where the association with CD151 is important for proper localization of laminin332-binding integrins during tumor cell–stromal cell interactions. In line with this, a CD151-knockdown carcinoma line revealed impaired motility, anomalously persistent adhesive contacts, and impaired integrin α3α1 internalization [270]. Further confirming the importance of CD151 for integrin traffic, expression of a CD151 molecule with a mutation of the sorting motif in the C-terminal domain markedly attenuates endocytosis of the CD151-associated integrins α3β1, α5β1, and α6β1 [147]. Thus, CD151 plays a critical role in integrin recycling as a mechanism to regulate tumor cell migration. Quantitative in vivo assays and intravital imaging using the chicken chorio-allantois membrane model confirmed the impact of CD151 on tumor cell migration. A CD151-specific antibody inhibits matrix-supported migration, but has no impact on extravasation. Migration inhibition is due to a failure to detach at the rear end. As migration of CD151-knockout cells was not affected, the authors suggested that—when present—CD151 might recruit partner molecules that control deadhesion, but this process is suppressed in the presence of the CD151 antibody [297].

Furthermore, CD151 contributes to pericellular activation of MMPs by associating with proMMP7, which results in activation of MMP7 [223]. In addition, CD151 has a positive effect on MMP9 expression through FAK, Src, p38, and JNK kinases activation. Signaling is initiated via CD151-associated integrins α3β1 or α6β1 and is stimulated by CD151 homophilic interactions [100, 285]. Reduced expression of MMP2, MMP7, and MMP9 in a CD151-knockdown carcinoma line confirmed the involvement of CD151 in MMP expression, complex formation, and colocalization at the leading edge of lamellipodia [223, 84]. CD151 also associates directly with MMP14/MT1-MMP [281]. MMP14 is a fundamental protease for matrix proteolysis. Besides degrading several components of the extracellular matrix, MMP14 is also a central activator of MMP2 [212]. The recruitment of MMP14 into TEM and thereby into the vicinity of CD151-associated laminin-binding integrins has been demonstrated to be crucial for focalized matrix degradation [281]. Thus, CD151 contributes to cell migration by associating with integrins and MMPs, by stimulating MMP production and by inducing integrin and MMP internalization and relocalization [92, 283].

Tspan8 overexpression mostly in gastrointestinal tumors as well as in melanoma correlates with metastasis [299, 20]. The few reports on the molecular mechanisms, whereby Tspan8 promotes metastasis, point toward similar mechanism as described for CD151. Tspan8-promoted tumor cell motility involves its association with α6β4 [94, 71]. Tspan8 associates with α6β4 only under stress conditions, when hemidesmosomes become dissolved and α6β4 becomes recruited into TEM. The association is accompanied by internalization of the Tspan8-α6β4 complex, which either rapidly recycles [103, 94] or becomes integrated into MVB and is released in exosomes [115]. Recycling is accompanied by increased motility [103, 94]. Continuing internalization to the endosomal compartment and rapid recycling back to the cell surface via a short loop recycling machinery under the control of rab4 has been described for several integrins [28] and may well account for the motility-promoting activity of Tspan8. Furthermore, Tspan8-expressing tumors interact with platelets, which could provide tumor cells with a shield supporting survival in the hostile environment encountered during metastatic spread [118], but also may induce platelet activation. Finally, Tspan8 associates with MMP14, which promote motility by matrix degradation (Rana et al. unpublished). It should be mentioned that, distinct to CD151, upregulated Tspan8 expression has also been described to promote proliferation [241]. We recently unraveled that this likely is a consequence of the association of Tspan8 with EpCAM and TACE. Upon PMA stimulation, EpCAM becomes cleaved by TACE. Subsequently, presenilin2 cleaves the intracellular domain of EpCAM, which is known to act as a cotranscription factor of β-catenin, promoting, besides others, c-myc transcription [154] PMA-induced EpCAM cleavage by TACE and presenilin essentially depends on EpCAM as well as TACE recruitment into TEM and association with Tspan8 (Thuma and Zöller submitted).

As already stated, TEM-located tetraspanin complexes are recovered in exosomes and all components are functionally active. As outlined, the tetraspanin integrin association is crucial for target cell selection and there is strong evidence that matrix binding by exosomal integrins is strengthened by the association with tetraspanins. Also, the collagenolytic and laminin-degradating activity of exosomes has repeatedly been described and demonstrated to facilitate angiogenesis and metastasis [208, 82, 76, 93,179, 186]. We are not aware of published evidence on the exosomal association of CD151 and Tspan8 with MMP14 or TACE, which, however, we have seen for both proteases in exosomes derived from a metastasizing rat pancreatic tumor line. Accordingly, exosomes degraded several matrix proteins, most strongly collagen IV and laminins (Rana et al. unpublished).

Last, not least, we want to discuss a possible engagement of exosomal tetraspanins in premetastatic niche formation. It has been hypothesized since more than 100 years that a given tumor type seeds in selected organs [240]. The seed and soil hypothesis has become reinforced with the description of the premetastatic niche, where it has been shown that tumors prepare defined organs for metastasizing cell settlement in advance of arrival [70, 119, 167]. Major contributors to premetastatic niche preparation have been suggested to be chemokines, chemokine receptors, adhesion molecules, and proteases [21, 27]. Our suggestion that rather than individual molecules, tumor-derived exosomes may play an important role in premetastatic niche preparation [115] has been confirmed by several groups [29, 101, 137, 145, 166, 186, 188, 196, 230]. Furthermore, there is strong evidence that premetastatic niche preparation via tumor-derived exosomes proceeds via exosomes directly affecting stromal elements in the premetastatic organ as well as by forcing hematopoietic progenitors to leave the bone marrow and contribute to preparation of the niche in the premetastatic organ [188].

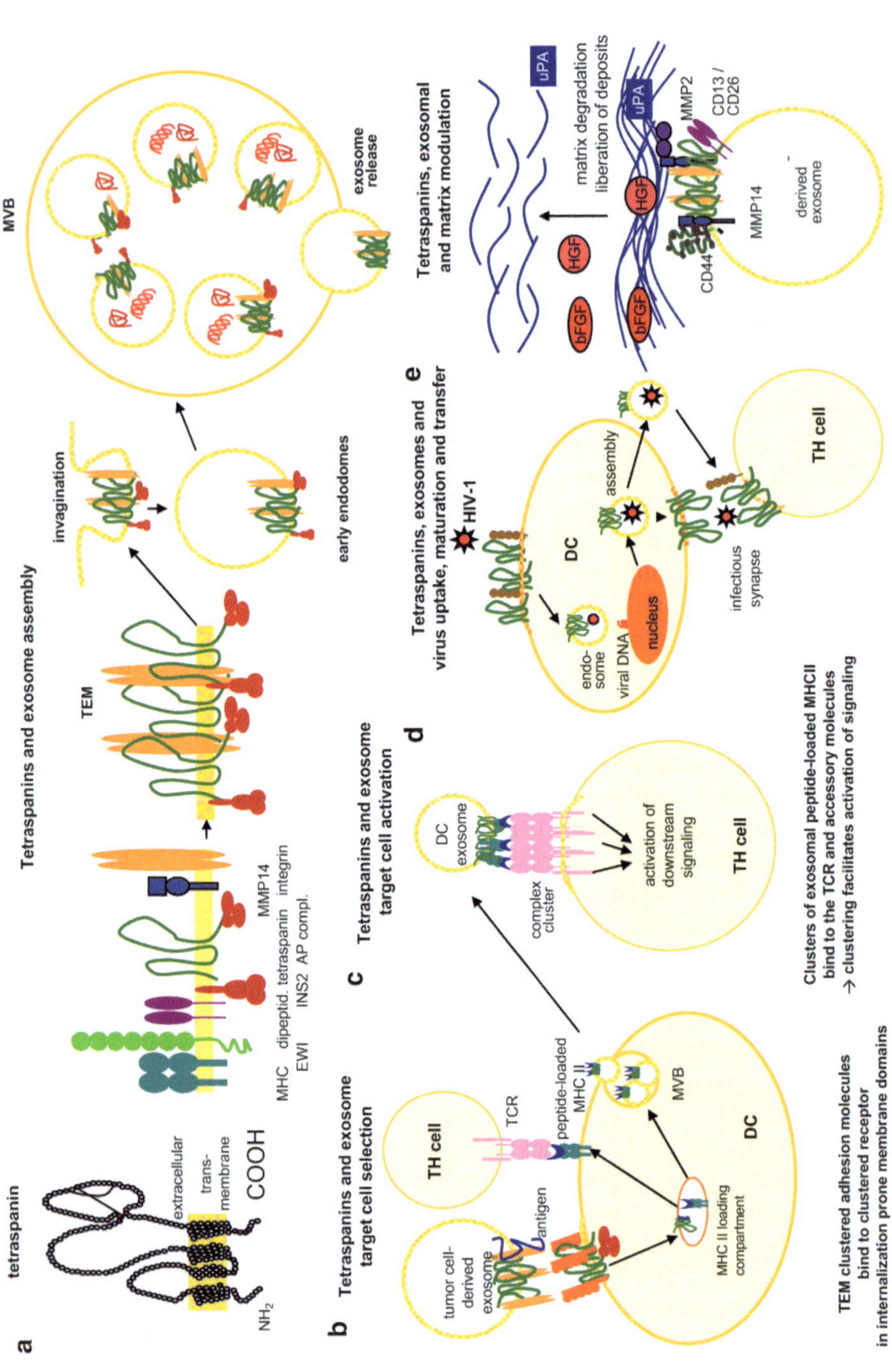
a
tetraspanin
extracellular
trans-
membrane
COOH
NH2
MHC
EWI
dipeptid.
INS2
tetraspanin
AP compl.
MMP14
integrin
Tetraspanins and exosome assembly
TEM
invagination
early endodomes
MVB
exosome
release
b
Tetraspanins and exosome
target cell selection
tumor cell-
derived
exosome
antigen
TH cell
TCR
peptide-loaded
MHC II
MVB
MHC II loading
compartment
DC
TEM clustered adhesion molecules
bind to clustered receptor
in internalization prone membrane domains
c
Tetraspanins and exosome
target cell activation
DC
exosome
complex
cluster
activation of
downstream
signaling
TH cell
Clusters of exosomal peptide-loaded MHCII
bind to the TCR and accessory molecules
→ clustering facilitates activation of signaling
d
Tetraspanins, exosomes and
virus uptake, maturation and transfer
HIV-1
DC
endo-
some
viral DNA
nucleus
assembly
infectious
synapse
TH cell
e
Tetraspanins, exosomal
and matrix modulation
uPA
bFGF
HGF
matrix degradation
liberation of deposits
MMP2
CD13 /
CD26
CD44
MMP14
derived
exosome

Fig. 4.1 Exosomal tetraspanins in intercellular communication. **a** *Tetraspanins* are 4-TM molecules located in glycolipid-enriched membrane microdomains prone for internalization. They can associate with many TM and intercellular molecules (examples: *MHC* molecules, *EWI, dipeptidases, MMP14, integrins*, as well as *INS2* and the *AP complex* both being engaged in tetraspanin internalization). By associating with tetraspanins, molecules are recruited into *TEM* and become internalized together with the tetraspanins into early endosomes, which integrate into *MVB*, fuse with the plasma membrane, and are released as exosomes. Accordingly, tetraspanin and tetraspanin-associated molecules are enriched in exosomes. **b** Exosomal tetraspanin complexes are engaged in target cell selection, where they bind to corresponding membrane microdomains, which facilitates their internalization (example: uptake of exosome by *DC*). **c** Clustering of tetraspanin-associated molecules facilitates synapse formation and activation of signaling in the target cell (example: *TH* activation by DC-derived exosomes). **d** Some viruses can bind to the large extracellular domain of tetraspanins or to tetraspanin-associated molecules, which facilitates their internalization and maturation in intracellular vacuoles as well as their release in exosomes (example: HIV binding to tetraspanin-associated CD4 in DC, HIV-1 assembly in tetraspanin-enriched intracellular vacuoles, and transfection via the tetraspanin-enriched infectious synapse or via released exosomes). **e** Tetraspanins associate with several matrix-binding proteins (examples: integrins, *CD44*) as well as matrix-degrading enzymes, particularly *MMP14*. Tetraspanin-adhesion molecule clusters in exosomes promote a strong interaction with the extracellular matrix, which facilitates degradation by associated MMPs and liberation of matrix-deposited growth factors and enzymes. In addition, many matrix protein degradation products exert biological activities

What is the role of exosomal tetraspanins in premetastatic niche preparation? It has been repeatedly demonstrated that tumor exosomes reach the premetastatic organ [115, 167] and, as outlined above, it likely is the TEM-localized tetraspanin-integrin complex that accounts for the selectivity of target cell binding, where exosomal CD9-αMβ2, -αLβ2, and -αXβ2 complexes preferentially target bone marrow cells, whereas Tspan8- and CD151-α3β1 and α6β4 complexes target lymph node stroma cells and in our model, where the tumor metastasizes to the lung [164], also lung fibroblasts [115, Rana et al. submitted]. In concern of the target cell modulation by tumor exosomes, changes in premetastatic stroma cells were mostly due to the transfer of miRNA rather than to protein or mRNA transfer (Rana et al. submitted), although we cannot exclude a contribution of tetraspanins to miRNA recruitment into endosomes/MVB.

Taken together, both metastasis-supporting tetraspanins CD151 and Tspan8 are abundantly recovered in tumor exosomes, where they are associated with α3β1, but more strongly with α6β4. The TEM clustering of the integrins strengthens the adhesion of exosomes to collagens and laminins and will facilitate matrix degradation by tetraspanin-associated MMP14 and possibly MMP2 and MMP9. Thereby tumor-derived exosomal TEM contribute to creating a path for migrating tumor cells and sprouting EC. Beyond this, matrix-degradation products and the liberation of matrix-embedded chemokines and cytokines will facilitate tumor cell and EC motility and the preparation of a premetastatic niche. With respect to the latter, it is again the tetraspanin-integrin complex that accounts for target cell selection, but exosomal tetraspanin complexes may not essentially contribute to target cell modulation.

4.5 Conclusion

Exosomes are enriched in tetraspanins, which are—as in the cell membrane—organized in TEM. It is this organization in glycolipid-enriched membrane domains, which are prone to internalization that accounts for the tetraspanin enrichment in exosomes. This also implies recovery of clusters of tetraspanins with associated molecules, reflecting the TEM composition in the cell membrane, in exosomes. As summarized in Fig. 4.1, these exosomal TEM are essential for several exosome activities:

1. Exosomal TEM are an essential component in target cell selection, mostly due to tetraspanin-integrin clusters.
2. The tetraspanin-associated internalization and sorting complexes facilitate exosome uptake by selected target cells.
3. The tetraspanin-TM protein complexes allow exosomes to fulfill activities similar to the target cell. Prominent examples are:

   (a) Tetraspanin-clustered MHC II molecules in DC-derived exosomes, which allow for T cell activation.

(b) Capture of selected viruses and transfer of viruses and other infectious organisms.
(c) Initiation of clotting and thrombus formation.
(d) Concomitant clustering of adhesion molecules and MMP, which facilitates ECM binding and degradation, which is important in tissue remodeling and metastasizing tumor cell motility and survival.

4. Though not yet proven, tetraspanins may be engaged in mRNA and miRNA recruitment. Irrespective of their direct engagement, by the tetraspanin engagement in exosome uptake by target cells, they become involved in the horizontal gene transfer, that might possibly be the most important activity of exosomes.

By all these means, tetraspanins become one of the most important parameters in intercellular communication via exosomes. In view of the exciting power of exosomes, it becomes demanding to explore the therapeutic potential of blocking undue exosomal activities via tetraspanins.

References

1. Abache T, Le Naour F, Planchon S, Harper F, Boucheix C, Rubinstein E (2007) The transferrin receptor and the tetraspanin web molecules CD9, CD81, and CD9P-1 are differentially sorted into exosomes after TPA treatment of K562 cells. J Cell Biochem 102:650–664
2. Admyre C, Bohle B, Johansson SM, Focke-Tejkl M, Valenta R, Scheynius A, Gabrielsson S (2007) B cell-derived exosomes can present allergen peptides and activate allergen-specific T cells to proliferate and produce TH2-like cytokines. J Allergy Clin Immunol 120:1418–1424
3. Admyre C, Johansson SM, Paulie S et al (2006) Direct exosome stimulation of peripheral human T cells detected by ELISPOT. Eur J Immunol 36:1772–1781
4. Albini A, Mirisola V, Pfeffer U (2008) Metastasis signatures: genes regulating tumor-microenvironment interactions predict metastatic behavior. Cancer Metastasis Rev 27:75–83
5. André F, Schartz NE, Chaput N, Flament C, Raposo G, Amigorena S, Angevin E, Zitvogel L (2002) Tumor-derived exosomes: a new source of tumor rejection antigens. Vaccine 20(Suppl 4):A28–A31
6. André M, Le Caer JP, Greco C et al (2006) Proteomic analysis of the tetraspanin web using LC-ESI-MS/MS and MALDI-FTICR-MS. Proteomics 6:1437–1449
7. Ardón-Alonso M, Yañez-Mó M, Barreiro O, Alvarez S, Muñoz-Fernández MA, Valenzuela-Fernández A, Sánchez-Madrid F (2006) Tetraspanins CD9 and CD81 modulate HIV-1-induced membrane fusion. J Immunol 177:5129–5137
8. Arduise C, Abache T, Li L, Billard M, Chabanon A, Ludwig A, Mauduit P, Boucheix C, Rubinstein E, Le Naour F (2008) Tetraspanins regulate ADAM10-mediated cleavage of TNF-alpha and epidermal growth factor. J Immunol 181:7002–7013
9. Artavanis-Tsakonas K, Kasperkovitz PV, Papa E, Cardenas ML, Khan NS, Van Der Veen AG, Ploegh HL, Vyas JM (2011) The tetraspanin CD82 is specifically recruited to fungal and bacterial phagosomes prior to acidification. Infect Immun 79:1098–1106
10. Atay S, Gercel-Taylor C, Taylor DD (2011) Human trophoblast-derived exosomal fibronectin induces pro-inflammatory Il-1β production by macrophages. Am J Reprod Immunol 66:259–69
11. Barreiro O, Yáñez-Mó M, Sala-Valdés M, Gutiérrez-López MD, Ovalle S, Higginbottom A, Monk PN, Cabañas C, Sánchez-Madrid F (2005) Endothelial tetraspanin microdomains regulate leukocyte firm adhesion during extravasation. Blood 105:2852–2861

12. Bartee E, Eyster CA, Viswanathan K, Mansouri M, Donaldson JG, Früh K (2010) Membrane-associated RING-CH proteins associate with Bap31 and target CD81 and CD44 to lysosomes. PLoS ONE 5:e15132
13. Bass R, Werner F, Odintsova E, Sugiura T, Berditchevski F, Ellis V (2005) Regulation of urokinase receptor proteolytic function by the tetraspanin CD82. J Biol Chem 280:14811–14818
14. Bauer M, Pelkmans L (2006) A new paradigm for membrane-organizing and -shaping scaffolds. FEBS Lett 580:5559–64
15. Berckmans RJ, Neiuwland R, Böing AN, Romijn FP, Hack CE, Sturk A (2001) Cell-derived microparticles circulate in healthy humans and support low grade thrombin generation. Thromb Haemost 85:639–646
16. Berditchevski F (2001) Complexes of tetraspanins with integrins: more than meets the eye. J Cell Sci 114:4143–4151
17. Berditchevski F, Odintsova E (2007) Tetraspanins as regulators of protein trafficking. Traffic 8:89–96
18. Berditchevski F, Tolias KF, Wong K, Carpenter CL, Hemler ME (1997) A novel link between integrins, transmembrane-4 superfamily proteins (CD63 and CD81), and phosphatidylinositol 4-kinase. J Biol Chem 272:2595–2598
19. Berditchevski F, Odintsova E, Sawada S, Gilbert E (2002) Expression of the palmitoylation-deficient CD151 weakens the association of alpha 3 beta 1 integrin with the tetraspanin-enriched microdomains and affects integrin-dependent signaling. J Biol Chem 277:36991–37000
20. Berthier-Vergnes O, Kharbili ME, de la Fouchardière A, Pointecouteau T, Verrando P, Wierinckx A, Lachuer J, Le Naour F, Lamartine J (2011) Gene expression profiles of human melanoma cells with different invasive potential reveal TSPAN8 as a novel mediator of invasion. Br J Cancer 104:155–165
21. Bidard FC, Pierga JY, Vincent-Salomon A, Poupon MF (2008) A "class action" against the microenvironment: do cancer cells cooperate in metastasis? Cancer Metastasis Rev 27:5–10
22. Brabletz T, Jung A, Spaderna S, Hlubek F, Kirchner T (2005) Opinion: migrating cancer stem cells—an integrated concept of malignant tumour progression. Nat Rev Cancer 5:744–749
23. Bradbury LE, Kansas GS, Levy S, Evans RL, Tedder TF (1992) The CD19/CD21 signal transducing complex of human B lymphocytes includes the target of antiproliferative antibody-1 and Leu-13 molecules. J Immunol 149:2841–2850
24. Brisson C, Azorsa DO, Jennings LK, Moog S, Cazenave JP, Lanza F (1997) Co-localization of CD9 and GPIIb-IIIa (alpha IIb beta 3 integrin) on activated platelet pseudopods and alpha-granule membranes. Histochemistry J 29:153–165
25. Burghoff S, Ding Z, Gödecke S, Assmann A, Wirrwar A, Buchholz D, Sergeeva O, Leurs C, Hanenberg H, Müller HW, Bloch W, Schrader J (2008) Horizontal gene transfer from human endothelial cells to rat cardiomyocytes after intracoronary transplantation. Cardiovasc Res 77:534–543
26. Bussolati B, Grange C, Camussi G (2011) Tumor exploits alternative strategies to achieve vascularization. FASEB J 25:2874–82
27. Carlini MJ, De Lorenzo MS, Puricelli L (2011) Cross-talk between Tumor Cells and the Microenvironment at the Metastatic Niche. Curr Pharm Biotechnol Apr 6. (Epub ahead of print)
28. Caswell P, Norman J (2008) Endocytic transport of integrins during cell migration and invasion. Trends Cell Biol 18:257–263
29. Chairoungdua A, Smith DL, Pochard P, Hull M, Caplan MJ (2010) Exosome release of β-catenin: a novel mechanism that antagonizes Wnt signaling. J Cell Biol 190:1079–91
30. Charrin S, Le Naour F, Oualid M, Billard M, Faure G, Hanash SM, Boucheix C, Rubinstein E (2001) The major CD9 and CD81 molecular partner. Identification and characterization of the complexes. J Biol Chem 276:14329–14337
31. Charrin S, Manié S, Oualid M, Billard M, Boucheix C, Rubinstein E (2002) Differential stability of tetraspanin/tetraspanin interactions: role of palmitoylation. FEBS Lett 516:139–144

32. Charrin S, Manié S, Thiele C, Billard M, Gerlier D, Boucheix C, Rubinstein E (2003) A physical and functional link between cholesterol and tetraspanins. Eur J Immunol 33:2479–2489
33. Cherukuri A, Shoham T, Sohn HW, Levy S, Brooks S, Carter R, Pierce SK (2004) The tetraspanin CD81 is necessary for partitioning of coligated CD19/CD21-B cell antigen receptor complexes into signaling-active lipid rafts. J Immunol 172:370–380
34. Claas C, Seiter S, Claas A, Savelyeva L, Schwab M, Zöller M (1998) Association between the rat homologue of CO-029, a metastasis-associated tetraspanin molecule and consumption coagulopathy. J Cell Biol 141:267–280
35. Claas C, Stipp CS, Hemler ME (2001) Evaluation of prototype transmembrane 4 superfamily protein complexes and their relation to lipid rafts. J Biol Chem 276:7974–7984
36. Claas C, Wahl J, Orlicky DJ, Karaduman H, Schnölzer M, Kempf T, Zöller M (2005) The tetraspanin D6.1A and its molecular partners on rat carcinoma cells. Biochem J 389:99–110
37. Classen A, Lloberas J, Celada A (2009) Macrophage activation: classical versus alternative. Methods Mol Biol 531:29–43
38. De Cicco M (2004) The prothrombotic state in cancer: pathogenic mechanisms. Crit Rev Oncol Hematol 50:187–196
39. de Parseval A, Lerner DL, Borrow P, Willett BJ, Elder JH (1997) Blocking of feline immunodeficiency virus infection by a monoclonal antibody to CD9 is via inhibition of virus release rather than interference with receptor binding. J Virol 71:5742–5749
40. Delaguillaumie A, Lagaudrière-Gesbert C, Popoff MR, Conjeaud H (2002) Rho GTPases link cytoskeletal rearrangements and activation processes induced via the tetraspanin CD82 in T lymphocytes. J Cell Sci 115:433–443
41. Delcayre A, Shu H, Le Pecq JB (2005) Dendritic cell-derived exosomes in cancer immunotherapy: exploiting nature's antigen delivery pathway. Expert Rev Anticancer Ther 5:537–547
42. Deneka M, Pelchen-Matthews A, Byland R, Ruiz-Mateos E, Marsh M (2007) In macrophages, HIV-1 assembles into an intracellular plasma membrane domain containing the tetraspanins CD81, CD9, and CD53. J Cell Biol 177:329–341
43. Deng J, Yeung VP, Tsitoura D, DeKruyff RH, Umetsu DT, Levy S (2000) Allergen-induced airway hyperreactivity is diminished in CD81-deficient mice. J Immunol 165:5054–5061
44. Denzer K, Kleijmeer MJ, Heijnen HF, Stoorvodel W, Geuze HJ (2000) Exosome: from internal vesicle of the multivesicular body to intercellular signaling device. J Cell Sci 113:3365–3374
45. Deregibus MC, Cantaluppi V, Calogero R, Lo Iacono M, Tetta C, Biancone L, Bruno S, Bussolati B, Camussi G (2007) Endothelial progenitor cell derived microvesicles activate an angiogenic program in endothelial cells by a horizontal transfer of mRNA. Blood 110:2440–2448
46. Desai M, Mirzay-Razzaz J, von Delft D, Sarkar S, Hamilton G, Seifalian AM (2010) Inhibition of neointimal formation and hyperplasia in vein grafts by external stent/sheath. Vasc Med 15:287–297
47. Dong LM, Potter JD, White E, Ulrich CM, Cardon LR, Peters U (2008) Genetic susceptibility to cancer: the role of polymorphisms in candidate genes. JAMA 299:2423–2436
48. Drummer HE, Boo I, Maerz AL, Poumbourios P (2006) A conserved Gly436-Trp-Leu-Ala-Gly-Leu-Phe-Tyr motif in hepatitis C virus glycoprotein E2 is a determinant of CD81 binding and viral entry. J Virol 80:7844–7853
49. Dumartin L, Quemener C, Laklai H Dumartin L, Quemener C, Laklai H (2010) Netrin-1 mediates early events in pancreatic adenocarcinoma progression, acting on tumor and endothelial cells. Gastroenterology 138:1595–1606
50. Engering A, Pieters J (2001) Association of distinct tetraspanins with MHC class II molecules at different subcellular locations in human immature dendritic cells. Int Immunol 13:127–134
51. Escola JM, Kleijmeer MJ, Stoorvogel W, Griffith JM, Yoshie O, Geuze HJ (1998) Selective enrichment of tetraspan proteins on the internal vesicles of multivesicular endosomes and on exosomes secreted by human B-lymphocytes. J Biol Chem 273:20121–20127

52. Escrevente C, Keller S, Altevogt P, Costa J (2011) Interaction and uptake of exosomes by ovarian cancer cells. BMC Cancer 11:108
53. Espenel C, Margeat E, Dosset P, Arduise C, Le Grimellec C, Royer CA, Boucheix C, Rubinstein E, Milhiet PE (2008) Single-molecule analysis of CD9 dynamics and partitioning reveals multiple modes of interaction in the tetraspanin web. J Cell Biol 182:765–776
54. Evergren E, Gad H, Walther K, Sundborger A, Tomilin N, Shupliakov O (2007) Intersectin is a negative regulator of dynamin recruitment to the synaptic endocytic zone in the central synapse. J Neurosci 27:379–390
55. Fanaei M, Monk PN, Partridge LJ (2011) The role of tetraspanins in fusion. Biochem Soc Trans 39:524–528
56. Fang Y, Wu N, Gan X Yan W, Morrell JC, Gould SJ (2007) Higher-order oligomerization targets plasma membrane proteins and HIV gag to exosomes. PLoS Biol 5:e158
57. Farquhar MJ, Harris HJ, McKeating JA (2011) Hepatitis C virus entry and the tetraspanin CD81. Biochem Soc Trans 39:532–536
58. Fevrier B, Raposo G (2004) Exosomes: endosomal-derived vesicles shipping extracellular messages. Curr Opin Cell Biol 16:415–421
59. Figdor CG, van Spriel AB (2010) Fungal pattern-recognition receptors and tetraspanins: partners on antigen-presenting cells. Trends Immunol 31:91–96
60. Fitter S, Tetaz T, Berndt MC, Ashman LK (1995) Molecular cloning of cDNA encoding a novel platelet, endothelial tetraspan antigen, PETA-3. Blood 86:1348–1355
61. Flaumenhaft R (2006) Formation and fate of platelet microparticles. Blood Cells Mol Dis 36:182–187
62. Folkman J (2004) Endogenous angiogenesis inhibitors. APMIS 112:496–507
63. Fomina AF, Deerinck TJ, Ellisman MH (2003) Regulation of membrane trafficking and subcellular organization of endocytic compartments revealed with FM1-43 in resting and activated human T cells. Exp Cell Res 291:150–166
64. Forman JD, Klein JT, Silver RF, Liu MC, Greenlee BM, Moller DR (1994) Selective activation and accumulation of oligoclonal V beta-specific T cells in active pulmonary sarcoidosis. J Clin Invest 94:1533–1542
65. Franchini M, Montagnana M, Targher G, Lippi G (2007) Reduced von Willebrand factor-cleaving protease levels in secondary thrombotic microangiopathies and other diseases. J Thromb Thrombolysis 24:29–38
66. Gambim MH, do Carmo Ade O, Marti L, Veríssimo-Filho S, Lopes LR, Janiszewski M (2007) Platelet-derived exosomes induce endothelial cell apoptosis through peroxynitrite generation: experimental evidence for a novel mechanism of septic vascular dysfunction. Crit Care 11:R107
67. Garcia E, Nikolic DS, Piguet V (2008) HIV-1 replication in dendritic cells occurs through a tetraspanin-containing compartment enriched in AP-3. Traffic 9:200–214
68. Garcia-España A, Chung PJ, Sarkar IN, Stiner E, Sun TT, Desalle R (2008) Appearance of new tetraspanin genes during vertebrate evolution. Genomics 91:326–334
69. Gartlan KH, Belz GT, Tarrant JM, Minigo G, Katsara M, Sheng KC, Sofi M, van Spriel AB, Apostolopoulos V, Plebanski M, Robb L, Wright MD (2010) A complementary role for the tetraspanins CD37 and Tssc6 in cellular immunity. J Immunol 185:3158–3166
70. Geiger TR, Peeper DS (2009) Metastasis mechanisms. Biochim Biophys Acta 1796:293–308
71. Gesierich S, Paret C, Hildebrand D, Weitz J, Zgraggen K, Schmitz-Winnenthal FH, Horejsi V, Yoshie O, Herlyn D, Ashman LK, Zöller M. (2005) Colocalization of the tetraspanins, CO-029 and CD151, with integrins in human pancreatic adenocarcinoma: impact on cell motility. Clin Cancer Res 11:2840–2852
72. Gesierich S, Berezovskiy I, Ryschich E, Zöller M (2006) Systemic induction of the angiogenesis switch by the tetraspanin D6.1A/CO-029. Cancer Res 66:7083–7094
73. Gibbings DJ, Ciaudo C, Erhardt M, Voinnet O (2009) Multivesicular bodies associate with components of miRNA effector complexes and modulate miRNA activity. Nat Cell Biol 11:1143–1149

74. Gordón-Alonso M, Yañez-Mó M, Barreiro O, Alvarez S, Muñoz-Fernández MA, Valenzuela-Fernández A, Sánchez-Madrid F (2006) Tetraspanins CD9 and CD81 modulate HIV-1-induced membrane fusion. J Immunol 177:5129–5137
75. Goschnick MW, Lau LM, Wee JL, Liu YS, Hogarth PM, Robb LM, Hickey MJ, Wright MD, Jackson DE (2006) Impaired "outside-in" integrin alphaIIbbeta3 signaling and thrombus stability in TSSC6-deficient mice. Blood 108:1911–1918
76. Grange C, Tapparo M, Collino F, Vitillo L, Damasco C, Deregibus MC, Tetta C, Bussolati B, Camussi G (2011) Microvesicles released from human renal cancer stem cells stimulate angiogenesis and formation of lung premetastatic niche. Cancer Res 71:5346–556
77. Griffith L, Slupsky J, Seehafer J, Boshkov L, Shaw AR (1991) Platelet activation by immobilized monoclonal antibody: evidence for a CD9 proximal signal. Blood 78:1753–1759
78. Grigorov B, Attuil-Audenis V, Perugi F, Nedelec M, Watson S, Pique C, Darlix JL, Conjeaud H, Muriaux D (2009) A role for CD81 on the late steps of HIV-1 replication in a chronically infected T cell line. Retrovirology 6:28
79. Gruenberg J, Stenmark H (2004) The biogenesis of multivesicular endosomes. Nat Rev Mol Cell Biol 5:317–323
80. Guo J, Wenk MR, Pellegrini L, Onofri F, Benfenati F, De Camilli P (2003) Phosphatidylinositol 4-kinase type IIalpha is responsible for the phosphatidylinositol 4-kinase activity associated with synaptic vesicles. Proc Natl Acad Sci U S A 100:3995–4000
81. Hakomori SI (2010). Glycosynaptic microdomains controlling tumor cell phenotype through alteration of cell growth, adhesion, and motility. FEBS Lett 584:1901–1906
82. Hakulinen J, Sankkila L, Sugiyama N, Lehti K, Keski-Oja J (2008) Secretion of active membrane type 1 matrix metalloproteinase (MMP-14) into extracellular space in microvesicular exosomes. J Cell Biochem 105:1211–1218
83. Hamamoto K, Ohga S, Nomura S, Yasunaga K (1994) Cellular distribution of CD63 antigen in platelets and in three megakaryocytic cell lines. Histochem J 26:367–375
84. Hasegawa M, Furuya M, Kasuya Y, Nishiyama M, Sugiura T, Nikaido T, Momota Y, Ichinose M, Kimura S (2007) CD151 dynamics in carcinoma-stroma interaction: integrin expression, adhesion strength and proteolytic activity. Lab Invest 87:882–892
85. Hashimoto A, Tarner IH, Bohle RM, Gaumann A, Manetti M, Distler O, Steinmeyer J, Ulfgren AK, Schulz A, Gay S, Müller-Ladner U, Neumann E (2007) Analysis of vascular gene expression in arthritic synovium by laser-mediated microdissection. Arthritis Rheum 56:1094–1105
86. Hassuna N, Monk PN, Moseley GW, Partridge LJ (2009) Strategies for targeting tetraspanin proteins: potential therapeutic applications in microbial infections. BioDrugs 23:341–359
87. Hasuwa H, Shishido Y, Yamazaki A, Kobayashi T, Yu X, Mekada E (2001) CD9 amino acids critical for upregulation of diphtheria toxin binding. Biochem Biophys Res Commun 289:782–790
88. He B, Liu L, Cook GA, Grgurevich S, Jennings LK, Zhang XA. (2005) Tetraspanin CD82 attenuates cellular morphogenesis through down-regulating integrin alpha6-mediated cell adhesion. J Biol Chem 280:3346–3354
89. Helle F, Dubuisson J (2008) Hepatitis C virus entry into host cells. Cell Mol Life Sci 65:100–112
90. Hemler ME (2001) Specific tetraspanin functions. J Cell Biol 155:1103–1107
91. Hemler ME (2003) Tetraspanin proteins mediate cellular penetration, invasion, and fusion events and define a novel type of membrane microdomain. Annu Rev Cell Dev Biol 19:397–422
92. Hemler ME (2005) Tetraspanin functions and associated microdomains. Nat Rev Mol Cell Biol 6:801–811
93. Hendrix A, Westbroek W, Bracke M, De Wever O (2010) An ex(o)citing machinery for invasive tumor growth. Cancer Res 70:9533–9537
94. Herlevsen M, Schmidt DS, Miyazaki K, Zöller M (2003) The association of the tetraspanin D6.1 A with the alpha6beta4 integrin supports cell motility and liver metastasis formation. J Cell Sci 116:4373–4390

95. Herrera MB, Fonsato V, Gatti S, Deregibus MC, Sordi A, Cantarella D, Calogero R, Bussolati B, Tetta C, Camussi G (2010) Human liver stem cell-derived microvesicles accelerate hepatic regeneration in hepatectomized rats. J Cell Mol Med 14:1605–1618
96. Hickey MM, Simon MC (2006) Regulation of angiogenesis by hypoxia and hypoxia-inducible factors. Curr Top Dev Biol 76:217–257
97. Hillen F, Griffioen AW (2007) Tumour vascularization: sprouting angiogenesis and beyond. Metastasis Rev 26:489–502
98. Ho SH, Martin F, Higginbottom A, Partridge LJ, Parthasarathy V, Moseley GW, Lopez P, Cheng-Mayer C, Monk PN (2006) Recombinant extracellular domains of tetraspanin proteins are potent inhibitors of the infection of macrophages by human immunodeficiency virus type 1. J Virol 80:6487–6496
99. Holderfield MT, Hughes CC (2008) Crosstalk between vascular endothelial growth factor, notch, and transforming growth factor-beta in vascular morphogenesis. Circ Res 102:637–652
100. Hong IK, Jin YJ, Byun HJ, Jeoung DI, Kim YM, Lee H (2006) Homophilic interactions of Tetraspanin CD151 up-regulate motility and matrix metalloproteinase-9 expression of human melanoma cells through adhesion-dependent c-Jun activation signaling pathways. J Biol Chem 281:24279–24292
101. Hood JL, San Roman S, Wickline SA (2011) Exosomes released by melanoma cells prepare sentinel lymph nodes for tumor metastasis. Cancer Res 71:3792–801
102. Horváth G, Serru V, Clay D, Billard M, Boucheix C, Rubinstein E (1998) CD19 is linked to the integrin-associated tetraspans CD9, CD81, and CD82. J Biol Chem 273:30537–30543
103. Huerta S, Harris DM, Jazirehi A, Bonavida B, Elashoff D, Livingston EH, Heber D (2003) Gene expression profile of metastatic colon cancer cells resistant to cisplatin-induced apoptosis. Int J Oncol 22:663–670
104. Hurley JH, Emr SD (2006) The ESCRT complexes: structure and mechanism of a membrane-trafficking network. Annu Rev Biophys Biomol Struct 35:277–298
105. Hwang I, Ki D (2011) Receptor-mediated T cell absorption of antigen presenting cell-derived molecules. Front Biosci 16:411–421
106. Iero M, Valenti R, Huber V, Filipazzi P, Parmiani G, Fais S, Rivoltini L (2008) Tumour-released exosomes and their implications in cancer immunity. Cell Death Differ 15:80–88
107. Israels SJ, McMillan-Ward EM (2010) Palmitoylation supports the association of tetraspanin CD63 with CD9 and integrin alphaIIbbeta3 in activated platelets. Thromb Res 125:152–158
108. Iwai K, Ishii M, Ohshima S, Miyatake K, Saeki Y (2007) Expression and function of transmembrane-4 superfamily (tetraspanin) proteins in osteoclasts: reciprocal roles of Tspan-5 and NET-6 during osteoclastogenesis. Allergol Int 56:457–463
109. Iwamoto R, Higashiyama S, Mitamura T, Taniguchi N, Klagsbrun M, Mekada E (1994) Heparin-binding EGF-like growth factor, which acts as the diphtheria toxin receptor, forms a complex with membrane protein DRAP27/CD9, which up-regulates functional receptors and diphtheria toxin sensitivity. EMBO J 13:2322–2330
110. Izquierdo-Useros N, Naranjo-Gómez M, Erkizia I, Puertas MC, Borràs FE, Blanco J, Martinez-Picado J (2010) HIV and mature dendritic cells: Trojan exosomes riding the Trojan horse? PLoS Pathog 6:e1000740
111. Janvier K, Bonifacino JS (2005) Role of the endocytic machinery in the sorting of lysosome-associated membrane proteins. Mol Biol Cell 16:4231–4242
112. Johnstone RM (2006) Exosomes biological significance: a concise review. Blood Cells Mol Dis 36:315–321
113. Jones EL, Demaria MC, Wright MD (2011) Tetraspanins in cellular immunity. Biochem Soc Trans 39:506–511
114. Jones MC, Caswell PT, Norman JC (2006) Endocytic recycling pathways: emerging regulators of cell migration. Curr Opin Cell Biol 18:549–557
115. Jung T, Castellana D, Klingbeil P, Cuesta Hernández I, Vitacolonna M, Orlicky DJ, Roffler SR, Brodt P, Zöller M (2009) CD44v6 dependence of premetastatic niche preparation by exosomes. Neoplasia 11:1093–1105

116. Kaji K, Oda S, Shikano T, Ohnuki T, Uematsu Y, Sakagami J, Tada N, Miyazaki S, Kudo A (2000) The gamete fusion process is defective in eggs of Cd9-deficient mice. Nat Genet 24:279–282
117. Kaji K, Takeshita S, Miyake K, Takai T, Kudo A (2001) Functional association of CD9 with the Fc gamma receptors in macrophages. J Immunol 166:3256–3265
118. Kanetaka K, Sakamoto M, Yamamoto Y, Takamura M, Kanematsu T, Hirohashi S et al (2003) Possible involvement of tetraspanin CO-029 in hematogenous intrahepatic metastasis of liver cancer cells. J Gastroenterol Hepatol 18:1309–1314
119. Kaplan RN, Rafii S, Lyden D (2006) Preparing the "soil": the premetastatic niche. Cancer Res 66:11089–11093
120. Karamatic Crew V, Burton N, Kagan A, Green CA, Levene C, Flinter F, Brady RL, Daniels G, Anstee DJ (2004) CD151, the first member of the tetraspanin (TM4) superfamily detected on erythrocytes, is essential for the correct assembly of human basement membranes in kidney and skin. Blood 104:2217–2223
121. Klein-Soyer C, Azorsa DO, Cazenave JP, Lanza F (2000) CD9 participates in endothelial cell migration during in vitro wound repair. Arterioscler Thromb Vasc Biol 20:360–369
122. Knobeloch KP, Wright MD, Ochsenbein AF, Liesenfeld O, Löhler J, Zinkernagel RM, Horak I, Orinska Z (2000) Targeted inactivation of the tetraspanin CD37 impairs T-cell-dependent B-cell response under suboptimal costimulatory conditions. Mol Cell Biol 20:5363–5369
123. Koh TW, Korolchuk VI, Wairkar YP, Jiao W, Evergren E, Pan H, Zhou Y, Venken KJ, Shupliakov O, Robinson IM, O'Kane CJ, Bellen HJ (2007) Eps15 and Dap160 control synaptic vesicle membrane retrieval and synapse development. J Cell Biol 178:309–322
124. Koh TW, Verstreken P, Bellen HJ (2004) Dap160/intersectin acts as a stabilizing scaffold required for synaptic development and vesicle endocytosis. Neuron 43:193–205
125. Kosaka N, Iguchi H, Yoshioka Y, Takeshita F, Matsuki Y, Ochiya T (2010) Secretory mechanisms and intercellular transfer of microRNAs in living cells. J Biol Chem 285:17442–17452
126. Kotha J, Zhang C, Longhurst CM, Lu Y, Jacobs J, Cheng Y, Jennings LK (2009) Functional relevance of tetraspanin CD9 in vascular smooth muscle cell injury phenotypes: a novel target for the prevention of neointimal hyperplasia. Atherosclerosis 203:377–386
127. Kovalenko OV, Yang X, Kolesnikova TV, Hemler ME (2004) Evidence for specific tetraspanin homodimers: inhibition of palmitoylation makes cysteine residues available for cross-linking. Biochem J 377:407–417
128. Kovalenko OV, Yang XH, Hemler ME (2007) A novel cysteine cross-linking method reveals a direct association between claudin-1 and tetraspanin CD9. Mol Cell Proteomics 6:1855–1867
129. Krementsov DN, Rassam P, Margeat E, Roy NH, Schneider-Schaulies J, Milhiet PE, Thali M (2010) HIV-1 assembly differentially alters dynamics and partitioning of tetraspanins and raft components. Traffic 11:1401–1414
130. Kropshofer H, Spindeldreher S, Röhn TA, Platania N, Grygar C, Daniel N, Wölpl A, Langen H, Horejsi V, Vogt AB (2002) Tetraspan microdomains distinct from lipid rafts enrich select peptide-MHC class II complexes. Nat Immunol 3:61–68
131. Lafleur MA, Xu D, Hemler ME (2009) Tetraspanin proteins regulate membrane type-1 matrix metalloproteinase-dependent pericellular proteolysis. Mol Biol Cell 20:2030–2040
132. Lakkaraju A, Rodriguez-Boulan E (2008) Itinerant exosomes: emerging roles in cell and tissue polarity. Trends Cell Biol 18:199–209
133. Lan RF, Liu ZX, Liu XC, Song YE, Wang DW (2005) CD151 promotes neovascularization and improves blood perfusion in a rat hind-limb ischemia model. J Endovasc Ther 12:469–478
134. Lau LM, Wee JL, Wright MD, Moseley GW, Hogarth PM, Ashman LK, Jackson DE (2004) The tetraspanin superfamily member CD151 regulates outside-in integrin alphaIIbbeta3 signaling and platelet function. Blood 104:2368–2375
135. Le Naour F, André M, Boucheix C, Rubinstein E (2006) Membrane microdomains and proteomics: lessons from tetraspanin microdomains and comparison with lipid rafts. Proteomics 6:6447–6454
136. Le Naour F, Zöller M (2008) The tumor antigen EpCAM: tetraspanins and the tight junction protein claudin-7, new partners, new functions. Front Biosci 13:5847–5865

137. Lee TH, D'Asti E, Magnus N, Al-Nedawi K, Meehan B, Rak J (2011) Microvesicles as mediators of intercellular communication in cancer-the emerging science of cellular 'debris'. Semin Immunopathol 33:455–467
138. Lee YS, Pressman S, Andress AP, Kim K, White JL, Cassidy JJ, Li X, Lubell K, Lim do H, Cho IS, Nakahara K, Preall JB, Bellare P, Sontheimer EJ, Carthew RW (2009) Silencing by small RNAs is linked to endosomal trafficking. Nat Cell Biol 11:1150–1156
139. Lefèvre B, Wolf JP, Ziyyat A (2010) Sperm-egg interaction: is there a link between tetraspanin(s) and GPI-anchored protein(s)? Bioessays 32:143–152
140. Lekishvili T, Fromm E, Mujoomdar M, Berditchevski F (2008) The tumour-associated antigen L6 (L6-Ag) is recruited to the tetraspanin-enriched microdomains: implication for tumour cell motility. J Cell Sci 121:685–694
141. Lermen D, Gorjup E, Dyce PW, von Briesen H, Müller P (2010) Neuro-muscular differentiation of adult porcine skin derived stem cell-like cells. PLoS ONE 5:e8968
142. Levental I, Grzybek M, Simons K (2010) Greasing their way: lipid modifications determine protein association with membrane rafts. Biochemistry 49:6305–6316
143. Levy S, Shoham T (2005) Protein-protein interactions in the tetraspanin web. Physiology (Bethesda) 20:218–224
144. Li XB, Zhang ZR, Schluesener HJ, Xu SQ (2006) Role of exosomes in immune regulation. J Cell Mol Med 10:364–375
145. Lima LG, Chammas R, Monteiro RQ, Moreira ME, Barcinski MA (2009) Tumor-derived microvesicles modulate the establishment of metastatic melanoma in a phosphatidylserine-dependent manner. Cancer Lett 283:168–75
146. Little KD, Hemler ME, Stipp CS (2004) Dynamic regulation of a GPCR-tetraspanin-G protein complex on intact cells: central role of CD81 in facilitating GPR56-Galpha q/11 association. Mol Biol Cell 15:2375–2387
147. Liu L, He B, Liu WM, Zhou D, Cox JV, Zhang XA (2007) Tetraspanin CD151 promotes cell migration by regulating integrin trafficking. J Biol Chem 282:31631–31642
148. Liu WF, Zuo HJ, Chai BL, Peng D, Fei YJ, Lin JY, Yu XH, Wang DW, Liu ZX (2011) Role of tetraspanin CD151-α3/α6 integrin complex: Implication in angiogenesis CD151-integrin complex in angiogenesis. Int J Biochem Cell Biol 43:642–650
149. Loewen CJ, Moritz OL, Tam BM, Papermaster DS, Molday RS (2003) The role of subunit assembly in peripherin-2 targeting to rod photoreceptor disk membranes and retinitis pigmentosa. Mol Biol Cell 14:3400–3413
150. Longo N, Yáez-Mó M, Mittelbrunn M, de la Rosa G, Muoz ML, Sánchez-Madrid F, Sánchez-Mateos P (2001) Regulatory role of tetraspanin CD9 in tumor-endothelial cell interaction during transendothelial invasion of melanoma cells. Blood 98:3717–3726
151. Lykke-Andersen S, Brodersen DE, Jensen TH (2009) Origins and activities of the eukaryotic exosome. J Cell Sci 122:1487–1494
152. Maecker HT, Levy S (1997) Normal lymphocyte development but delayed humoral immune response in CD81-null mice. J Exp Med 185:1505–1510
153. Maecker HT, Todd SC, Levy S (1997) The tetraspanin superfamily: molecular facilitators. FASEB J 11:428–442
154. Maetzel D, Denzel S, Mack B, Canis M, Went P, Benk M, Kieu C, Papior P, Baeuerle PA, Munz M, Gires O (2009) Nuclear signalling by tumour-associated antigen EpCAM. Nat Cell Biol 11:162–171
155. Mallardo M, Deitinghoff A, Müller J, Goetze B, Macchi P, Peters C, Kiebler MA (2003) Isolation and characterization of Staufen-containing ribonucleoprotein particles from rat brain. Proc Natl Acad Sci U S A 100:2100–2105
156. Mangin PH, Kleitz L, Boucheix C, Gachet C, Lanza F (2009) CD9 negatively regulates integrin alphaIIbbeta3 activation and could thus prevent excessive platelet recruitment at sites of vascular injury. J Thromb Haemost 7:900–902
157. Mantegazza AR, Barrio MM, Moutel S, Bover L, Weck M, Brossart P, Teillaud JL, Mordoh J (2004) CD63 tetraspanin slows down cell migration and translocates to the endosomal-lysosomal-MIICs route after extracellular stimuli in human immature dendritic cells. Blood 104:1183–1190

158. Marks MS, Ohno H, Kirchnausen T, Bonracino JS (1997) Protein sorting by tyrosine-based signals: adapting to the Ys and wherefores. Trends Cell Biol 7:124–128
159. Marsh M, Theusner K, Pelchen-Matthews A (2009) HIV assembly and budding in macrophages. Biochem Soc Trans 37:185–189
160. Martin F, Roth DM, Jans DA, Pouton CW, Partridge LJ, Monk PN, Moseley GW (2005) Tetraspanins in viral infections: a fundamental role in viral biology? J Virol 79:10839–10851
161. Martinez FO, Helming L, Gordon S (2009) Alternative activation of macrophages: an immunologic functional perspective. Annu Rev Immunol 27:451–483
162. Masciopinto F, Giovani C, Campagnoli S, Galli-Stampino L, Colombatto P, Brunetto M, Yen TS, Houghton M, Pileri P, Abrignani S (2004) Association of hepatitis C virus envelope proteins with exosomes. Eur J Immunol 34:2834–2842
163. Mathivanan S, Lim JW, Tauro BJ, Ji H, Moritz RL, Simpson RJ (2010) Proteomic analysis of A33-immunoaffinity-purified exosomes released from the human colon tumor cell line LIM1215 reveals a tissue-specific protein signature. Mol Cell Proteomics 9:197–208
164. Matzku S, Komitowski D, Mildenberger M, Zöller M (1983) Characterization of BSp73, a spontaneous rat tumor and its in vivo selected variants showing different metastasizing capacities. Invasion Metastasis 3:109–123
165. Mazurov D, Heidecker G, Derse D (2007) The inner loop of tetraspanins CD82 and CD81 mediates interactions with human T cell lymphotrophic virus type 1 Gag protein. J Biol Chem 282:3896–3903
166. McCready J, Sims JD, Chan D, Jay DG (2010) Secretion of extracellular hsp90alpha via exosomes increases cancer cell motility: a role for plasminogen activation. BMC Cancer 10:294
167. McGovern M, Voutev R, Maciejowski J, Corsi AK, Hubbard EJ (2009) A "latent niche" mechanism for tumor initiation. Proc Natl Acad Sci U S A 106:11617–11622
168. Meyer-Wentrup F, Figdor CG, Ansems M, Brossart P, Wright MD, Adema GJ, van Spriel AB (2007) Dectin-1 interaction with tetraspanin CD37 inhibits IL-6 production. J Immunol 178:154–162
169. Mitchell PJ, Welton J, Staffurth J, Court J, Mason MD, Tabi Z, Clayton A (2009) Can urinary exosomes act as treatment response markers in prostate cancer? J Transl Med 7:4
170. Mittelbrunn M, Yáñez-Mó M, Sancho D, Ursa A, Sánchez-Madrid F (2002) Cutting edge: dynamic redistribution of tetraspanin CD81 at the central zone of the immune synapse in both T lymphocytes and APC. J Immunol 169:6691–6695
171. Miura Y, Kainuma M, Jiang H, Velasco H, Vogt PK, Hakomori S (2004) Reversion of the Jun-induced oncogenic phenotype by enhanced synthesis of sialosyllactosylceramide (GM3 ganglioside). Proc Natl Acad Sci U S A 101:16204–16209
172. Montpellier C, Tews BA, Poitrimole J, Rocha-Perugini V, D'Arienzo V, Potel J, Zhang XA, Rubinstein E, Dubuisson J, Cocquerel L (2011) Interacting regions of CD81 and two of its partners, EWI-2 and EWI-2wint, and their effect on hepatitis C virus infection. J Biol Chem 286:13954–13965
173. Morel O, Hugel B, Jesel L, Mallat Z, Lanza F, Douchet MP, Zupan M, Chauvin M, Cazenave JP, Tedgui A, Freyssinet JM, Toti F (2004) Circulating procoagulant microparticles and soluble GPV in myocardial infarction treated by primary percutaneous transluminal coronary angioplasty. A possible role for GPIIb-IIIa antagonists. J Thromb Haemost 2:1118–1126
174. Morelli AE, Larregina AT, Shufesky WJ, Sullivan ML, Stolz DB, Papworth GD, Zahorchak AF, Logar AJ, Wang Z, Watkins SC, Falo LD Jr, Thomson AW (2004) Endocytosis, intracellular sorting, and processing of exosomes by dendritic cells. Blood 104:3257–3266
175. Murayama Y, Shinomura Y, Oritani K, Miyagawa J, Yoshida H, Nishida M, Katsube F, Shiraga M, Miyazaki T, Nakamoto T, Tsutsui S, Tamura S, Higashiyama S, Shimomura I, Hayashi N (2008) The tetraspanin CD9 modulates epidermal growth factor receptor signaling in cancer cells. J Cell Physiol 216:135–143
176. Muro Y, Okabe M (2011) Mechanisms of fertilization—a view from the study of gene-manipulated mice. J Androl 32:218–225

177. Nazarenko I, Rana S, Baumann A, McAlear J, Hellwig A, Trendelenburg M, Lochnit G, Preissner KT, Zöller M (2010) Cell surface tetraspanin Tspan8 contributes to molecular pathways of exosome-induced endothelial cell activation. Cancer Res 70:1668–1678
178. Ng T, Shima D, Squire A et al (1999) PKCalpha regulates beta1 integrin-dependent cell motility through association and control of integrin traffic. EMBO J 18:3909–3923
179. Nieuwland R, Van Der Post JA, Lok CA, Kenter G, Sturk A (2010) Microparticles and exosomes in gynecologic neoplasias. Semin Thromb Hemost 36:925–929
180. Nydegger S, Khurana S, Krementsov DN, Foti M, Thali M (2006) Mapping of tetraspanin-enriched microdomains that can function as gateways for HIV-1. J Cell Biol 173:795–807
181. Obregon C, Rothen-Rutishauser B, Gerber P, Gehr P, Nicod LP (2009) Active uptake of dendritic cell-derived exovesicles by epithelial cells induces the release of inflammatory mediators through a TNF-alpha-mediated pathway. Am J Pathol 175:696–705
182. Odintsova E, Butters TD, Monti E, Sprong H, van Meer G, Berditchevski F (2006) Gangliosides play an important role in the organization of CD82-enriched microdomains. Biochem J 400:315–325
183. Orlowski E, Chand R, Yip J, Wong C, Goschnick MW, Wright MD, Ashman LK, Jackson, DE (2009) Platelet tetraspanin superfamily member, CD151 is required for regulation of thrombus stability in vivo. J Thrombosis Haemostasis 7:2074–2084
184. Pan BT, Teng K, Wu C, Adam M, Johnstone RM (1985) Electron microscopic evidence for externalization of the transferrin receptor in vesicular form in sheep reticulocytes. J Cell Biol 101:942–948
185. Pap E, Pállinger E, Pásztói M, Falus A (2009) Highlights of a new type of intercellular communication: microvesicle-based information transfer. Inflamm Res 58:1–8
186. Park JE, Tan HS, Datta A, Lai RC, Zhang H, Meng W, Lim SK, Sze SK (2010) Hypoxia modulates tumor microenvironment to enhance angiogenic and metastastic potential by secretion of proteins and exosomes. Mol Cell Proteomics 9:1985–1099
187. Parthasarathy V, Martin F, Higginbottom A, Murray H, Moseley GW, Read RC, Mal G, Hulme R, Monk PN, Partridge LJ (2009) Distinct roles for tetraspanins CD9, CD63 and CD81 in the formation of multinucleated giant cells. Immunology 127:237–248
188. Peinado H, Lavotshkin S, Lyden D (2011) The secreted factors responsible for pre-metastatic niche formation: old sayings and new thoughts. Semin Cancer Biol 21:139–146
189. Percherancier Y, Planchenault T, Valenzuela-Fernandez A, Virelizier JL, Arenzana-Seisdedos F, Bachelerie F (2001) Palmitoylation-dependent control of degradation, life span, and membrane expression of the CCR5 receptor. J Biol Chem 276:31936–31944
190. Petersen SH, Odintsova E, Haigh TA, Rickinson AB, Taylor GS, Berditchevski F (2011) The role of tetraspanin CD63 in antigen presentation via MHC class II. Eur J Immunol 41:2556–2561
191. Pileri P, Uematsu Y, Campagnoli S, Galli G, Falugi F, Petracca R, Weiner AJ, Houghton M, Rosa D, Grandi G, Abrignani S (1998) Binding of hepatitis C virus to CD81. Science 282:938–941
192. Pols MS, Klumperman J (2009) Trafficking and function of the tetraspanin CD63. Exp Cell Res 315:1584–1592
193. Potolicchio I, Carven GJ, Xu X, Stipp C, Riese RJ, Stern LJ, Santambrogio L (2005) Proteomic analysis of microglia-derived exosomes: metabolic role of the aminopeptidase CD13 in neuropeptide catabolism. J Immunol 175:2237–2243
194. Qazi KR, Torregrosa Paredes P, Dahlberg B, Grunewald J, Eklund A, Gabrielsson S (2010) Proinflammatory exosomes in bronchoalveolar lavage fluid of patients with sarcoidosis. Thorax 65:1016–1024
195. Raimondo F, Morosi L, Chinello C, Magni F, Pitto M (2010) Advances in membranous vesicle and exosome proteomics improving biological understanding and biomarker discovery. Proteomics 11:709–720
196. Rak J (2010) Microparticles in cancer. Semin Thromb Hemost 36:888–906
197. Rana S, Zöller M (2011) Exosome target cell selection and the importance of exosomal tetraspanins: a hypothesis. Biochem Soc Trans 39:559–562

198. Rana S, Claas C, Kretz CC, Nazarenko I, Zöller M (2011) Activation-induced internalization differs for the tetraspanins CD9 and Tspan8: impact on tumor cell motility. Int J Biochem Cell Biol 43:106–119
199. Ratajczak J, Miekus K, Kucia M, Zhang J, Reca R, Dvorak P, Ratajczak MZ (2006) Embryonic stem cell-derived microvesicles reprogram hematopoietic progenitors: evidence for horizontal transfer of mRNA and protein delivery. Leukemia 20:847–856
200. Razmara M, Hu H, Masquelier M, Li N (2007) Glycoprotein IIb/IIIa blockade inhibits platelet aminophospholipid exposure by potentiating translocase and attenuating scramblase activity. Cell Mol Life Sci 64:999–1008
201. Re F, Belyanskaya SL, Riese RJ, Cipriani B, Fischer FR, Granucci F, Ricciardi-Castagnoli P, Brosnan C, Stern LJ, Strominger JL, Santambrogio L (2002) Granulocyte-macrophage colony-stimulating factor induces an expression program in neonatal microglia that primes them for antigen presentation. J Immunol 169:2264–2273
202. Rocha-Perugini V, Lavie M, Delgrange D, Canton J, Pillez A, Potel J, Lecoeur C, Rubinstein E, Dubuisson J, Wychowski C, Cocquerel L (2009) The association of CD81 with tetraspanin-enriched microdomains is not essential for Hepatitis C virus entry. BMC Microbiol 9:111
203. Ronquist KG, Ronquist G, Larsson A, Carlsson L (2010) Proteomic analysis of prostate cancer metastasis-derived prostasomes. Anticancer Res 30:285–290
204. Rous BA, Reaves BJ, Ihrke G, Briggs JA, Gray SR, Stephens DJ, Banting G, Luzio JP (2002) Role of adaptor complex AP-3 in targeting wild-type and mutated CD63 to lysosomes. Mol Biol Cell 13:1071–1082
205. Rubinstein E (2011) The complexity of tetraspanins. Biochem Soc Trans 39:501–505
206. Rubinstein E, Ziyyat A, Wolf JP, Le Naour F, Boucheix C (2006) The molecular players of sperm-egg fusion in mammals. Semin Cell Dev Biol 17:254–263
207. Ruf W, Mueller BM (2006) Thrombin generation and the pathogenesis of cancer. Semin Thromb Hemost 32(Suppl 1):61–68
208. Runz S, Keller S, Rupp C, Stoeck A, Issa Y, Koensgen D, Mustea A, Sehouli J, Kristiansen G, Altevogt P (2007) Malignant ascites-derived exosomes of ovarian carcinoma patients contain CD24 and EpCAM. Gynecol Oncol 107:563–571
209. Sachs N, Kreft M, Van Den Bergh Weerman MA, Beynon AJ, Peters TA, Weening JJ, Sonnenberg A (2006) Kidney failure in mice lacking the tetraspanin CD151. J Cell Biol 175:33–39
210. Sala-Valdés M, Ursa A, Charrin S, Rubinstein E, Hemler ME, Sánchez-Madrid F, Yáñez-Mó M (2006) EWI-2 and EWI-F link the tetraspanin web to the actin cytoskeleton through their direct association with ezrin-radixin-moesin proteins. J Biol Chem. 281:19665–19675
211. Sangaletti S, Colombo MP (2008) Matricellular proteins at the crossroad of inflammation and cancer. Cancer Lett 267:245–253
212. Sato H, Takino T (2010) Coordinate action of membrane-type matrix metalloproteinase-1 (MT1-MMP) and MMP-2 enhances pericellular proteolysis and invasion. Cancer Sci 101:843–847
213. Saunderson SC, Schuberth PC, Dunn AC, Miller L, Hock BD, MacKay PA, Koch N, Jack RW, McLellan AD (2008) Induction of exosome release in primary B cells stimulated via CD40 and the IL-4 receptor. J Immunol 180:8146–8152
214. Sawada S, Yoshimoto M, Odintsova E, Hotchin NA, Berditchevski F (2003) The tetraspanin CD151 functions as a negative regulator in the adhesion-dependent activation of Ras. J Biol Chem 278:26323–26326
215. Scherberich A, Moog S, Haan-Archipoff G, Azorsa DO, Lanza F, Beretz A (1998) Tetraspanin CD9 is associated with very late-acting integrins in human vascular smooth muscle cells and modulates collagen matrix reorganization. Arterioscler Thromb Vasc Biol 18:1691–1697
216. Schorey JS, Bhatnagar S (2008) Exosome function: from tumor immunology to pathogen biology. Traffic 9:871–881
217. Schröder J, Lüllmann-Rauch R, Himmerkus N, Pleines I, Nieswandt B, Orinska Z, Koch-Nolte F, Schröder B, Bleich M, Saftig P (2009) Deficiency of the tetraspanin CD63 associated with kidney pathology but normal lysosomal function. Mol Cell Biol 29:1083–1094

218. Seigneuret M (2006) Complete predicted three-dimensional structure of the facilitator transmembrane protein and hepatitis C virus receptor CD81: conserved and variable structural domains in the tetraspanin superfamily. Biophys J 90:212–227
219. Sharma C, Yang XH, Hemler ME (2008) DHHC2 Affects Palmitoylation, Stability, and Functions of Tetraspanins CD9 and CD151. Mol Biol Cell 19:3415–3425
220. Sheng KC, van Spriel AB, Gartlan KH, Sofi M, Apostolopoulos V, Ashman L, Wright MD (2009) Tetraspanins CD37 and CD151 differentially regulate Ag presentation and T-cell co-stimulation by DC. Eur J Immunol 39:50–55
221. Shet AS, Aras O, Gupta K, Hass MJ, Rausch DJ, Saba N, Koopmeiners L, Key NS, Hebbel RP (2003) Sickle blood contains tissue factor-positive microparticles derived from endothelial cells and monocytes. Blood 102:2678–2683
222. Shi GM, Ke AW, Zhou J, Wang XY, Xu Y, Ding ZB, Devbhandari RP, Huang XY, Qiu SJ, Shi YH, Dai Z, Yang XR, Yang GH, Fan J (2010) CD151 modulates expression of matrix metalloproteinase 9 and promotes neoangiogenesis and progression of hepatocellular carcinoma. Hepatology 52:183–196
223. Shiomi T, Inoki I, Kataoka F, Ohtsuka T, Hashimoto G, Nemori R, Okada Y (2005) Pericellular activation of proMMP-7 (promatrilysin-1) through interaction with CD151. Lab Invest 85:1489–1506
224. Silvie O, Rubinstein E, Franetich JF, Prenant M, Belnoue E, Rénia L, Hannoun L, Eling W, Levy S, Boucheix C, Mazier D (2003) Hepatocyte CD81 is required for *Plasmodium falciparum* and *Plasmodium yoelii* sporozoite infectivity. Nat Med 9:93–96
225. Simpson F, Hussain NK, Qualmann B, Kelly RB, Kay BK, McPherson PS Schmid SL (1999) SH3-domain-containing proteins function at distinct steps in clathrin-coated vesicle formation. Nat Cell Biol 1:119–124
226. Simpson RJ, Lim JW, Moritz RL, Mathivanan S (2009) Exosomes: proteomic insights and diagnostic potential. Expert Rev Proteomics 6:267–283
227. Sincock PM, Fitter S, Parton RG, Berndt MC, Gamble JR, Ashman LK (1999) PETA-3/CD151, a member of the transmembrane 4 superfamily, is localised to the plasma membrane and endocytic system of endothelial cells, associates with multiple integrins and modulates cell function. J Cell Sci 112:833–844
228. Singethan K, Müller N, Schubert S, Lüttge D, Krementsov DN, Khurana SR, Krohne G, Schneider-Schaulies S, Thali M, Schneider-Schaulies J (2008) CD9 clustering and formation of microvilli zippers between contacting cells regulates virus-induced cell fusion. Traffic 9:924–235
229. Smalheiser NR (2007) Exosomal transfer of proteins and RNAs at synapses in the nervous system. Biol Direct 2:35
230. Spoden G, Freitag K, Husmann M, Boller K, Sapp M, Lambert C, Florin L (2008) Clathrin- and caveolin-independent entry of human papillomavirus type 16-involvement of tetraspanin-enriched microdomains (TEMs). PLoS ONE 3:e3313
231. Sridhar SC, Miranti CK (2006) Tetraspanin KAI1/CD82 suppresses invasion by inhibiting integrin-dependent crosstalk with c-Met receptor and Src kinases. Oncogene 25:2367–2378
232. Stahl PD, Barbieri MA (2002) Multivesicular bodies and multivesicular endosomes: the "ins and outs" of endosomal traffic. Sci STKE 2002(141):PE32
233. Sterk LM, Geuijen CA, Oomen LC, Calafat J, Janssen H, Sonnenberg A (2000) The tetraspan molecule CD151, a novel constituent of hemidesmosomes, associates with the integrin alpha6beta4 and may regulate the spatial organization of hemidesmosomes. J Cell Biol 149:969–982
234. Stipp CS, Kolesnikova TV, Hemler ME (2001a) EWI-2 is a major CD9 and CD81 partner and member of a novel Ig protein subfamily. J Biol Chem 276:40545–40554
235. Stipp CS, Orlicky D, Hemler ME (2001b) FPRP, a major, highly stoichiometric, highly specific CD81- and CD9-associated protein. J Biol Chem 276:4853–4862
236. Stipp CS, Kolesnikova TV, Hemler ME (2003) Functional domains in tetraspanin proteins. Trends Biochem Sci 28:106–112

237. Stoorvogel W, Kleijmeer MJ, Geuze HJ, Raposo G (2002) The biogenesis and functions of exosomes. Traffic 3:321–330
238. Tait JF, Smith C (1999) Phosphatidylserine receptors: role of CD36 in binding of anionic phospholipid vesicles to monocytic cells. J Biol Chem 274:3048–3054
239. Takeda Y, Kazarov AR, Butterfield CE, Hopkins BD, Benjamin LE, Kaipainen A, Hemler ME (2007) Deletion of tetraspanin Cd151 results in decreased pathologic angiogenesis in vivo and in vitro. Blood 109:1524–1532
240. Talmadge JE, Fidler IJ (2010) AACR centennial series: the biology of cancer metastasis: historical perspective. Cancer Res 70:5649–5669
241. Tanaka F, Hori N, Sato K (2002) Identification of differentially expressed genes in rat hepatoma cell lines using subtraction and microarray. J Biochem 131:39–44
242. Tarrant JM, Groom J, Metcalf D, Li R, Borobokas B, Wright MD, Tarlinton D, Robb L (2002) The absence of Tssc6, a member of the tetraspanin superfamily, does not affect lymphoid development but enhances in vitro T-cell proliferative responses. Mol Cell Biol 22:5006–5018
243. Taylor DD, Gercel-Taylor C (2011) Exosomes/microvesicles: mediators of cancer-associated immunosuppressive microenvironments. Semin Immunopathol 33:441–54
244. Testa JE, Brooks PC, Lin JM, Quigley JP (1999) Eukaryotic expression cloning with an antimetastatic monoclonal antibody identifies a tetraspanin (PETA-3/CD151) as an effector of human tumor cell migration and metastasis. Cancer Res 59:3812–3820
245. Thali M (2011) Tetraspanin functions during HIV-1 and influenza virus replication. Biochem Soc Trans 39:529–531
246. Tian T, Wang Y, Wang H, Zhu Z, Xiao Z (2010) Visualizing of the cellular uptake and intracellular trafficking of exosomes by live-cell microscopy. J Cell Biochem 111:488–496
247. Tlsty TD, Coussens LM (2006) Tumor stroma and regulation of cancer development. Annu Rev Pathol 1:119–150
248. Todeschini AR, Dos Santos JN, Handa K, Hakomori SI (2008) Ganglioside GM2/GM3 complex affixed on silica nanospheres strongly inhibits cell motility through CD82/cMet-mediated pathway. Proc Natl Acad Sci U S A 105:1925–1930
249. Todeschini RA, Hakomori SI (2008) Functional role of glycosphingolipids and gangliosides in control of cell adhesion, motility, and growth, through glycosynaptic microdomains. Biochim Biophys Acta 1780:421–433
250. Tsopanoglou NE, Maragoudakis ME (2007) Inhibition of angiogenesis by small-molecule antagonists of protease-activated receptor-1. Semin Thromb Hemost 33:680–687
251. Tsukita S, Yonemura S (1999) Cortical actin organization: lessons from ERM (ezrin/radixin/moesin) proteins. J Biol Chem 274:34507–34510
252. Tsukita S, Furuse M (2000) The structure and function of claudins, cell adhesion molecules at tight junctions. Ann N Y Acad Sci 915:129–135
253. Tumne A, Prasad VS, Chen Y, Stolz DB, Saha K, Ratner DM, Ding M, Watkins SC, Gupta P (2009) Noncytotoxic suppression of human immunodeficiency virus type 1 transcription by exosomes secreted from CD8+ T cells. J Virol 83:4354–4364
254. Unternaehrer JJ, Chow A, Pypaert M, Inaba K, Mellman I (2007) The tetraspanin CD9 mediates lateral association of MHC class II molecules on the dendritic cell surface. Proc Natl Acad Sci U S A 104:234–239
255. Valadi H, Ekström K, Bossios A, Sjöstrand M, Lee JJ, Lötvall JO (2007) Exosome-mediated transfer of mRNAs and microRNAs is a novel mechanism of genetic exchange between cells. Nat Cell Biol 9:654–659
256. Vallhov H, Gutzeit C, Johansson SM, Nagy N, Paul M, Li Q, Friend S, George TC, Klein E, Scheynius A, Gabrielsson S (2011) Exosomes containing glycoprotein 350 released by EBV-transformed B cells selectively target B cells through CD21 and block EBV infection in vitro. J Immunol 186:73–82
257. van Niel G, Porto-Carreiro I, Simoes S, Raposo G (2006) Exosomes: a common pathway for a specialized function. J Biochem 140:13–21
258. van Spriel AB, Figdor CG (2010) The role of tetraspanins in the pathogenesis of infectious diseases. Microbes Infect 12:106–112

259. van Spriel AB, Puls KL, Sofi M, Pouniotis D, Hochrein H, Orinska Z, Knobeloch KP, Plebanski M, Wright MD (2004) A regulatory role for CD37 in T cell proliferation. J Immunol 172:2953–2961
260. van Spriel AB, Sofi M, Gartlan KH, Van Der Schaaf A, Verschueren I, Torensma R, Raymakers RA, Loveland BE, Netea MG, Adema GJ, Wright MD, Figdor CG (2009) The tetraspanin protein CD37 regulates IgA responses and anti-fungal immunity. PLoS Pathog 5:e1000338
261. Veenbergen S, van Spriel AB (2011) Tetraspanins in the immune response against cancer. Immunol Lett 138:129–136
262. Vlodavsky I, Korner G, Ishai-Michaeli R, Bashkin P, Bar-Shavit R, Fuks Z (1990) Extracellular matrix-resident growth factors and enzymes: possible involvement in tumor metastasis and angiogenesis. Cancer Metastasis Rev 9:203–226
263. Vogt AB, Spindeldreher S, Kropshofer H (2002) Clustering of MHC-peptide complexes prior to their engagement in the immunological synapse: lipid raft and tetraspan microdomains. Immunol Rev 189:136–151
264. Walker JW (2008) Protein scaffolds, lipid domains and substrate recognition in protein kinase C function: implications for rational drug design. Handb Exp Pharmacol 186:185–203
265. Wang H, Min G, Glockshuber R, Sun TT, Kong XP (2009) Uropathogenic E. coli adhesin-induced host cell receptor conformational changes: implications in transmembrane signaling transduction. J Mol Biol 392:352–361
266. Wang HX, Li Q, Sharma C, Knoblich K, Hemler ME (2011) Tetraspanin protein contributions to cancer. Biochem Soc Trans 39:547–552
267. Wells A, Grandis JR (2003) Phospholipase C-gamma1 in tumor progression. Clin Exp Metastasis 20:285–290
268. Whiteside TL, Mandapathil M, Szczepanski M, Szajnik M (2011) Mechanisms of tumor escape from the immune system: Adenosine-producing Treg, exosomes and tumor-associated TLRs. Bull Cancer 98:25–31
269. Willett BJ, Picard L, Hosie MJ, Turner JD, Adema K, Clapham PR (1997) Shared usage of the chemokine receptor CXCR4 by the feline and human immunodeficiency viruses. J Virol 71:6407–6415
270. Winterwood NE, Varzavand A, Meland MN, Ashman LK, Stipp CS (2006) A critical role for tetraspanin CD151 in alpha3beta1 and alpha6beta4 integrin-dependent tumor cell functions on laminin-5. Mol Biol Cell 17:2707–2721
271. Witteveldt J, Evans MJ, Bitzegeio J, Koutsoudakis G, Owsianka AM, Angus AG, Keck ZY, Foung SK, Pietschmann T, Rice CM, Patel AH (2009) CD81 is dispensable for hepatitis C virus cell-to-cell transmission in hepatoma cells. J Gen Virol 90:48–58
272. Wittmann J, Jäck HM (2010) Serum microRNAs as powerful cancer biomarkers. Biochim Biophys Acta 1806:200–207
273. Wright MD, Geary SM, Fitter S, Moseley GW, Lau LM, Sheng KC, Apostolopoulos V, Stanley EG, Jackson DE, Ashman LK (2004) Characterization of mice lacking the tetraspanin superfamily member CD151. Mol Cell Biol 24:5978–5988
274. Wright MD, Moseley GW, van Spriel AB (2004) Tetraspanin microdomains in immune cell signalling and malignant disease. Tissue Antigens 64:533–542
275. Wubbolts R, Leckie RS, Veenhuizen PT, Schwarzmann G, Möbius W, Hoernschemeyer J, Slot JW, Geuze HJ, Stoorvogel W (2003) Proteomic and biochemical analyses of human B cell-derived exosomes. Potential implications for their function and multivesicular body formation. J Biol Chem 278:10963–10972
276. Xie B, Zhou G, Chan SY, Shapiro E, Kong XP, Wu XR, Sun TT, Costello CE (2006) Distinct glycan structures of uroplakins Ia and Ib: structural basis for the selective binding of FimH adhesin to uroplakin Ia. J Biol Chem 281:14644–14653
277. Yalaoui S, Zougbédé S, Charrin S, Silvie O, Arduise C, Farhati K, Boucheix C, Mazier D, Rubinstein E, Froissard P (2008) Hepatocyte permissiveness to Plasmodium infection is conveyed by a short and structurally conserved region of the CD81 large extracellular domain. PLoS Pathog 4:e1000010

278. Yamabhai M, Hoffman NG, Hardison NL, McPherson PS, Castagnoli L, Cesareni G, Kay BK (1998) Intersectin, a novel adaptor protein with two Eps15 homology and five Src homology 3 domains. J Biol Chem 273:31401–31407
279. Yamamoto Y, Grubisic K, Oelgeschläger M (2007) Xenopus Tetraspanin-1 regulates gastrulation movements and neural differentiation in the early Xenopus embryo. Differentiation 75:235–245
280. Yáñez-Mó M, Alfranca A, Cabañas C, Marazuela M, Tejedor R, Ursa MA, Ashman LK, de Landázuri MO, Sánchez-Madrid F (1998) Regulation of endothelial cell motility by complexes of tetraspan molecules CD81/TAPA-1 and CD151/PETA-3 with alpha3 beta1 integrin localized at endothelial lateral junctions. J Cell Biol 141:791–804
281. Yáñez-Mó M, Barreiro O, Gonzalo P, Batista A, Megías D, Genís L, Sachs N, Sala-Valdés M, Alonso MA, Montoya MC, Sonnenberg A, Arroyo AG, Sánchez-Madrid F (2008) MT1-MMP collagenolytic activity is regulated through association with tetraspanin CD151 in primary endothelial cells. Blood 112:3217–3226
282. Yáñez-Mó M, Barreiro O, Gordon-Alonso M, Sala-Valdés M, Sánchez-Madrid F (2009) Tetraspanin-enriched microdomains: a functional unit in cell plasma membranes. Trends Cell Biol 19:434–446
283. Yáñez-Mó M, Gutiérrez-López MD, Cabañas C (2011) Functional interplay between tetraspanins and proteases. Cell Mol Life Sci 68:3323–3335
284. Yang J, Weinberg RA (2008) Epithelial-mesenchymal transition: at the crossroads of development and tumor metastasis. Dev Cell 14:818–829
285. Yang XH, Richardson AL, Torres-Arzayus MI, Zhou P, Sharma C, Kazarov AR, Andzelm MM, Strominger JL, Brown M, Hemler ME (2008) CD151 accelerates breast cancer by regulating alpha 6 integrin function, signaling, and molecular organization. Cancer Res 68:3204–3213
286. Yates AJ, Rampersaud A (1998) Sphingolipids as receptor modulators. An overview. Ann N Y Acad Sci 845:57–71
287. Yauch RL, Berditchevski F, Harler MB, Reichner J, Hemler ME (1998) Highly stoichiometric, stable, and specific association of integrin alpha3beta1 with CD151 provides a major link to phosphatidylinositol 4-kinase, and may regulate cell migration. Mol Biol Cell 9:2751–2765
288. Yauch RL, Hemler ME (2000) Specific interactions among transmembrane 4 superfamily (TM4SF) proteins and phosphoinositide 4-kinase. Biochem J 351:629–637
289. Yunta M, Lazo PA (2003) Tetraspanin proteins as organisers of membrane microdomains and signalling complexes. Cell Signal 15:559–564
290. Zakharova L, Svetlova M, Fomina AF (2007) T cell exosomes induce cholesterol accumulation in human monocytes via phosphatidylserine receptor. J Cell Physiol 212:174–181
291. Zhang F, Kotha J, Jennings LK, Zhang XA (2009) Tetraspanins and vascular functions. 83:7–15
292. Zhang XA, Bontrager AL, Hemler ME (2001) Transmembrane-4 superfamily proteins associate with activated protein kinase C (PKC) and link PKC to specific beta(1) integrins. J Biol Chem 276:25005–25013
293. Zhang XA, Kazarov AR, Yang X, Bontrager AL, Stipp CS, Hemler ME (2002) Function of the tetraspanin CD151-alpha6beta1 integrin complex during cellular morphogenesis. Mol Biol Cell 13:1–11
294. Zheng Z, Liu Z (2006) CD151 gene delivery activates PI3K/Akt pathway and promotes neovascularization after myocardial infarction in rats. Mol Med 12:214–220
295. Zheng ZZ, Liu ZX (2007) Activation of the phosphatidylinositol 3-kinase/protein kinase Akt pathway mediates CD151-induced endothelial cell proliferation and cell migration. Int J Biochem Cell Biol 39:340–348
296. Zhou B, Liu L, Reddivari M, Zhang XA (2004) The palmitoylation of metastasis suppressor KAI1/CD82 is important for its motility- and invasiveness-inhibitory activity. Cancer Res 64:7455–7463
297. Zijlstra A, Lewis J, Degryse B Stuhlmann H, Quigley JP (2008) The inhibition of tumor cell intravasation and subsequent metastasis via regulation of in vivo tumor cell motility by the tetraspanin CD151. Cancer Cell 13:221–234

298. Zöller M (2006) Gastrointestinal tumors: metastasis and tetraspanins. Z Gastroenterol 44:573–586
299. Zöller M (2009) Tetraspanins: push and pull in suppressing and promoting metastasis. Nat Rev Cancer 9:40–55
300. Zomer A, Vendrig T, Hopmans ES, van Eijndhoven M, Middeldorp JM, Pegtel DM (2010) Exosomes: fit to deliver small RNA. Commun Integr Biol 3:447–450

Chapter 5
The Effects of Exosomes and Related Vesicles on Cancer Development, Progression, and Dissemination

Huang-Ge Zhang and William E. Grizzle

Abstract Exosomes are small bilayer-membrane-bound nanoparticles that are released from normal, diseased, and neoplastic cells and are present in blood and other bodily fluids. Exosomes contain a variety of signal molecules including signal peptides, mRNA, microRNA (miRNA), and lipids and have characteristic molecular features on their external surfaces (e.g., CD9 and CD81). Exosomes can function to export from cells unneeded endogenous molecules and therapeutic drugs. When exosomes and similar types of vesicles are secreted from cells and are taken up by other cells, they may act locally to provide autocrine or paracrine signals or distantly as a newly described nanoparticle-based endocrine system. Specifically, mRNA transferred to cells by exosomes can result in the production of new and novel proteins. In cancer, signals via exosomes affect the immune system via inhibition of the functions of T cells and natural killer (NK) cells, by inhibiting the differentiation of precursors to mature antigen-presenting cells, e.g., dendritic cells, and by increasing the number and/or activity of immune suppressor cells including myeloid-derived suppressor cells, T regulatory cells, and CD14, HLA-DR$^{-/\text{low}}$ cells. Exosomes from neoplastic lesions also act to provide a fertile environment in which neoplastic lesions can develop, progress, and metastasize, especially via the stimulation of angiogenesis. Exosomes have multiple potential clinical uses including the development of vaccines for targeting tumors; also, tumor-derived exosomes may be useful in diagnosis/early detection, risk assessment and prediction, in determining prognosis, and as surrogate endpoints in evaluating therapeutic and preventive approaches to cancer.

5.1 Introduction

Exosomes are small bilayer-membrane-bound nanoparticles that typically are flattened ovoid-like structures, which have been described as cup-shaped (blood) or doughnut-shaped (saliva), depending upon how and from where exosomes are

W. E. Grizzle (✉)
Department of Pathology, University of Alabama at Birmingham, Zeigler Research Building, ZRB 408 703 South 19th Street, Birmingham, AL 35294-0007, USA
e-mail: wgrizzle@uab.edu

H.-G. Zhang
Department of Microbiology and Immunology, University of Louisville, Louisville, KY, USA

H.-G. Zhang (ed.), *Emerging Concepts of Tumor Exosome–Mediated Cell–Cell Communication,* DOI 10.1007/978-1-4614-3697-3_5,

isolated. They are released from normal and diseased cells via fusion of intracellular multivesicular bodies (MVBs) with cellular membranes and subsequent extracellular release of these nanovesicles from MVBs [1–5]. After their release from cells, a proportion of exosomes enters the lymphatic-vascular system and are transported to distant sites [6]. Thus, both normal and diseased individuals have exosomes present in their blood and other bodily fluids. Care should be taken when referring to exosomes in that the terminology, "cytoplasmic exosomes" or sometimes just "exosomes," also is used to describe the 3′–5′ exoribonuclease complex, which functions in mRNA and apoptotic DNA degradation and other pathways [7]. The 3′– 5′ exoribonuclease complex will not be discussed.

5.2 Immune Regulation by Exosomes

Exosomes are an important regulatory feature of the normal function of the immune system and in diseases in which exosomes may be involved in dysregulation of immunity [8, 9]. Many, if not all, cells of the immune system secrete exosomes including B and T lymphocytes, DCs, and mast cells [8–13]. Normally, specific forms of exosomes facilitate how antigen-presenting cells (APCs), especially DCs, present antigens to T lymphocytes and natural killer (NK) cells. For example, exosomes, which express MHC class II antigen-peptide complexes, stimulate T cells much more efficiently if the exosomes have been absorbed by functional APCs [10–13]. $CD4^+$ T cells can be stimulated to proliferate by exosomes carrying the intact stimulating antigen or the antigen-peptide in an MHC class II complex; however, the overall process of stimulating these T cells requires at least two separate subpopulations of DCs. One population of mature DCs carrying an MHC class II complex interacts with the antigen and secretes exosomes typically carrying an antigen-peptide MHC class II complex. A second CD-$8\alpha^-$, $CD80^+$, $CD86^+$ subpopulation of DCs, which may be MHC class II negative, can then interact with these exosomes to present the exosome antigen-peptide class II complexes for efficient stimulation of $CD4^+$ T cells. In addition, the activation response may be amplified by the transfer of the peptide-antigen MHC class II complexes among DCs [10]. Other pathways involving APCs and their effects on the responses of T cells to antigens are important in exosomal mediation of peripheral immune tolerance [11–14]. Similarly, MHC class I-restricted $CD8^+$ T cells are stimulated by exosomes only when the exosomes have interacted with mature DCs. Such stimulation is facilitated by adjuvant molecules including ligands for Toll-like receptors 3 and 9 [13]. Other studies have reported that exosomes from DCs can produce an antibody response and that functional B cells are needed for efficient stimulation of T cells by exosomes from DCs [14].

Exosomes participate in maternal-fetal tolerance that is important for survival of the fetus [15–18]. Specifically, the placenta releases exosomes, which contain Fas ligand (FasL), which in various models has been shown to reduce inflammation [19–22]. In addition, exosomes released from syncytiotrophoblasts contain surface ligands, i.e., ULBP1–5, to the NKG2D receptor that is found on NK, CD8, and $\gamma\delta$ T cells [15]. Release of these exosomes leads to a reduction in NKG2D receptors and subsequent decrease in cytotoxicity of these cells in vitro [15–17]. Also, exosome-like particles have been reported to convert $CD4^+$, $CD25^-$ T cells into $CD4^+$, $CD25^+$, $Foxp3^+$ T regulatory cells (Tregs). Tregs normally suppress autoreactive T cells and hence, inhibit abnormal immune responses by their actions to prevent immune reactions to normal tissues [22].

There are numerous other interactions among cells involved in immunity via intercellular communication mediated by exosomes. For example, exosomes from mast cells can activate B and T lymphocytes and DCs. Exosomes from mast cells stimulated in vivo (mice) the maturation of DCs and facilitated their capability to present antigens to T lymphocytes. The presence of heat shock proteins (HSP) 60 and 70 in the exosomes were reported to be critical to the ability of DCs to acquire the functions necessary for antigen presentation.

Normal immune pathways may be dysregulated in autoimmune diseases. Exposure of DCs to interleukin-10 (IL-10) has been reported to increase the release of exosomes from DCs and to reduce the inflammation as well as the extent of the arthritis caused by injections of collagen [23]. Fibroblasts from patients with rheumatoid arthritis (RA) release exosomes, which contain TNFα, which can kill specific immune cells. In addition, these exosomes activate Akt via its phosphorylation causing increased phenotypic expression of NF-κB, which may increase the severity of RA [23, 24]. In addition, exosomes released from salivary glands contain autoantigens that may induce autoimmunity [25]. Exosomes also are involved in inflammation of the liver induced by fatty diets [26].

5.3 Production of Exosomes by Neoplastic Lesions

There have been many reports of exosomes being produced by a wide range of various types of tumors. Such exosomes are frequently referred to as tumor-derived (TD)-exosomes. TD-exosomes are separated primarily from the physiological fluids (e.g., urine) of patients with neoplastic lesions and spent or conditioned media from cellular cultures. Malignant lesions that have been reported to be associated with TD-exosomes include most cancers including those of the bladder, brain, breast, colorectum, kidney, lung, oral cavity, ovary, and prostate. In addition, lymphomas and melanomas have been reported to produce exosomes [8, 9].

The exosomes in the bodily fluids of patients with malignant lesions would be a combination of exosomes including those typical of normal individuals, those secondary to comorbid conditions, as well as TD-exosomes. This is the same situation that is found as to molecules circulating in blood of cancer patients in that these molecules also are from normal cells, nonneoplastic diseased cells, and tumors. Thus, the molecular characteristics of TD-exosomes used in translational research are potentially just as specific and just as sensitive as the characteristics of biomarkers in bodily fluids that are outside exosomes [8, 9, 27–29]. Actually, the molecular features of TD-exosomes may be more sensitive and specific for certain questions in translational research because selected molecules characteristic of tumors may be more concentrated in TD-exosomes than the same molecules in the matching bodily fluids. In addition, for some malignancies, the amount of TD-exosomes correlates with the extent of the tumor.

TD-exosomes are potentially useful in translational research because they usually contain molecules characteristic of the malignant tissues from which the TD-exosomes are derived. Specifically, TD-exosomes from bodily fluids from the patients with a tumor mirror the molecular characteristics of the matching tumors [30–41]. For example, patients with melanomas have exosomes in blood, which contain Melan A/Mart 1 and other melanoma-specific molecules [3, 37]. Of interest, TD-exosomes from the blood of patients with tumors of the central nervous system (CNS) with blood-brain barriers assumed to be intact, have been reported to contain L1-NCAM/CD171, a neural molecule that was also found in TD-exosomes isolated from the matching tumors of the brain, but not in exosomes isolated from the blood from controls [30, 38–41]; however, the blood of both patients with tumors of the CNS as well as controls without tumors contained HSP70 [30, 38]. In contrast, TD-exosomes from tumors of the CNS have been reported to variably contain EGFR, EGFRvIII, HSPs27, 60, 72, 73, 80, and 90 [30, 38–41].

5.4 Nonimmune Effects of Exosomes on Tumor Progression

The growth of tumors depends upon the balance between proliferation and cell death, processes, which are regulated by multiple molecular mechanisms. TD-exosomes are involved in the growth of specific cancers via autocrine, paracrine, and endocrine regulation of the local tumor environment and hence the growth of tumors. As demonstrated in Fig. 5.1, TD-exosomes generate a fertile environment locally, supporting the growth of the primary tumor [39–46]. The enrichment of local and metastatic environments are frequently secondary to TD-exosomes whose contents are enriched in mRNAs generating proteins associated with stimulating the cell cycle of stromal cells including endothelial cells. Also, TD-exosomes may contain proteins that stimulate angiogenesis.

TD-exosomes have been demonstrated to stimulate the proliferation of endothelial cells and thus aid progression of tumors by increasing angiogenesis [39–47]. For example, paracrine stimulation by TD-exosomes from melanomas when compared with a control, i.e., autocrine exosomes derived from endothelial cells, selectively increased spheroid formation by endothelial cells and budding of spheroids as well as levels of proangiogenic cytokines including TGFβ and VEGF [42]. Similarly, Skog et al. [38] isolated microvesicles from the conditional media from short-term cultures of cells derived from human samples of glioblastomas as well as from sera from these patients and control patients. These vesicles were isolated by the typical preparations that are used to isolate exosomes, but CD9 and CD81 were not evaluated. When the source cells were examined by EM, they were described as being "covered with microvesicles" that ranged from 50 to 500 nm in diameter. Similar to "exosome" from glioblastoma patients described previously, the EGFR

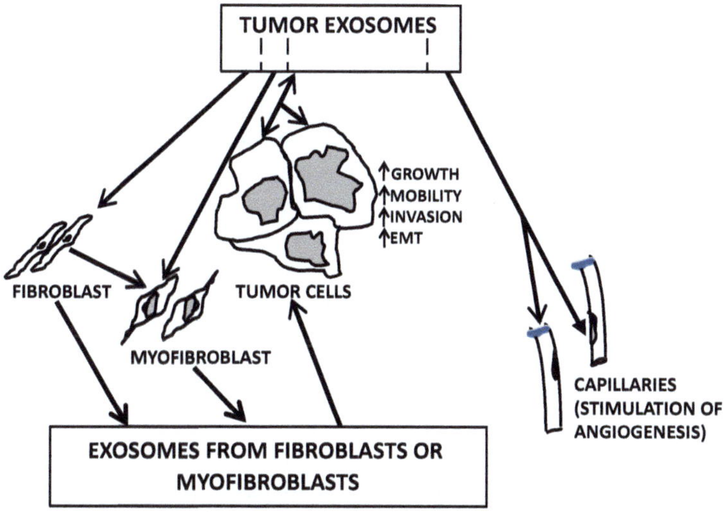

Fig. 5.1 This figure represents a model via which malignant cells interact via autocrine and paracrine activities to prepare an environment, which supports local growth of the tumor, mobility and invasion of malignant cells, induction of myofibroblasts, increased angiogenesis, and epithelial mesenchymal transition (EMT). Various responses from different compartments (subpopulations) of tumor-derived exosomes (TD-exosomes) are demonstrated by "!"

VIII receptor was found in these vesicles including vesicles from the sera of 9 out of 25 patients with glioblastomas. Also, additional mRNA mutants or variants as well as microRNAs that are typical of glioblastomas were identified in such vesicles. In these studies, they also found that microvesicles isolated by this method contained proteins that stimulate angiogenesis (e.g., angiogenin, VEGF, IL-6, and IL-8). These angiogenic proteins were enriched in the TD-vesicles compared with the contents of glioblastoma-source cells. When the TD-vesicles were internalized by microvascular endothelial cells of the brain, they stimulated an increase in tubule length that was similar to that caused by proteins that stimulate angiogenesis. The microvesicles also stimulated proliferation of the U87 glioma cell line that was derived from a glioblastoma. This suggests, in addition to stimulation of angiogenesis, autocrine stimulation by these TD-vesicles [39]. Similarly, a hypoxic environment in a primary tumor may cause the release of exosomes that are enriched in proteins that stimulate angiogenesis. This has been demonstrated with the A-431 cell line in which VEGF and other cytokine were increased by hypoxia. In addition, hypoxia reduced adhesion and increased invasion. Of note, E-cadherin was decreased and its negative regulator, snail, was increased—two proteins involved in EMT [44].

As described, TD-exosomes and related vesicles may support the growth and dissemination of tumors via providing autocrine signals to the tumor cells; however, the reports on the autocrine effects of specific cancer cell lines has been variable among different types of tumors. Specifically, TD-exosomes have been reported to increase apoptosis of well-differentiated, but not poorly differentiated pancreatic cell lines because of the Notch pathway. Such autocrine responses would indicate that for some pancreatic tumors, TD-exosomes might inhibit local growth [48, 49]. In contrast, TD-exosomes are reported to increase proliferation via an autocrine signal in the breast cancer cell line BT-474 [50]. In addition, growth of the glioblastoma-derived cell line, U87, was increased by exosomes derived from primary tumors of the glioblastoma type. The increased growth was assumed to be due to increased proliferation; however, decreased apoptosis was not excluded [39]. Similarly, a cervical cancer cell line, C33A, incubated with exosomes from the C33A cell line transfected with the protein LAMP-1, had increased phosphorylation of Akt and ERK [51]. Of note, the activation of the Akt pathway is associated with increased proliferation and other tumor-promoting downstream molecules.

Exosomes from a gastric cancer cell line stimulated the proliferation of that cell line as well as a second gastric carcinoma cell line. This stimulation of proliferation also was via increased expression of the Akt pathway [52].

As described, tumor growth may be influenced via exosomal modulation of apoptosis. The antiapoptotic protein survivin and its associated proteins, HSP70 and HSP90, may be packaged in and on exosomes and their concentrations can be increased by cellular stresses. Of note, survivin also increases cellular proliferation and invasion, so exosomal survivin is a potent microenvironmental stimulus for the growth and dissemination of tumors [53].

Establishing a supportive environment to facilitate local growth as well as progression and dissemination of a primary tumor not only depends upon TD-exosomes, but also other microvesicles released from malignant cells as well as exosomes and other microvesicles from nonmalignant cells of the local environment of a tumor. For example, the human prostate cancer cell line (PC-3) has been reported to secrete microvesicles that are thought not to be exosomes and yet activate human fibroblasts; these microvesicles were separated after a 45-min centrifugation at 14.000 g without a nanofiltration step and without pelleting by ultracentrifugation, yet these vesicles were important in intercellular communication. The activated fibroblasts also demonstrated increased microvesicular shedding and these vesicles increased the migration and invasion of PC-3 cells. These autocrine and paracrine responses, which acted as a signal loop did not occur in the LNCaP prostate cancer cell line, which is much less metastatic. For the much more metastatic PC-3 cell line, the effects of the microvesicles on chemotaxis were attributed to CX3CL1 and its associated receptor, CX3CLR1 [43]. The differences between LNCaP and the more aggressive cell line PC-3 might be explained by the signal molecules on the surfaces of exosomes. For example, TGF-β and its associated binding complex, type III receptor betaglycan, is very strongly expressed on the surface of PC-3 cells, but not LNCaP cells [54].

TD-exosomes secreted locally may induce the activation of fibroblasts/myofibroblasts, which in turn may degrade the local extracellular matrix, increase cellular motility and invasion, and epithelial mesenchymal transitions (EMT) of neoplastic cells. Also, TD-exosomes can induce expression of α-smooth muscle actin in fibroblasts as an indication of their differentiation to myofibroblasts. Differentiation into myofibroblasts can lead to the modification of the tumor microenvironment via the increased production of extracellular matrix such as pericellular hyaluronic acid [54]. TD-exosomes may also facilitate the growth of tumors by transfer of phenotypic properties via molecular signals to local cells, including limited oncogenic features, and may induce local cells to produce or modulate their phenotypic and molecular characteristics in order to produce phenotypic changes that support the development and dissemination of the tumor [27–29, 41, 47]. Sometimes such changes are complex, requiring sequential interactions of TD-exosomes with varying molecular characteristics [55]. In turn, local changes in the tumor microenvironment such as hypoxia, pH, cellular detachment, and stresses such as heat may affect the production, secretion, and molecular contents of exosomes [54–57].

Exosomes are efficient in transferring molecular features of cellular membranes, cytoplasm, and nuclei of the source cells to target cells and when these molecules produce oncogenic phenotypic features, TD-exosomes and/or other similar functioning vesicles have been designated as "oncosomes." For example, the variant, constitutively activated EGFR-related receptor, EGFRvIII, is oncogenic via its continual activation of the MAPK and Akt pathway and production of VEGF. Vesicles that can be obtained in vivo from glioblastoma tumors (SCID xenografts) can transfer EGFRvIII in cell culture to less aggressive glioma cells, which lack the EGFRvIII surface molecule. Once transferred, the expression of EGFRvIII remained relatively stable in these neoplastic cells so this transfer of EGFRvIII could represent a mechanism by which more aggressive phenotypes can spread horizontally among neoplastic cells. Continuing exposure of endothelial cells to EGFRvIII from exosomes also was able to transfer an EGFRvIII phenotype to endothelial cells, which stimulated their growth and hence, would stimulate angiogenesis. Note, nonneoplastic endothelial cells were not stably transformed and lost the expression of EGFRvIII when the exposure to the TD-exosomes was stopped [40, 41].

Another example of the transfer of an oncogenic phenotype by the release of vesicles from cancer cells has been described by Antonyak et al. who studied vesicles released from a breast cancer cell line (MDA-MB-231 cells) and separately from a glioma cell line (U87) that were retained on a 450 nm filter. These larger vesicles contained tissue transglutaminase (tTG) that is an oncogenic enzyme that cross-links proteins. They reported that such vesicles containing tTG plus the tTG substrate, fibronectin, could transfer to NIH/3T3 fibroblast or to human benign breast epithelial cells, MCF10A, some of the specific features of cancer cells including growth in low-serum medium and anchorage—independent growth in agar. As with other studies, some of these features could not be transferred to benign cells upon a single exposure; however, transfer of some oncogenic features to benign cells could be accomplished by continuous exposure to these vesicles [47].

5.5 Exosomes and Metastases

Metastases of tumors are controlled by many molecular pathways including genes and pathways associated with stimulating as well as inhibiting metastases [58]. In addition, metastasis of tumors is facilitated by TD-exosomes secreted from the primary

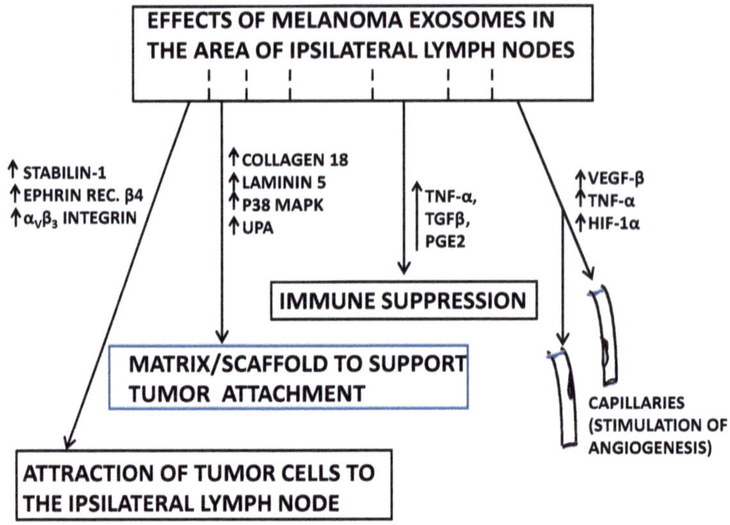

Fig. 5.2 This figure demonstrates a model of how exosomes from melanomas facilitate metastates to local ipsilateral lymph nodes via developing a fertile environment in the lymph node. Data are based on analysis at the mRNA level as adapted from Ref. [81]. The potential variability of TD-exosomes is emphasized in the cartoon by different compartments or subpopulations of TD-exosomes. *HIF* hypoxia induced factor, *MAPK* map kinase, *REC* receptor, *TNF* tumor necrosis factor, *TGFβ* transforming growth factor β, *PGE2* prostaglandin E2, *VEGF* vascular endothelial growth factor

tumor. These TD-exosomes can induce factors at the metastatic site, which attract tumors, build scaffolds for the attachment of metastatic cells, aid in the survival of metastatic tumor cells, suppress immunity, and stimulate angiogenesis [42–50, 59].

Exosomes from melanomas have been reported to travel to sentinel lymph nodes and aid in establishing an environment favorable for metastases. Melanoma cells were reported to be recruited to the sentinel lymph node by exosomal stimulated production of stabilin 1, ephrin receptor β4, and integrin $\alpha_v\beta_3$. The matrix to support the growth of melanoma cells also was found to be modulated by exosomes via the increased expression of collagen 18, laminin 5, map kinase (p38), and urokinase plasminogen activator protease. In addition, the exosomes stimulated angiogenesis by the production of vascular endothelial growth factor β (VEGFβ), TNFα, and hypoxia inducible factor 1α (HIF-1α). As in Fig. 5.2, this molecular priming of the lymph node ipsilateral to the primary site of the melanoma causes selective metastases to the ipsilateral lymph node [59]. In addition, exosomes induce local changes in melanoma that may aid in dissemination [42].

In a rat model of pancreatic cancer, conditioned media from a cell line of a syngenic tumor was separated into a soluble and exosomal fraction. The exosomal fraction was most effective in supporting metastatic spread to lymph nodes and the lung; however, the combination of soluble plus exosomal fractions were much more efficient at supporting these metastases. This effect was greatly reduced in cell lines in which there was knockdown of splice variants of CD44, especially CD44v6, on which the induction of an extracellular matrix was reported to depend; the extracellular matrix contained c-Met and urokinase-type plasminogen activator receptor. Other exosomes from pancreatic tumors independent of their aggressiveness could interact with the soluble matrix to facilitate tumor attachment and subsequent growth [55].

Many of these studies have been based upon the responses of cells to TD-exosomes or other vesicles isolated from culture media of cell lines. Much more work is required to characterize autocrine, paracrine, and endocrine interactions of exosomes in the growth, progression, and metastases of tumors in vivo.

5.6 Effects of Exosomes on Immune Surveillance of Neoplasia

As discussed, the immune system should recognize the development, progression, and dissemination of neoplastic lesions and should theoretically inhibit the growth of such lesions via cellular and other immune reactions. Unexpectedly, the presence of tumors actually suppresses immunity permitting the more rapid growth of tumors [60–62]. This, plus the involvement of exosomes in immune regulation, suggested to investigators that neoplastic lesions might release TD-exosomes, which could provide signals to inhibit immunity. Thus, it was hypothesized that signals provided via TD-exosomes were involved in the partial suppression of immune surveillance and that this facilitated the progression of neoplasia. This potential mechanism is supported by our report that transplanted tumors demonstrated increased growth in mice after injection of exosomes isolated from the same tumors [61]. In subsequent studies, it has been shown that TD-exosomes act to suppress immune surveillance via multiple pathways and multiple phenotypic effects. These include reduced cytotoxicity and decreased interleukin-2- mediated proliferation of NK and T cells; APCs are decreased by TD-exosomal signals and cells that suppress immunity are increased [60–67]. Specifically, the observed changes in the function of NK cells caused by TD-exosomes is associated with decreased release of perforin, inhibition of Jak-3, and decreased expression of cyclin D3 [61]. Based upon these other studies, it has been concluded that the decrease in immune surveillance induced by tumors is, in part, caused by the release of exosomes [8, 9, 19, 61–73].

The molecular signals provided by TD-exosomes to immune cells are likely to be diverse among the various categories of tumors. For example, ovarian epithelial and oral neoplastic lesions have been reported to secrete exosomes that contain FasL, TRAIL, and/or related molecules (e.g., TNFα) via the TRAIL/FasL pathways that can target and cause apoptotic cellular death of activated T cells [19, 21, 24, 63]. Specifically, TRAIL or FasL can suppress CD-3ζ and Jak-3 in activated T lymphocytes and induce apoptosis [63]. However, in our prior study of the effects of TD-exosomes on the in vivo growth of matching tumors, the TD-exosomes contained neither TRAIL nor Fas ligand [61]. In addition, TD–exosomes, which contain TGF-β1 down regulate the expression of the NKG2D receptor, which is a receptor involved in activation of CD8^{+}T and NK cells [64]. These observations support the concept that TD-exosomes may inhibit immune surveillance via multiple pathways, which directly inhibit the ability of T and NK cells to target and destroy neoplastic cells.

Exosomes also inhibit the immune system and increase immune tolerance by reducing the activity and/or numbers of APCs, especially dendritic cells [65–70]. Normally, exosomes from DCs are necessary to stimulate T and NK cells to kill tumor cells and hence APCs should aid in the reduction of the growth of tumors [65–73]. However, TD-exosomes inhibit immune surveillance by reducing the number of mature DCs in a time- and dose-dependent manner. This occurs via signals from exosomes, which increase the phenotypic expression in DCs of IL-6 and the phosphorylation of Stat 3 [70]. TD-exosomes also directly reduce the numbers of APCs via inhibiting the differentiation of $CD14^+$ monocytes to mature APCs. This inhibition results in a shift of $CD14^+$ cells to an immunosuppressive subset of $CD14^+HLA\text{-}DR^{-/low}$ cells, which may secrete TGFβ and thus directly inhibit T cells [68, 69].

In addition to the direct targeting of T and NK cells and the indirect effects on immunity via reducing the number and effectiveness of APCs, TD-exosomes may cause the suppression of the immune system by increasing subpopulations of cells, which inhibit immunity such as $CD11b^+Gr\text{-}1^+$, myeloid-derived suppressor cells (MDSCs) [74–82]. MDSCs have been noted to be increased in tumors as well as in blood and spleens of mice with transplanted syngenic tumors and MDSCs can be decreased in the blood and spleens of these mice by removing the implanted tumors [75–77]. Of note, increased numbers of MDSCs in humans with specific tumors have been associated with increased progression of tumors and decreased patient survival [76–79]. This may be via the action of MDSCs to inhibit $CD4^+$ and $CD8^+$ lymphocytes as well as NK cells; the direct contact of MDSCs with NK cells decreases activation of NK cells [74, 75, 83]. The release from tumor cells of soluble molecular mediators such as GM-CSF, in addition to exosomes, may also increase the number of MDSCs. TD-exosomes that contain TGFβ and prostaglandin E2 can increase MDSCs [78, 80]. Also, exosomes with HSP72 may modulate receptors to MyD88 as well as Toll-like receptor2 of MDSCs, causing MDSCs to be activated by increases in IL-6 and Stat 3 phosphorylation [80]. Of note, the study of the interaction of TD-exosomes with TLR pathways and MyD88 should be studied more extensively in cells isolated from in vivo tumors because of phenotypic changes in cells induced by long-term cultures [8, 81, 83].

Immune surveillance of tumors also can be reduced by TD-exosomes, which can cause $CD4^+$, $CD25^-$ T cells to differentiate into $CD4^+$, $CD25^+$, $Foxp3^+$ Tregs via signals resulting in the phosphorylation of SMAD2/3 and Stat3 [82].

In summary, immune surveillance is reduced and avoided by tumors because TD-exosomes can inhibit the activity of T and NK cells, reduce the maturation of precursor cells to APCs especially dendritic cells and thus indirectly inhibit the functions of APCs, as well as increase cellular subpopulations, which suppresses immunity including myeloid-derived suppressor cells, Treg cells, and $CD14^+\text{-}HLA\text{-}DR^{-/low}$ cells. The effects of TD-exosomes on immunity are demonstrated in Fig. 5.3.

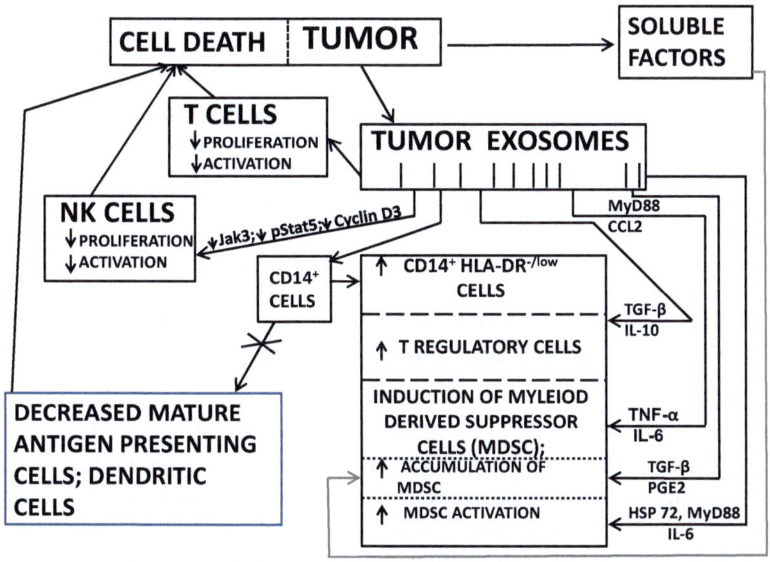

Fig. 5.3 This figure demonstrates a murine model of the multiple actions of TD-exosomes released from neoplastic cells to suppress immune surveillance. TD-exosomes reduce the number of antigen-presenting cells (e.g., blocking the differentiation of myeloid precursor cells into dendritic cells) as well as the activity and/or proliferation of NK and T cells. In addition, cells, which suppress the immune system, including Treg cells, MDSCs, and $CD14^{+}$, HLA-$DR^{-/low}$ cells are increased in number, distribution, and/or activity via the actions of TD-exosomes. Overall, these actions reduce the cytotoxicity of the immune system against malignant cells. In this cartoon, the variability of TD-exosomes is emphasized by the individual compartments or subpopulations of the TD-exosomes, each of which produces different effects on the immune system. *MDSC* myeloid-derived suppressor cells, *NK* natural killer, *Treg* T regulatory cell, *HSP* heat shock protein, *PGE* prostaglandin E, *IL* interleukin, *TGF* transforming growth factor, *TNF* tumor necrosis factor

5.7 Effects of TD-Exosomes on the Resistance of Tumors to Therapy

Exosomes provide a pathway via which unneeded and other molecules are exported from cells [2, 8, 9, 84–87]. Specific chemotherapeutic and other drugs may be eliminated via this pathway and TD-exosomes may be important to the development and maintenance of chemoresistance [8, 9, 85–88]. Doxorubicin and cis-platinum, but not 5-fluorouracil, are efficiently exported by TD-exosomes and thus TD-exosomes may provide a pathway of chemoresistance of specific malignant cells [86–87]. Shedden et al. [86] combined genes associated with the shedding of exosomes into a "vesicle shedding index," which in the "NCI 60-cell-line panel" positively correlated with the 50 % growth inhibition (GI_{50}) for most of the 171 compounds of the NCI "Standard Anticancer Agent Database." Also, when the actual shedding of vesicles from six cell lines were measured, the shedding rate correlated positively with resistance to doxorubicin, a drug that is exported in vesicles; however, there was not a correlation with resistance to the drug, 5-fluorouracil, which is not efficiently exported in vesicles [86]. The export of drugs such as doxorubicin is thought to involve the vacuolar protein sorting 4a (VPS4a), which is a protein essential for the formation of multivesicular bodies and thus the secretion of exosomes. Specifically, in a model using an erythroleukemic cell line, K562, disruption of the VP54a pathway inhibited the formation of exosomes and the efflux of doxorubicin [88].

5.8 The Uses of Exosomes in Detection, Diagnosis, Prognosis, Prediction, and Treatment of Neoplasia

TD-exosomes affect many aspects of the development, progression, and dissemination of their neoplastic cells of origin, typically mirroring the molecular features of their originating tumors. TD-exosomes may contain enriched concentrations of biomarkers, hence molecular assays of TD-exosomal preparations may be useful in translational research and/or in making clinical decisions. Uses include early detection, diagnosis, and risk assessment as well as in prediction of therapeutic efficacy, in determining prognosis, and as surrogate endpoints in evaluating chemotherapeutic and novel therapies as well as approaches to prevention of neoplasia [27–29]. Of importance in translational research is that exosomes can be picked up and carried in blood as well as shed into multiple biological fluids such as urine, ascites, and pleural fluids, biological fluids that can be obtained easily for clinical uses. The measurements of biomarkers in the exosomal compartments of biological fluids are likely to be just as sensitive and specific as measurements of the same biomarkers in the matching fluids [8, 9, 27–29]. For specific biomarkers, which may be even more concentrated in exosomes than in the biofluid from which exosomes have been removed, the measurement of a biomarker or biomarker panel in the exosomal fraction may be more sensitive and specific for answering specific clinical questions [8, 9, 27–29]. This must be determined for each clinical use by translational research for each separate biomarker or biomarker panel [27–29]. Such multiple uses have been proposed [89–98]. For example, Claudin 4, a biomarker of ovarian epithelial tumors, was increased in exosomes separated from the blood of patients with ovarian carcinoma [89]. MicroRNAs that have been proposed to be useful in the diagnosis of lung cancer have been found in TD-exosomes from patients with lung cancer [90, 91]. Similarly, δ-catenin, PCA-3, and TMPRSS2:ERG, which are putative molecular biomarkers useful in the early diagnosis of prostate cancer, have been detected in TD-exosomes in urine of patients with prostatic cancer [92–95]; such biomarkers in exosomes also may be useful in determining therapeutic responses. Similarly, the presence in biological fluids of TD-exosomes containing biomarkers characteristic of tumors of the bladder, brain, colorectum, and other tumors such as melanomas and renal cell carcinomas have been reported [3, 6, 30, 37, 39, 96–101]. Thus, the presence of increased molecular biomarkers in the biological fluids from individuals may be useful in clinical decisions as well as in translational research based on measurements of specific biomarkers.

As TD-exosomes may contain increased concentrations of specific tumor antigens, exosomes could be used in stimulating cytotoxic T-lymphocyte responses (CTL). Specifically, TD-exosomes could be used to stimulate DCs [3–5] and the in vitro generation by the DCs of exosomes could avoid the negative therapeutic effects of the use of TD-exosomes (e.g., inhibition of immune surveillance). To date, trials with this approach have been safe, but responses have been only modest [57, 62, 63, 67, 102].

Alternatively, due to the actions of TD-exosomes in suppressing immunity in patients with neoplastic lesions, the direct targeting of TD-exosomes could be used to reduce the suppression of the immune system in patients with tumors. Several strategies have been proposed to reduce the inhibitory effects of TD-exosomes on immune surveillance in patients with cancer. One approach would be to prevent or reduce the transfer into the TD-exosomes of molecules, which may suppress immunity such as FasL. Alternatively, molecules, which might selectively stimulate immune cells could be introduced into TD-exosomes in vitro. This would permit targeting tumors selectively [8, 9]. In addition, factors that decrease the release of TD-exosomes such as cellular pH might be utilized [56, 103] as could methods, which can specifically remove TD-exosomes from the circulation. For example, TD-exosomes might be removed from blood using some form of "hemopurifier" that selectively removes TD-exosomes via fixed antibodies that specifically bind TD-exosomes and thus, reduce the inhibitory effects of exosomes on immune surveillance [8, 9].

Exosomes have been reported to function in selectively exporting specific drugs (e.g., doxorubicin and cisplatinum) from neoplastic cells [86, 87]. Thus, TD-exosomes involved in exporting drugs from cells might be targeted to make specific drugs more efficacious. Such targeting to decrease the removal of therapeutic drugs via exosomes might be via reducing the production or release of TD-exosomes from neoplastic cells. This might be accomplished via local changes in pH or by selectively targeting the VPS4a signaling pathway [56, 86–88, 103].

As TD-exosomes can selectively target malignant cells via autocrine interactions, they could serve as a method of delivery of drugs, small molecules, and agents of gene therapy to selected tumors. For example, curcumin, a polyphenol with antiinflammatory and antitumor activities, has been used to reduce the effects of TD-exosomes in inhibiting NK cytotoxicity [104]. Unfortunately, solubility and bioavailability has limited the effectiveness of curcumin therapeutically, although its usefulness can be improved by its delivery via encapsulation in lipid preparations such as liposomes [105]. In comparison, the delivery and bioavailability of curcumin can be improved greatly by the incorporation of curcumin into exosomes. Delivery of curcumin by exosomes was found to be much more effective than delivery by liposomes in preventing septic shock in a murine model [106]. While not directly related to cancer therapeutics, this approach shows how specific drugs could be targeted to neoplastic cells by exosomes.

Developments in this area are being advanced by several commercial as well as academic organizations. The potential use of exosomes in clinical medicine has been the focus of many patents including the uses of exosomes to increase immune reactions to tumors, to facilitate the specific delivery of therapeutic molecules to tumors, to block or reduce the export of drug and molecules utilized in cancer therapy, and to remove molecules by "hemopurifier". This work is being performed on exosomal as well as nonexosomal vesicles. For example, vesicles from lower organisms such as ameba are being proposed as vehicles to transfer a wide range of molecules to eukaryocytic cells (e.g., patent application to I. Tatischeff et al.). In addition, vesicles have been constructed from erythrocytic ghosts and these vesicles were filled with doxorubicin coupled to folate. The preparation was stable and demonstrated increased cytotoxicity compared with doxorubicin or doxorubicin-loaded vesicles [107]. Of interest, mesenchymal stem cells have been selectively immortalized by a lentiviral transfection of the myc gene. The purpose of this cell line was to provide a stable, uniform source of exosomes for potential therapeutic uses [108]. Thus, approaches using vesicles for the selective delivery of drugs, bioactive molecules, preventive agents, and gene therapy are in active development.

5.8.1 Exosomes should be Defined or Subclassified Based on the Combination of Size and their Biological Properties

Lastly, we would like to express our concerns as to the current terminology and "definitions" used in the field of exosomes and the general field of extracellular vesicles. Why not extend the definition of exosomes into nanoparticles of various sizes instead of restricting their size to 30–100 nm? Most bodily fluids, including blood (serum and plasma), breast milk, saliva, pleural fluid, amniotic fluid, ascites, urine, cerebrospinal fluid, semen, bronchoalveolar lavage fluid, and synovial fluid contain exosomes [99–101] and other membrane-bound vesicles. Exosomes may have different molecular, morphological, and biophysical properties based on the tissue or biofluid from which they are isolated. For example, the exosome-like particles from semen were reported to be round to egg-shaped, to contain internal vesicles and to range in size from 50 to 500 nm [100]; however, exosomes isolated from ascites were noted to be almost identical in size (30–100 nm) and ultrastructure to exosomes from matching samples of blood [101]. Exosomes isolated from saliva were described as < 100 nm in diameter and doughnut-shaped with a height to width ratio of about 0.04 [34]. The vesicles recovered from cell culture media have been reported to be larger and in some cases their sizes have been described as biphasic, with most being 100–200 nm and 400–1,000 nm [86] in largest dimension.

The "sizes" of exosomes become an important issue based on past and current studies, which have used the terms "microvesicles," "oncosomes," or "exosomal-like particles" instead of exosomes to include together with exosomes, vesicles that bud directly from the surface lipid rafts of cellular membranes without involvement of MVBs [40–50, 109]. Usually, such studies vary as to how membrane-bound particles are separated from fluids, especially from spent or conditioned media from cell cultures, as well as the surface biomarkers that are used to identify subgroups of "vesicles." One possibility that may explain the larger vesicles found in spent media is that cells in two-dimensional (2D) cultures produce exosomes with different sizes, functions, molecular contents, and/or molecular surface markers than classic "exosomes" isolated from blood. For example, interactions of cell lines of tumors grown in contact with other cells (leukocytes) may secrete exosomes with different contents and functions than cells grown in isolated 2D culture environments [83, 110]. Similarly, exosomes obtained from short-term cultures of cells derived from xenografts function differently than exosomes from cells maintained in longer term cultures [83]. In addition, the actual biological functions of extracellular vesicles that do not meet the typical definition of exosomes may have similar functions in general intercellular communications. If the basic structures and functions of particles are the same, an arbitrary size separation may not be that important. Similarly, using a group of surface biomarkers to define the exosome subcomparment in some cases may be too restrictive in that the exact surface markers may vary, depending upon the specific physiological actions of exosomes (e.g., a different surface marker or lack of a surface marker might target exosomes to different subgroups of cells). Also, even a preparation used in isolation of vesicles that meets the standards of a classically defined exosomal preparation (e.g., CD9 and CD81 molecules on Western blots) does not ensure that all of the membrane-bound particles in that preparation meet the standard molecular characteristics of "exosomes." In addition, just because most of the vesicles in an exosomal preparation meet typical definitions of an exosome, does not mean that those exosomes creating the observed responses are from the subpopulation expressing classical exosomal markers. It is likely that each cell produces many different subpopulations of exosomes as well as different, nonexosomal vesicles; thus, one cell may provide multiple signals to different cellular groups both locally and distantly. Each population of exosomes secreted by one cell type would likely contain multiple distinct subgroups of exosomes/vesicles based both on their surface molecules (interaction signals) and molecular contents (cellular modulation signals). In addition, the sizes of exosomes may be dynamically changed over time from the point of initial release to their reaching their final destination. To test this hypothesis, we need to develop in vivo imaging technology to keep trafficking of an individual exosomes.

A major problem in the analysis of exosomal functions is that one must read carefully the details of the methods of isolation of each study in order to understand which subgroups of membrane-bound vesicles are being studied; this includes understanding their biochemical and physiological characteristics and functions and how studies of specific vesicles relate to those of classic exosomes. Specifically, some studies use a 200–220 nm filter to exclude larger particles from a preparation; alternatively, this filter or a larger pore filter (450 nm) may be used to retain the particles to be evaluated. Similarly, it is frequently unclear as to how the preparations are tested or not tested as to the expression of surface markers such as tetraspanins (e.g., especially CD9 and CD81). Publications are becoming chaotic as to studies of exosomes or related vesicles and must in the future be clearer as to which subpopulations of "vesicles" are being evaluated and how experimental results relate to those elicited by classic exosomal preparations. Unfortunately, methods are not always clear or the methods may be relegated to supplementary information and thus, are not an integral component of the manuscript. Even manuscripts, which indicate that the "vesicles" studied are similar to other studies may have studied different vesicles. Specifically, studies of "exosomes" may refer to the results of manuscripts that were focused on different types of "vesicles" and indicate that these prior results were obtained from "exosomes." The molecular, functional, and morphological characteristics of different subgroups of vesicular nanoparticles secreted/excreted by cells are an area that should be clarified by future research.

Specifically, molecules on the surface of exosomes are frequently used to define and to demonstrate that exosomal preparations contain isolated exosomes [37, 111–116]. These include selected tetraspanins, CD-9, CD-49d, CD-63, CD-81, and CD-82, which have been described as markers of the external surfaces of exosomes. Other molecules proposed to be characteristic of the surface molecules of exosomes include the integrin $\alpha_v\beta_3$, (CD51, CD61), transferin receptor (CD71), Alix, CD80, CD86, externalized phosphatidylserine, milk fat globule-E8/lactadhern (MFG-E8), ICAM-1, CD-96, Rab-5b, and MHC class I and II complexes . Exosomes secreted by specific types of tumors such as melanomas may express specific molecules such as caveolin-1 [3–5, 8, 9, 37, 115] while other molecules, e.g., CD11a and its ligand, CD54 and CD11b, are more specific for inflammatory cells. In addition, even the tetraspanins may not always be expressed on every type of exosome; however, the surface molecules used most commonly to define exosomes from all cell types include CD9 and CD81.

Similar to nanoparticles, the sizes of exosomes may also have effect on exosomes trafficking. The exosomes circulating in blood and/or present in some bodily fluids may be influenced by the sizes of vesicles initially released from cells, in that smaller exosomes may penetrate the spaces of lymphatics and capillaries more readily and selectively than larger exosomes/vesicles; hence, smaller exosomes may be represented more extensively in blood. It remains to be determined if smaller exosomes actually penetrate more readily into the vascular-lymphatic system or into specific cells and/or if larger exosomes remaining in the interstitial space might be more important in modulating local cellular functions via autocrine, paracrine, and "other-crine" interactions.

5.9 Summary

Exosomes are an important component of a newly identified method of local (autocrine, paracrine, and other-crine) and distant (endocrine) intra- and intercellular communication that functions in normal cells. Tumors have adapted or hijacked this approach to intercellular communication to aid in their growth, progression, and dissemination by signals, which aid in the production of a fertile environment of the primary tumor as well as the potential metastatic sites of tumors. TD-exosomes also inhibit the immune system in humans and other animals with tumors and TD-exosomes facilitate the resistance of neoplasia to specific drugs by exporting selective chemotherapeutic agents and other potentially therapeutic molecules from cells.

Thus, the clinical impact of TD-exosomes is that TD-exosomes aid tumors in avoiding immune surveillance, in aiding neoplastic lesions to progress and disseminate more rapidly, and as a pathway of therapeutic resistance. Because of the deleterious effects of TD-exosomes in blood and other bodily fluids, TD-exosomes might be selective targets for therapeutic interventions via their removal; exosomes and other vesicles might, in contrast, serve as vehicles for selective delivery of drugs, small molecules, preventive agents, or agents used in gene therapy to neoplastic lesions. Additional clinical uses might focus on using TD-exosomes as a subcompartment of biological fluids in which biomarkers are measured to aid in the early detection and diagnosis of diseases for determining prognosis, for prediction of therapeutic efficacy, and to measure therapeutic responses to therapy. Such uses would rely on the molecular contents of TD-exosomes, which should be different than exosomes from either associated diseased controls or from normal individuals. All these potential uses of exosomes are currently active areas of translational research.

Acknowledgments This work is supported in part by grants from the Susan G. Komen for the Cure to Huang-Ge Zhang (BCTR0707323), NIH grants to Huang-Ge Zhang (RO1CA116092, RO1CA107181, RO1AT004294, R01CA137037, BX000962), and grants of William E. Grizzle, Susan G. Komen for the Cure (BCTR0600484); the Breast (5P50CA089019), Pancreatic (2P50CA101955) and Cervical (5P50CA098252) SPORES at UAB, the DOD Prostate Cancer grant (PC093309)' and the U54 MSM/TU/UAB Comprehensive Cancer Center Partnership (2U54CA118948).

References

1. Trams EG, Lauter CJ, Salem N Jr, Heine U (1981) Exfoliation of membrane ecto-enzymes in the form of microvesicles. Biochim Biophys Acta 645:63–70
2. Johnstone RM, Adam JR, Hammond LO, Turbide C (1987) Vesicle formation during reticulocyte maturation. Association of plasma membrane activities with released vesicles (exosomes). J Biol Chem 262:9412–9420
3. Wolfers J, Lozier A, Raposo G, Regnault A, Thery C, Masurier C., et al. (2001) Tumor-derived exosomes are a source of shared tumor rejection antigens for CTL cross-priming. Nat Med 7:297–303
4. Clayton A, Court J, Navabi H, Adams M, Mason MD, Hobot JA et al (2001) Analysis of antigen presenting cell derived exosomes, based on immuno-magnetic isolation and flow cytometry. J Immunol Methods 247:163–174
5. Thery C, Zitvogel L, Amigorena S (2002) Exosomes: composition, biogenesis and function. Nat Rev Immunol 2:569–579
6. Denzer K, Kleijmeer MJ, Heijnen HFG, Stoorvogel W, Geuze HJ (2000) Exosome: from internal vesicle of the multivesicular body to intercellular signaling device. J Cell Sci 113:3365–3374
7. Van Dijk EL, Schilders G, Pruijn JM (2007) Human cell growth requires a functional cytoplasmic exosome, which is involved in various mRNA decay pathways. RNA 13:1027–1035
8. Zhang H-G, Grizzle WE (2011) Exosomes and cancer: a newly described pathway of immune suppression. Clin Cancer Res 17(5):1–6
9. Zhang H-G, Liu Y, Deng Z-B, Liu C, Xiang X, Grizzle WE (2012) Exosomes and immune surveillance of neoplastic lesions: a review. Biotech Histochem 87(3):161–168

10. Théry C, Duban L, Seguar E, Véron P, Lantz O, Amigorena S (2002) Indirect activation of naïve CD4$^+$ T cells by dendritic cell-derived exosomes. Nat Immunol 3(12):1156–1162
11. Vincent-Schneider H, Stumptner-Cuvelette P, Lankar D, Pain S, Raposo G, Benaroch P et al (2002) Exosomes bearing HLA-DR1 molecules need dendritic cells to efficiently stimulate specific T cells. Int Immunol 14:713–722
12. Andre F, Chaput N, Schartz NEC Flament C, Aubert N, Bernard J, Lemonnier F, Raposo G, Escudier B, Hsu D-H, Tursz T, Amigorena S, Angevin E, Zitvogel L (2004) Exosomes are potent cell-free peptide-based vaccine. I. dendritic cell-derived exosomes tranfer functional MHC class 1/peptide complexes to dendritic cells. Blood 172:2126–2136
13. Chaput N, Schartz NEC, André F, Taïeb J, Novault S, Bonnaventure P, Aubert N, Bernard J, Lemonnier F, Merad M, Adema G, Adams M, Ferrantini M, Carpentier AF, Escudier B, Tursz T, Angevin E, Zitvogel L (2004) Exosomes as potent cell-free peptide-based vaccine. II. Exosomes in CpG adjuvants efficiently prime naïve Tc1 lymphocytes leading to tumor rejection. J Immunol 172:2137–2146
14. Qazo KR, Gehrmann U, Jordö ED, Karlsson MCI, Gabrielsson S (2009) Antigen-loaded exosomes alone induce Th 1-type memory thruogh a B cell-dependent mechanism. Blood 113:2673–2683
15. Hedlund M, Stenqvist A-C, Nagaeva O, Kjellberg L, Wulff M, Baranov V, Mincheva-Nilsson L (2009) Human placenta expresses and secretes NKG2D ligands via exosomes that down-modulate the cognate receptor expression: evidence for immunosuppressive function. J Immunol 183(1):340–351
16. Mincheva-Nilsson L (2006) Immune cells and molecules in pregnancy: friends or foes to the fetus? Exp Rev Clin Immunol 2:457–470
17. Sabapatha A, Gerçel-Taylor C, Taylor DD (2006) Specific isolation of placenta-derived exosomes from the circulation of pregnant women and their immunoregulatory consequences. Am J Reprod Immunol 56:345–355
18. Taylor DD, Akyol S, Gerçel-Taylor C (2006) Pregnancy-associated exosomes and their modulation of T cell signaling. J Immunol 176:1534–1542
19. Kim JW, Wieckowski E, Taylor DD, Reichert TE, Watkins S, Whiteside TL (2005) Fas ligand-positive membranous vesicles isolated from sera of patients with oral cancer induce apoptosis of activated T lymphocytes. Cancer Res 11:1010–1020
20. Kim SH, Bianco N, Menon R, Lechman ER, Shufesky WJ, Morelli AE, Robbins PD (2006) Exosomes derived from genetically modified DC expressing FasL are anti-inflammatory and immunosuppressive. Mol Therapy 13(2):289–300
21. Abusamra AJ, Zhong Z, Zheng X, Li M, Ichim TE, Chin JL et al (2005) Tumor exosomes expressing Fas ligand mediate CD8$^+$ T-cell apoptosis. Blood Cells Mol Dis 35:169–173
22. Wang GJ, Liu Y, Qin A, Shah SV, Deng ZB, Xiang X, Cheng Z, Liu C, Wang J, Zhang L, Grizzle WE, Zhang HG (2008) Thymus exosomes-like particles induce regulatory T cells. J Immunol 181:5242–5248
23. Kim SH, Lechman ER, Bianco N, Menon R, Keravala A, Nash J et al (2004) Exosomes derived from IL-10-treated dendritic cells can suppress inflammation and collagen-induced arthritis. J Immunol 174:6440–6448
24. Zhang HG, Liu C, Su K, Yu S, Zhang L, Zhang S, Wang J, Cao X, Grizzle W, Kimberly RP (2006) A membrane form of TNF-alpha presented by exosomes delays T cell activation-induced cell death. J Immunol 176:7385–7393
25. Kapsogeorgou EK, Abu-Helu RF, Moutsopoulos HM, Manoussakis MN (2005) Salivary gland epithelial cell exosomes: a source of autoantigenic ribonucleoproteins. Arthritis Rheum 52:1517–1521
26. Deng ZB, Liu Y, Liu C, Xiang X, Wang J, Cheng Z, Shah SV, Zhang S, Zhang L, Zhuang X, Michalek S, Grizzle WE, Zhang HG (2009) Immature myeloid cells induced by a high-fat diet contribute to liver inflammation. Hepatology 50(5):1412–1420 (PMCID: PMC2852608)
27. Grizzle WE, Srivastava S, Manne U (2012) Translational Pathology of Cancer. In: S Srivastava, WE Grizzle (eds) Translational pathology of early cancer. IOS Press, Amsterdam (in press)

28. Grizzle WE, Srivastava S, Manne U (2012) The biology of incipient, pre-invasive or intraepithelial neoplasia. In: S Srivastava, WE Grizzle (eds) Translational pathology of early cancer. IOS Press, Amsterdam (in press)
29. Grizzle WE, Srivastava S, Manne U (2012) The genetics of early cancer. In: S Srivastava, WE Grizzle (eds) Translational pathology of early cancer. IOS BV, Amsterdam (in press)
30. Graner MW, Cumming RI, Bigner DD (2007) The heat shock response and chaperones/heat shock proteins in brain tumors: surface expression, release, and possible immune consequences. J Neurosci 27:11214–11227
31. Lässer C, Alikhani VS, Ekström K, Eldh M, Paredes PT, Bossios A, Sjöstrand M, Gabrielsson S, Lötvall J, Valadi H (2011) Human saliva, plasma and breast milk exosomes contain RNA: uptake by macrophages. J Transl Med 9:9
32. Harrington MG, Fonteh AH, Oberins E, Lise P, Cowan RP, McComb G, Chavez JN, Rush J, Biringer RG, Hühmer AF (2009) The morphology and biochemistry of nanostructures provide evidence for synthesis and signaling functions in human cerebrospinal fluid. Cerebrospinal Fluid Res 6:10
33. Admyre C, Grunewald J, Thyberg S, Gripenbäck S, Tornling G, Eklund A, Scheynius A, Gabrielsson S (2003) Exosomes with major histocompatibility complex class II and co-stimulatory molecles are present in human BAL fluid. Eur Respir J 22:578–583
34. Palanisamy V, Sharma S, Deshpande A, Zhou H, Gimzewski J, Wong DT (2010) Nanostructural and transcriptomic analyses of human salive derived exosomes. PLoS ONE 5(1):38577. doi:10.1371/journalo.pone.008577
35. Skriner K, Adolph K, Jungblut PR, Burmester GR (2006) Association of citrullinated proteins with synovial exosomes. Arthritis Rheum 54:3809–3814
36. Wolfers J, Lozier A, Raposo G, Regnault A, Thery C, Masurier C et al (2001) Tumor-derived exosomes are a source of shared tumor rejection antigens for CTL cross-priming. Nat Med 7:297–303
37. Logozzi M, de Milito A, Lugini L, Borghi M, Calabrò L, Spada M, Perdicchio M, Marino ML, Federici C, Lessi E, Brambilla D, Venturi G, Lozopone F, Santinami M, Huber V, Maio M, Rivoltini L, Fais S (2009) High levels of exosomes expressing CD63 and caveolin-1 in plasma of melanoma patients. PLoS ONE 4(4):e5219. doi:10.1371/journal.pone.0005219
38. Graner MW, Alzate O, Dechkovskaia AM, Keene JD, Sampson JH, Mitchell DA, Bigner DD (2009) Proteomic and immunologic analyses of brain tumor exosomes. FASEB J 23:1541–1557
39. Skog J, Würdinger T, van Rijn S, Meijer DH, Gainche L, SenaEsteves M, Curry WT Jr, Carter BS, Krichevsky AM, Breakefield XO (2008) Glioblastoma microvesicles transport RNA and proteins that promote tumour growth and provide diagnostic biomarkers. Nat Cell Biol 10(12):1470–1476
40. Al-Nedawi K, Meehan B, Kerbel RS, Allison AC, Rak J (2009) Endothelial expression of autocrine VEGF upon the uptake of tumor-derived microvesicles containing oncogenic EGFR. Proc Natl Acad Sci U S A 106(10)L:3794–3799
41. AL-Nedawi K, Meehan B, Micallef J, Lhotak V, May L, Guha A, Rak J (2008) Intercellular transfer of the oncogenic receptor EGFRvIII by microvesicles derived from tumour cells. Nat Cell Biol 10(5):619–624
42. Hood JL, Pan H, Lanza GM, Wickline SA (2009) Consortium for translational research in advanced imaging and nanomedicine (C-TRAIN). Paracrine induction of endothelium by tumor exosomes. Lab Invest 89:1317–1328
43. Castellana D, Zobairi F, Martinez MC, Panaro MA, Mitolo V, Freyssinet JM, Kunzelmann C (2009) Membrane microvesicles as actors in the establishment of a favorable prostatic tumoral niche: a role for activated fibroblasts and CX3CL1-CX3CR1 axis. Cancer Res 69(3):785–793
44. Park JE, Tan HS, Datta A. Lai RC, Zhang H, Meng W, Lim SK, Sze SK (2010) Hypoxic tumor cell modulates its microenvironment to enhance angiogenic and metastatic potential by secretion of proteins and exosomes. Mol Cell Proteomics 9:1085–1099

45. Hong BS, Cho J-H, Kim H, Choi E-J, Rho S, Kim J, Kim JH, Choi D-S, Kim Y-K, Hwang D, Ghoo YS (2009) Colorectal cancer cell-derived microvesicles are enriched in cell cycle-related mRNAs that promote proliferation of endothelial cells. BMC Genomics 10:556. doi:10.1186/1471-2164-10-556
46. Gesierich S, Berezovskly I, Ryschich E, Zöller M (2006) Systemic induction of the angiogenesis switch by the tetraspanin D6.1A/CO-029. Cancer Res 66:7083–7094
47. Antonyak MA, Li B, Boroughs LK, Johnson JL, Druso JE, Bryant KL, Holowks DA, Cerione RA (2011) Cancer cell-derived microvesicles induce transformation by tranferring tissue transglutaminase and fibronectin to recipient cells. Proc Natl Acad Sci U S A 12:4852–4857
48. Ristorcelli E, Beraud E, Verrando P, Villard C, Lafitte D, Sbarra V, Lombardo D, Verine A (2008) Human tumor nanoparticles induce apoptosis of pancreatic cancer cells. FASEB J 22(9):3358–3369
49. Ristorcelli E, Beraud E, Mathieu S, Lombardo D, Verine A (2009) Essential role of Notch signaling in apoptosis of human pancreatic tumoral cells mediated by exosomal nanoparticles. Int J Cancer 125(5):1016–1026
50. Koga K, Matsumoto K, Akuyoshi T, Kubo M, Yamanaka N, Tasaki A, Nakashima H, Nakamura M, Kuroki S, Tanaka M, Katano M (2005) Purification, characterization and biological significance of tumor-derived exosomes. Anticancer Res 25(6A):3703–3707
51. Meckes DG Jr, Shair KHY, Marquitz AR, Kung C-P, Edwards RH, Raab-Traub N (2010) Human tumor virus utilizes exosomes for intercellular communication. PNAS 107(47):20370–20375
52. Qu JL, Qu XJ, Zhao MF, GTeng JE, Zhang Y, Hou KZ, Jiang YH, Yang XH, Liu YP (2009) Gastric cancer exosomes promote tumour cell proliferation through P13K/Akt and MAPK/ERK activation. Dig Liver Dis 41(12):875–880
53. Khan S, Jutzy JMS, Aspe JR, McGregor DW, Neidigh JW, Wall NR (2011) Survivin is released from cancer cells via exosomes. Apoptosis 16:1–12
54. Webber J, Steadman R, Mason MD, Tabi Z, Clayton A (2010) Cancer exosomes trigger fibroblast to myofibroblast differentiation. Cancer Res 70(23):9621–9630
55. Jung T, Castellana D, Klingbell P, Hernåndez IC, Vitacolonna M, Orlicky DJ, Roffler SR, Brodt P, Zöller M (2009) CD44v6 dependence of premetastatic niche preparation by exosomes. Neoplasia 11(10):1093–1105
56. Parolini I, Federici C, Raggi C, Lugini L, Palleschi S, de Milito A, Coscia C, Iessi E, Logozzi M, Molinari A, Colone M, Tatti M, Sargiacomo M, Fais S (2009) Microenvironment pH is a key factor for exosomes traffic in tumor cells. J Biol Chem 284(49):34211–34222
57. Clayton A, Turkes A, Navabi H, Mason MD, Tabi Z (2005) Induction of heat shock proteins in B-cell exosomes. J Cell Sci 118:3631–3638
58. McNally LR, Welch DR, Beck BH, Stafford LJ, Long JW, Sellers JC, Huang ZQ, Grizzle WE, Stockard CR, Nash KT, Buchsbaum DJ (2010) KISS1 over-expression suppresses metastasis of pancreatic adenocarcinoma in a xenograft mouse model. Clin Exp Metastasis 27(8):591–600
59. Hood JL, San RS, Wickline SA (2011) Exosomes released by melanoma cells prepare sentinel lymph nodes for tumor metastasis. Cancer Res 71:3792. doi:10.1158/0008–5472.CAN-10-4455
60. Zhang HG, Grizzle WE (2003) Aging, immunity, and tumor susceptibility. Immunol Allergy Clin North Am 23:83–102
61. Liu C, Yu S, Zinn K, Wang J, Zhang L, Yujiang J, Kappes JC, Barnes S, Kimberly RP, Grizzle WE, Zhang H-G (2006) Murine mammary carcinoma exosomes promote tumor growth by suppression of NK cell function. J Immunol 176:1375–1385
62. Iero M, Valenti R, Huber V, Filipazzi P, Parmiani G, Fais S et al (2008) Tumour-released exosomes and their implications in cancer immunity. Cell Death Differ 15:80–88
63. Taylor DD, Gerçel-Taylor C, Lyons KS, Stanson J, Whiteside TL (2003) T-cell apoptosis and suppression of T-cell receptor/CD3-ζ by fas ligand-containing membrane vesicles shed from ovarian tumors. Clin Cancer Res 9:5113–5119

64. Clayton A, Mitchell JP, Court J, Linnane S, Mason MD, Zsuzsanna T (2008) Human tumor-derived exosomes down-modulate NKG2D expression. J Immunol 180:7249–7258
65. Zitvogel L, Regnault A, Lozier A, Wolfers J, Flament C, Tenza D et al (1998) Eradication of established murine tumors using a novel cell-free vaccine: dendritic cell-derived exosomes. Nat Med 4:594–600
66. Andre F, Shartz NE, Movassagh M, Flament C, Pautier P, Morice P et al (2002) Malignant effusions and immunogenic tumour-derived exosomes. Lancet 360:295–305
67. Andre F, Escudier B, Angevin E, Tursz T, Zitvogel L (2004) Exosomes for cancer immunotherapy. Ann Oncol 15:141–144
68. Valenti R, Huber V, Filipazzi P, Pilla L, Sovena G, Villa A, Corbelli A, Fais S, Parmiani G, Rivoltini L (2006) Human tumor-released microvesicles promote the differentiation of myloid cells with transforming growth factor-β-mediated suppressive activity on T lymphocytes. Cancer Res 66:9290–9298
69. Valenti R, Huber V, Iero M, Filipazzi P, Parmiani G, Rivoltini L (2007) Tumor-released microvesicles as vehicles of immunosuppression. Cancer Res 67:2912–2915
70. Yu S, Liu C, Su K, Wang J, Liu Y, Zhang L et al (2007) Tumor exosomes inhibit differentiation of bone marrow dendritic cells. J Immunol 178:6867–6875
71. Luketic L, Delanghe J, Sobol PT, Yang P, Frotten E, Mossman KL et al (2007) Antigen presentation by exosomes released from peptide-pulsed dendritic cells is not suppressed by the presence of active CTL. J Immunol 179:5024–5032
72. Viaud S, Terme M, Flament C, Taieb J, Andre F, Novault S et al (2009) Dendritic cell-derived exosomes promote natural killer cell activation and proliferation: a role for NKG2D ligands and IL-15Ralpha. PLoS ONE 4:e4942
73. Van Doormaal FF, Kleinjan A, Di Nisio M, Büller HR, Nieuwland R (2009) Cell-derived microvesicles and cancer. J Med 67(7):266–273
74. Xiang X, Poliakov A, Liu C, Liu Y, Deng ZB, Wang J et al (2009) Induction of myeloid-derived suppressor cells by tumor exosomes. Int J Cancer 124:2621–2633
75. Liu C, Yu S, Kappes J, Wang J, Grizzle WE, Zinn KR et al (2007) Expansion of spleen myeloid suppressor cells represses NK cell cytotoxicity in tumor-bearing host. Blood 109:4336–4342
76. Serafini P, Borrello I, Bronte V (2006) Myeloid suppressor cells in cancer: recruitment, phenotype, properties, and mechanisms of immune suppression. Sem Cancer Biol 16:53–65
77. Gabrilovich DI, Nagaraj S (2009) Myeloid-derived suppressor cells as regulators of the immune system. Nat Rev Immunol 9:162–174
78. Ochoa AC, Zea AH, Hernandez C, Rodriguez PC (2007) Arginase, prostaglandins and myeloid-derived suppressor cells in renal cell carcinoma. Clin Cancer Res 13:721–736
79. Diaz-Montero CM, Diaz-Montero CM, Salem ML, Nishimura MI, Garrett-Mayer E, Cole DJ, Montero AJ (2009) Increased circulating myeloid-derived suppressor cells correlate with clinical cancer stage, metastatic tumor burden and doxorubicin-cyclophosphamide chemotherapy. Cancer Immunol Immunother 58:49–59
80. Chalmin F, Ladoire S, Mignot G, Vincent J, Bruchard M, Remy-Martin J-P, Boireau W, Rouleau A, Simon B, Lanneau D, De Thonel A, Multhoff G, Hamman A, Martin F, Chauffert B, Solary E, Zitvogel L, Garrido C, Ryffel B, Borg C, Apetoh L, Rébé C, Ghiringhelli F (2010) Membrane-associated Hsp72 from tumor-derived exosomes mediates STAT3-dependent immunosuppressive function of mouse and human myeloid-derived suppressor cells. J Clin Invest 120:457–471
81. Liu Y, Xiang X, Zhang X, Zhang S, Liu C, Cheng Z, Michalek SM, Grizzle WE, Zhang HG (2010) Contribution of MyD88 to the tumor exosome-mediated induction of myeloid derived suppressor cells. Am J Pathol 176:2490–2499
82. Szajnik M, Czystowska M, Szczepanski MJ, Mandapathil M, Whiteside TL (2010) Tumor-derived microvesicles induce, expand and up-regulate biological activities of human regulatory T cells (Treg). PLoS ONE 5:e11149. doi:10.1371/journal.pone.0011469
83. Xiang X, Liu Y, Zhuang X, Zhang S, Michalek S, Taylor DD, Grizzle W, Zhang H-G (2010) TLR2-mediated expansion of MDSCs is dependent on the source of tumor exosomes. Am J Pathol 177(4):1606–1610

84. Blanc L, Barres C, Bette-Bobillo P, Vidal M (2007) Reticulocyte-secreted exosomes bind natural IgM antibodies: involvement of a ROS-activatable endosomal phospholipase. Blood 110:3407–3416
85. Blanc L, De Gassart A, Geminard C, Bette-Bobillo P, Vidal M (2005) Exosome release by reticulocytes—an integral part of the red blood cell differentiation system. Blood Cells Mol Dis 35:21–26
86. Shedden K, Xie XT, Chandaroy P, Chang YT, Rosania GR (2003) Expulsion of small molecules in vesicles shed by cancer cells: association with gene expression and chemosensitivity profiles. Cancer Res 63:4331–4337
87. Safaei R, Larson BJ, Cheng TC, Gibson MA, Otani S, Naerdemann W, Howell SB (2005) Abnormal lysosomal trafficking and enhanced exosomal export of cisplatin in drug-resistant human ovarian carcinoma cells. Mol Cancer Ther 4:1595–1604
88. Chen YY, Posada MM, Blazer LL, Zhao T, Rosania GR (2006) The role of the VPS4A-exosome pathway in the intrinsic egress route of a DNA-binding anticancer drug. Pharm Res 23(8):1687–1695
89. Li J, Sherman-Baust CA, Tsai-Turton M, Bristow RE, Roden RB, Morin P (2009) Claudin-containing exosomes in the peripheral circulation of women with ovarian cancer. BMC Cancer 9:244
90. Rabinowits G, Gercel-Taylor C, Day JM, Taylor DD, Kloecker GH (2009) Exosomal microRNA: a diagnostic marker for lung cancer. Clin Lung Cancer 10:42–46
91. Mallick R, Patnaik SK, Yendamuri S (2010) MicroRNAs and lung cancer: biology and applications in diagnosis and prognosis. J Carcinog 9:8
92. Lu Q, Zhang J, Allison R, Gay H, Yang W-X, Bhowmick NA, Frelix G, Shappell S, Chen Y-H (2009) Identification of extracellular δ-Catenin accumulation for prostate cancer detection. Prostate 69:411–418
93. Nilsson J, Skog J, Nordstrand A, Baranov V, Mincheva-Nilsson L, Breakefield XO, Widmark A (2009) Prostate cancer-derived urine exosomes: a novdel approach to biomarkers for prostate cancer. Br J Cancer 100:1603–1607
94. Mitchell PJ, Welton J, Staffurth J, Court J, Mason MD, Tabi Z, Clayton A (2009) Can urinary exosomes act as treatment response markers in prostate cancer? J Transl Med 7:4
95. Jansen FH, Krijgsveld J, van Rijswijk A, van den Bernd G-J, van den Berg MS, van Weerden WM, Willemsen R, Dekker LJ, Luider TM, Jenster G (2009) Exosomal secretion of cytoplasmic prostate cancer xenograft-derived proteins. Mol Cell Proteomics 8:1192–1205
96. Klein-Scory S, Kübler S, Diehl H, Eilert-Micus C, Reinacher-Schick A, Stuhler K, Warscheid B, Meyer HE, Schmiegel W, Schwarte-Waldhoff I (2010) Immunoscreening of the extracellular proteome of colorectal cancer cells. BMC Cancer 10:70. doi:10.1186/1471-2407-10-70
97. Morrissey JJ, London AN, Luo J, Kharasch ED (2010) Urinary biomarkers for the early diagnosis of kidney cancer. Mayo Clin Proc 85(5):413–421
98. Grebe SK, Erickson LA (2010) Screening for kidney cancer: is there a role for aquaporin-1 and adipophilin? Mayo Clin Proc 85(5):410–412
99. Welton JL, Khanna S, Giles PJ, Brennan P, Brewis IA, Staffurth J, Mason MD, Clayton A (2010) Proteomics analysis of bladder cancer. Mol Cell Proteomics 9:1324–1338
100. Poliakov A, Spilman M, Dokland T, Amling CL, Mobley JA (2009) Structural heterogeneity and protein composition of exosome-like vesicles (prostasomes) in human semen. Prostate 69:159–167
101. Taylor DD, Gerçel-Taylor C (2005) Tumour-derived exosomes and their role in cancer-associated T-cell signalling defects. Br J Cancer 92:305–311
102. Ichim TE, Zhong Z, Kaushal S, Zheng X, Ren X, Hao X, Joyce JA, Hanley HH, Riordan NH, Koropatnick J, Bogin V, Miney BR, Min WP, Tullis RH (2008) Exosomes as a tumor immune escape mechanism: possible therapeutic implications. J Transl Med 6:37
103. Iessi E, Marino ML, Lozupone F, Fais S, de Milito A (2008) Tumor acidity and malignancy: novel aspects in the design of anti-tumor therapy. Cancer Therapy 6:55–66
104. Zhang HG, Kim H, Liu C, Yu S, Wang J, Grizzle WE et al (2007) Curcumin reverses breast tumor exosomes mediated immune suppression of NK cell tumor cytotoxicity. Biochim Biophys Acta 1773:1116–1123

105. Narayanan NK, Nargi D, Randolph C, Narayanan BA (2009) Liposome encapsulation of curcumin and resveratrol in combination reduces prostate cancer incidence in PTEN knockout mice. Int J Cancer 125:1–8
106. Sun D, Zhuang X, Xiang X, Liu Y, Zhang S, Liu C, Grizzle W, Miller D, Zhang H-G (2010) A novel nanoparticle drug delivery system. The anti-inflammatory activity of curcumin is enhanced when encapsulated in exosomes. Mol Therapy (in press)
107. Mishra PR, Jain NK (2003) Folate conjugated Doxorubicin-loaded membrane vesicles for improved cancer therapy. Drug Deliv 10(4):277–282
108. Chen TS, Arsian F, Yin Y, Tan SS, Lai RC, Choo ABH, Padmanabhan J, Lee CN, de Kleijn DPV, Lim SK (2011) Enabling a robust scalable manufacturing process for therapeutic exosomes through ocogenic immortalization of human ESC-derived MSCs. J Transl Med 9:47
109. Camussi G, Deregibus M-C, Brunpo S, Grange C, Fonsato V, Tetta C (2011) Exosome/microvesicle-mediated epigenetic reprogramming of cells. Am J Cancer Res 1(1):98–110
110. Deng Z, Cheng Z, Xiang X, Yan J, Zhuang X, Lkiu C, Jiang H, Ju S, Zhang L, Grizzle W, Mobley J, Roman J, Miller D, Zhang HG (2012) Tumor cell cross talk with tumor-associated leukocytes leads to induction of tumor exosomal fibronectin and promotes tumor progression. Am J Pathol 180(1):390–398
111. Thery C, Boussac M, Veron P, Ricciardi-Castagnoli P, Raposo G, Garin J, Amigorena S (2001) Proteomic analysis of dendritic cell-drived exosomes: a secreted subcellular compartment distinct form apoptotic vesicles. J Immunol 166:7309–7318
112. Van Niel G, Mallegol J, Bevilacqua C, Candalh C, Brugière S, Tomaskovic-Crook E, Heath JK, Cerf-Besussan N, Heyman M (2003) Investinal epithelial exosomes carry MHC class II/peptides able to inform the immune system in mice. GUT 52:1690–1697
113. Nazarenko I, Rana S, Baumann A, McAlear J, Hellwig A, Trendelenburg M, Lochnit G, Preissner KL, Zöller M (2010) Cell surface tetraspanin tspan8 contributes to molecular pathways of exosome-induced endothelial cell activation. Cancer Res 70:1668. doi:10.1158/0008-5472.CAN-09-2470
114. Morelli AE, Larregina AT, Shufesky WJ, Sullivan MMLG, Sullivan DBS, Papworth GD, Zahorchak AF, Logar AJ, Wang Z, Watkins SC, Falo, LD Jr, Thomson AW (2004) Endocytosis, intracellular sorting, and processing of exosomes by dendritic cells. Blood 104:3257–3266
115. Caby M-P, Lankar D, Vincendeau-Scherrer Raposo G, Bonnerot C (2005) Exosomal-like vesicles are present in human blood plasma. Int Immunol 17(7):879–887
116. Koumangoye RB, Sakwe AM, Goodwin JS, Patel T, Ochieng J (2011) Detachment of breast tumor cells induces rapid secretion of exosomes which subsequently mediate cellular adhesion and spreading. PLoS ONE 6(9):e24234. doi:10.1371/journal.pone.0024234

Chapter 6
Circulating Cell-derived Vesicles Mediate Tumor Progression

Douglas D. Taylor and Cicek Gercel-Taylor

Abstract Tumor cells have been demonstrated to release membranous structures, which are termed as microvesicles or exosomes depending on specific characteristics, including size and composition. These cell-derived vesicles can exhibit an array of proteins, lipids, and nucleic acids derived from the originating tumor. It is now recognized that these vesicular components are critical conveyers of intercellular communication and mediated many of the pathological conditions associated with cancer development, progression, and therapeutic failures. Through the expression of components responsible for angiogenesis promotion, stromal remodeling, signaling pathway activation through growth factor/receptor transfer, chemoresistance, and genetic intercellular exchange, tumor exosomes/microvesicles could represent a central mediator of the tumor microenvironment.

6.1 Historical Overview

The release of nanosized membranous vesicles into biological fluids by viable tumor cells was initially described by us over three decades ago [1] and has since been demonstrated multiple cell types and systems. In vivo, exosomes released by tumor cells accumulate in biologic fluids including blood, urine, ascites, and pleural fluids [2]. The release and accumulation of exosomes/microvesicles appear to be important in the malignant transformation process. These membranous vesicles have been identified by various terms including high molecular weight complexes, membrane fragments, microvesicles, microparticles, and exosomes. While more recently restrictive definitions have been applied to these cell-derived vesicles, significant overlap exists between structures identified as exosomes and microvesicles. In many studies, uncharacterized cell-derived vesicles are termed as "microvesicles." In contrast, numerous investigators define "exosomes" solely based on density.

D. D. Taylor (✉)
Division of Gynecologic Oncology, James Graham Brown Cancer Center,
505 South Hancock Street, CTRB 316, Louisville, KY 40202, USA
e-mail: ddtaylor@louisville.edu

C. Gercel-Taylor
Department of Obstetrics, Gynecology, and Women's Health, University of Louisville
School of Medicine, Louisville, KY 40202, USA

H.-G. Zhang (ed.), *Emerging Concepts of Tumor Exosome–Mediated Cell–Cell Communication,* DOI 10.1007/978-1-4614-3697-3_6,

These overlaps in vesicle properties and terms suggest these distinctions may not be clear-cut.

The term, "exosome" was coined by Trams et al. in 1981 [3] for "exfoliated membrane vesicles with 5′-nucleotidase activity." The "exosome" term originated from a discovery of the secretion of neoplastic cell line-derived exfoliated vesicles (within larger vesicles), which mirrored the 5′-nucleotidase activity of the parent cells [3]. Subsequently, the canonical pathway of "vesicle" release upon multivesicular endosome fusion with the cell surface was demonstrated in cultured sheep [4] and rat [5] reticulocytes. After purification by ultracentrifugation, the sedimented microvesicles were found to contain the transferrin receptor, which was also found in native reticulocytes [6]. These microvesicles were redefined as "exosomes" to differentiate them from "endosomes." The disparate nature of these studies is reflected in the various names that were proposed and which are still used to identify the cell-surface-released and endocytic microvesicles of different origins.

The appearance of circulating tumor-derived vesicles was initially demonstrated in ovarian cancer patients [1, 7, 8]. These gynecologic cancer patients exhibited intact membranous vesicles within their peripheral circulation and malignant effusions (ascites and cyst fluids). These tumor-derived vesicles were found to express molecular markers that were generally linked with tumor plasma membranes, including placental-type alkaline phospohatase and mdr-1 [9–11]; however, proteins not generally associated with plasma membranes, such as p53, GRP78, and nucleophosmin, could be associated and identified [12, 13].

Microvesicles/exosomes have been subsequently demonstrated to be released by a variety of noncancerous cells, particularly cells of the immune system, including dendritic cells, macrophages, B cells, T cells and NK cells [14], as well as embryonic cells [15]. Exosomes have been demonstrated to be key mediators/regulators of normal immune regulation [14]. By an analogous mechanism, exosomes released by the tumor may elicit a tolerogenic response and participate in other immune mechanisms, such as platelet activation, mast cell degranulation, germinal center reaction, and potential engulfment of apoptotic cells. The aberrant release of exosomes by tumors may allow them to circumvent these immunoregulatory antigen delivery pathways and evade immunosurveillance [2, 10].

6.2 Cell-derived Vesicle Composition and Characterization

Over the past three decades, these shed tumor-derived vesicles have been demonstrated not to be an exact replicate of the plasma membrane or other membranous compartments of the originating tumor cells, but they represent "micromaps" with enhanced expression of tumor antigens, as well as other macromolecules, including major histocompatibility antigens [2]. Exosomes versus microvesicles have been defined based on size (30–100-nm lipid bilayer vesicles), density (1.12 g/ml for B cell derived to 1.19 g/ml for intestinal cell derived), and expression of specific biomarkers (including tetraspanins) [16].

Biophysically, exosomes are cytoplasm enclosed in a lipid layer with the external domains of transmembrane proteins exposed to the extracellular environment. Electron microscopic studies have demonstrated the fusion of the limiting membrane of multivesicular bodies (MVB) with the plasma membrane as well as release of ILVs in different cell types of hematopoietic origin, such as Epstein-Barr virus (EBV)-transformed B cells [17], mastocytes [18], dendritic cells [19], platelets [20], macrophages [21], and cells of nonhematopoietic origin like neurons and epithelial cells [22].

The protein composition of cell-derived vesicles has been extensively analyzed by various techniques including Western blotting, fluorescence-activated cell sorting, immunoelectron microscopy and mass spectrometry. Their composition is dependent on the cell type of origin. Nevertheless, exosomes contain a number of common protein components [23]. The cytosolic proteins present on exosomes include Rabs, which promote exosome docking and the membrane fusion events [23]. The annexins, including annexin I, II, V, and VI, may regulate membrane cytoskeleton dynamics and membrane fusion events [24].

Recently, we have analyzed the patient-derived exosomal proteome using ion trap mass spectrometry. This study identified 232 unique proteins. These proteins were classified as percent of the identified total proteins into molecular chaperones (8.5 %), vesicle fusion (8.5 %), cytoskeletal proteins and proteins involved in the assembly/disassembly of the cytoskeletal networks (17.6 %), anionic and cationic ion transport channels (3.7 %), proteins involved in lipid (6.9 %), carbohydrate (3.2 %) and amino acids (2.1 %) metabolisms, proteins involved in DNA replication (6.9 %), mRNA splicing (5.3 %), transcription/translation (5.3 %), posttranscriptional protein modification (13.8 %), and signal transduction (2.7 %). Our studies demonstrated that cytosolic proteins were highly represented and we observed a diverse array of cytoskeletal constituents (actin, α-actinin-1, cofilin, filamin-A-B-C, tubulins, gelsolin, profilin-1, spectrin, symplekin, talin, vinculin, myosins). We identified that transmembrane proteins were also abundant, including multiple integrins (β1, α3, αv), intercellular adhesion molecule 1 (ICAM-1), and mucin-4. A variety of channels were observed, such as the voltage-dependent anion-selective channel protein 2 and 3, chloride intracellular channel protein 1, sodium-/potassium-transporting ATPase subunit β-3, long of sodium-/potassium-transporting ATPase subunit α-1, and transitional endoplasmic reticulum ATPase. In line with their endocytic origin, exosomal proteins belonging to the ESCRT complex that are important protein complexes involved in ubiquitin-dependent exosome biogenesis have also been observed [25]. These ESCRT-associated proteins include vacuolar protein sorting-associated protein 35 (VPS-35), Alix, ubiquitin-like modifier-activating enzyme and ubiquitin carboxyl-terminal hydrolases. We demonstrated that proteins involved in membrane trafficking and fusion processes were enriched (annexin A2, A5, A6, clathrin heavy chain 1/2, coatomer subunit β, Rab1b, Rab2a, and Rab7a). A group of markers of endosomes and lysosomes were also detected (cathepsin-C, -D, -EH domain-containing protein 1, and β-hexoaminidase), and several chaperonnes were identified (HSP70, HSP90, HSC70, HSPA4, 8, 9, HSPA1 A/B, HSPB1, HSP47, HSPA5, HSPβ1, HSPD1, HSP90AB1, B1, AA1; T-complex protein 1, endoplasmin, and protein disulfide-isomerase A3, A4, A6) [25, 26].

6.3 Formation of Cell-Derived Vesicles

The exact mechanisms of exosome/microvesicle release remain unclear; however, the release is modulated by extracellular signals [27]. Three mechanisms have been proposed for the release of cellular components into the extracellular space. These include: (1) exocytic fusion of MVBs resulting in exosomes, (2) budding of vesicles directly from the plasma membranes resulting in shed "microvesicles," and (3) cell death leading to apoptotic bodies. The first two mechanisms are properties of viable cells and are energy-requiring events. Most isolation protocols readily exclude apoptotic bodies; however, these methods do not differentiate exosomes from the endocytic pathway from shed "microvesicles" from the plasma membrane. As a result, most studies on these extracellular vesicle populations include a mixture of exosomes and microvesicles, which may confuse interpretation of biochemical data. Exosomes/microvesicles derived from the extracellular environment of tumors, either in vitro or in vivo, exhibit overlapping similarities in size (defined by dynamic light scattering), morphology (defined by electron microscopy), density (defined by sucrose gradient centrifugation), and protein markers of both the endosomes and plasma membranes (defined by Western immunoblotting and mass spectrometry) [22, 28]. We have compared the extracellular vesicle populations obtained from biologic fluids of ovarian cancer patients by both the techniques described to isolate exosomes and our original chromatographic method isolating "microvesicles" [7]. This comparative study demonstrated that these in vivo-derived vesicles from both techniques isolated cup-shaped vesicles, with a density between 1.13 and 1.17 g/ml, a diameter between 50 and 100 nm, and expressing CD63, Alix, VPS35, galectin 3, HSP90, fibronectin, and placental alkaline phosphatase [28]. While patient-derived circulating extracellular vesicles fit within the definition of exosomes, the contribution of these two populations is unclear. Since both populations exist within the peripheral circulation of cancer patients, this distinction between exosomes and shed microvesicles may not be critical to understand the biologic activities of these vesicles as they can interact with target cells of the host as a mixture.

The most common process is the release of large biomolecules through the plasma membrane by a process termed exocytosis, which has regulatory and signaling functions. Exocytosis can be either constitutive (non-calcium-triggered) or regulated (calcium-triggered) [28]. Constitutive exocytosis occurs in all cells and serves to secrete extracellular matrix components or to incorporate newly synthesized proteins into the plasma membrane following fusion with transport vesicles. Regulated exocytosis is critical to events, such as neurologic signaling, as synaptic vesicles fuse with the membrane at the synaptic cleft. The formation of these endosomes is initiated cell surface-mediated invagination to generate endocytic vesicles that migrate and fuse with the early endosome [29]. Exosomes appear to form by invagination and budding from the limiting membrane of late endosomes, resulting in vesicles that contain cytosol and that expose the extracellular domain of transferring receptors at their surface. Electron microscopic studies have shown fusion profiles of multivesicular endosomes with the plasma membrane, leading to the release of the internal vesicles into the extracellular environment [29].

Since the formation of these membrane vesicles has an endocytic origin, this mechanism is a secretion process of the endosomal system, including endocytic vesicles, early endosomes, late endosomes and lysosomes. These endocytic vesicles form through clathrin- or non-clathrin-mediated endocytosis at the plasma membrane and are transported to early endosomes [30]. The late endosomes develop from early endosomes via acidification, changes in their protein content, and their ability to fuse with vesicles or other cellular membranes. Early endosomes are localized near the outer margin of the cells and exhibit a tubular appearance, in contrast late endosomes are localized proximal to the nucleus and are spherical in shape. A critical step in the formation of MVBs from late endosomes is reversed budding. In this step, limiting membranes of late endosomes "bud" into their lumen resulting in a continuous enrichment of internal luminal vesicles [31]. MVBs have been demonstrated to be involved in the exocytic fusion of their external membrane with the plasma membrane of the cell, resulting in release of their segregated vesicles to the extracellular space.

The increased release of exosomes/microvesicles and their accumulation appear to be important in the malignant transformation process. Although extracellular shedding of exosomes occurs in other types of cells under specific physiological conditions, membranous vesicles from tumor cells accumulate in biologic fluids, including sera, urine, ascites, and pleural fluids. Exosome release and its accumulation appear to be important features of the malignant transformation . Recently, we have analyzed circulating vesicles of normal individuals, patients with benign ovarian disease, and patients with ovarian cancer using the Nanoparticle Tracking Analysis system (NanoSight). Our study demonstrates the presence of circulating vesicular materials in all individuals (Fig. 6.1); however, ovarian cancer patients exhibit approximately threefold more vesicular material. In these cancer patients, the size range of these vesicles was between 50 and 250 nm, with the major peak at 98–99 nm.

6.4 Exosomes/Microvesicles Are Active Components of Intercellular Communication

Exosomes released from tumor cells have gained increased recognition as "vehicles" for intercellular communication with extensive autocrine/paracrine functions. By exposing cell-type specific adhesion receptors or ligands, exosomes can interact with specific cells and deliver their "signals," including bioactive lipids, cytokines, growth factors, receptors, and genetic materials. Thus, the microvesicle/exosomal pathway may constitute a mechanism for local and systemic intercellular transfer of information, with a complexity superior to that of secreted soluble factors, but similar to that observed with direct cell–cell contact. Exosomes provide stable conformational conditions for their protein content (due to maintenance of their 3-dimensional transmembrane structure), conserve bioactivity of their proteins (based on the protective membrane structure), improve biodistribution (based on their capacity to circulate

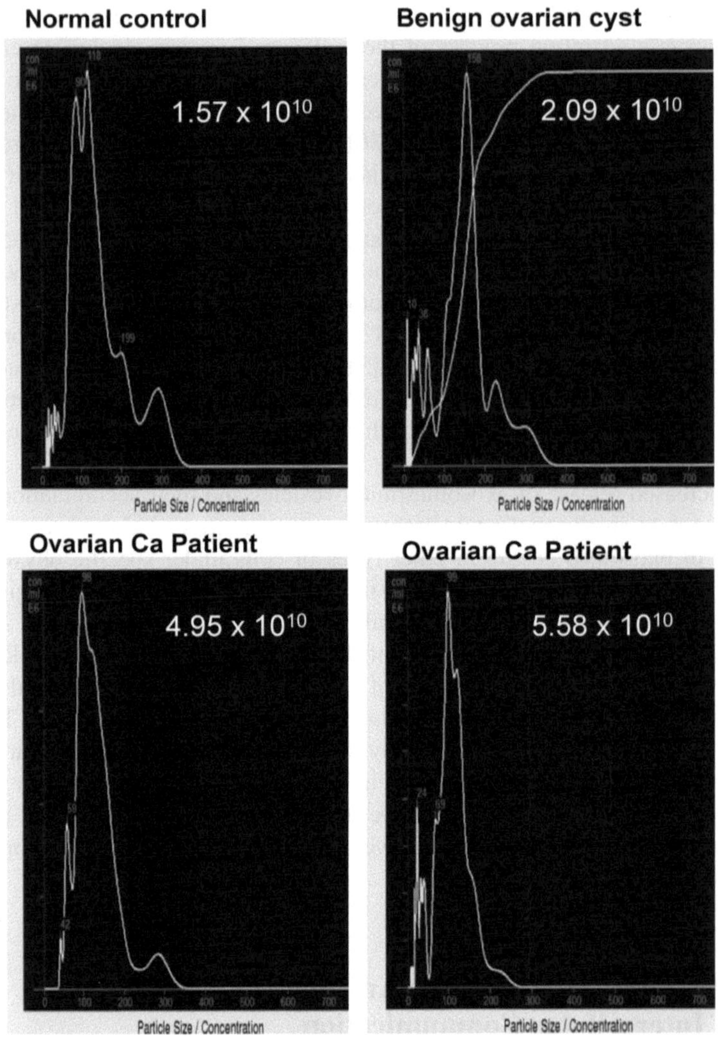

Fig. 6.1 Size distribution of total circulating membraneous vesicles in a patient with benign ovarian adenoma and two patients with advanced ovarian cancer, compared to a normal, nonpregnant female volunteer. Plasma from EDTA-treated blood specimens were diluted in PBS and injected into the laser chamber of a NanoSight LM10 and the number and size distribution of the circulating vesicles were defined using Nanoparticle Tracking software. The inset numbers indicate the total number of vesicles in 1 ml of plasma

in biologic fluids and migrate to secondary sites), and support an efficient interaction with target cells (due to the fusogenic properties of exosomes). Due to these features, the tumor-derived exosomes are efficient platforms for the in vivo transfer of cross-talk signals.

The multiplicity of bioactive molecules associated with exosomes suggests that they exhibit a central role in generating the tumor microenvironment [32, 33]. Exosomes have the ability to transfer specific proteins to homologous and heterologous target cells for the delivery of signaling pathways [34, 35]. The presence of tumor-derived exosomes can increase matrix metalloproteinase (MMP) secretion and VEGF expression in target cells through the expression of proangiogenic molecules, such as members of the tetraspanin family, thereby promoting neoangiogenesis even at secondary metastatic sites [36]. The released matrix metalloproteinases can digest the extracellular matrices (ECM) where they arise. This degradation is enhanced when MMPs are coreleased with exosome-associated extracellular matrix metalloproteinase inducer (EMMPRIN) [37].

Studies have shown that cancer ascites-derived exosomes carry extracellular matrix-remodeling enzymes, such as metalloproteinases 2 and 9 (MMP-2, MMP-9) [38, 39] and urokinase plasminogen activator [40], leading to an increase in extracellular matrix degradation. The expression of matrix-remodeling enzymes increases the tumor's invasive phenotype and promotes metastasis. The presence of proangiogenic factors supports neovascularization of the developing tumor. A common cellular component of the tumor microenvironment is the monocyte/macrophage. Within the microenvironment, these tumor-associated macrophages have been shown to assist in tumor progression by angiogenesis, growth, metastasis, and immunosuppression [41].

When shed vesicles fuse with their target cells, they can transfer important membrane components, such as receptors and ligands. The transferring of receptors between exosomes and target cells was demonstrated by the observation that bystander B cells acquire antigen receptors from activated B cells by membrane transfer [42]. This transfer allows the amplified expansion of antigen-binding B cells with the ability to present a specific antigen to CD4 T cells. Exosomes can transfer the adhesion molecule CD41 from platelets to endothelial cells or to tumor cells, conferring proadhesive properties to the target cell [43]. Exosome-mediated transfer of Fas ligand from tumor cells induces apoptosis of activated T cells favoring tumor-immune escape [44]. Exosomes can also be protective for cells that remove the potentially harmful molecules, such as Fas or the membrane attack complex, from their membranes to the extracellular compartment.

Exosomes have also been postulated to contribute to the spread of infective agents, such as human immunodeficiency virus type 1 [45]. In macrophages receiving chemokine receptors, this can induce an increased risk of HIV infection together with resistance to apoptosis. The transfer of the chemokine (CXC motif) receptor 4 and the chemokine (CC motif) receptor 5 chemokine coreceptors for human immunodeficiency virus type I by released exosomes can enhance the entry of the virus into cell types other than the lymphohemopoietic lineage [46]. In addition to transferring receptors, exosomes can transfer viruses, contained within exosomes, by the "Trojan exosome hypothesis" involving direct delivery [47].

In human gliomas, only a fraction of the cells, exhibiting a transformed phenotype, expressed the truncated epidermal growth factor receptor, EGFRvIII, associated with dysregulated tumor growth [48]. Al-Nedawi et al. [49] demonstrated transfer of the oncogenic EGFRvIII from human glioma cancer cells expressing the receptor to glioma cells without the EGFRvIII via the fusion of exosomes. After transfer, the glioma cells, lacking the receptor, were transformed to express EGFRvIII-regulated genes, including VEGF, Bcl-x_L, and p27 [50]. Subsequent studies demonstrated that the oncogenic EGFRvIII from human squamous cell carcinoma cells was transferred via exosomes to tumor-associated endothelial cells to activate MAPK and Akt cell signaling pathways and promote endothelial VEGF expression [51].

The occurrence of epigenetic changes has been frequently reported in cancer. Epigenetic regulation of gene transcription, mediating cell proliferation, differentiation, and survival are additional targets in tumor progression, resulting in genomic instability [52]. One explanation for this genomic instability lies in the mediation by microvesicular horizontal transfer. Horizontal transfer via microvesicles has been validated in a number of tumor-associated cells including: gliomas, monocytes, mast cells, and T cells [53]. One explanation of this phenomenon is the transfer of genetic information between cells. It has been shown that tumor-derived exosomes may transfer not only surface determinants but also mRNA of tumor cells to monocytes. Janowska-Wieczorek et al. [54] demonstrated that exosomes derived from murine embryonic stem cells (ESCs) could induce epigenetic reprogramming of target cells. ESC-derived exosomes were shown to improve survival of hematopoietic stem/progenitor cells, to induce upregulation of early pluripotent and early hematopoietic markers, and to induce phosphorylation of mitogen-activated protein kinase p42/44 and Akt. ESC-derived exosomes were shown to express mRNAs for several pluripotent transcription factors that can be delivered to target cells and translated to their corresponding proteins [55]. As RNase-treatment inhibited their exosome-mediated biological effect, the involvement of mRNA in the observed biological effects was suggested. Yuan et al. [56] have shown that in addition to mRNA, the exosomes can transfer microRNA to target cells. They demonstrated that exosomes derived from ESCs contain abundant microRNA and that they can transfer a subset of microRNAs to mouse embryonic fibroblasts in vitro. Since microRNAs are regulators of protein translation, this observation opens the possibility that stem cells can alter the expression of genes in neighboring cells by transferring exosomal microRNAs. When shed vesicles fuse with their target cells, the portion of cytosol segregated within their lumen is discharged to and integrates with the cytosol of the target cell. Because this transfer can also include transmission of specific mRNAs, it can ultimately contribute to the epigenetic and proteomic properties of target cells.

It has been suggested that tumor cell progression could use multiple forms of exosome-/microvesicle-mediated communication to simultaneously affect multiple effector populations, based on release of tissue factors, immunoregulators, and oncogenic molecules. Thus, the signals transferred to neighboring cells via exosomes may mirror the transcriptional status of the parent cell; however, due to the exosomal messenger RNA and microRNA being transferred, their consequences on the translational machinery of the target cells are extensive.

6.5 Exosomes/Microvesicles Serve as Regulators of the Immune Response

The initial identification of exosome release by tumor cells was envisioned as the discovery of a new cell-free source of tumor antigens for in vivo immune priming or tumor vaccine design. Cancer cells are postulated to modulate components of the microenvironment and affect immune system function, primarily through pathways involving cell-to-cell contact and the release of suppressive soluble factors. However, a unique alternative mechanism has emerged that involves the active release of immunosuppressive microvesicles/exosomes by tumor cells [57, 58]. As tumor-derived microvesicles/exosomes are abundant in the blood and malignant effusions derived from cancer patients [59, 60], their release appears to be an important feature of intercellular communication. Evidence supports the concept that tumors constitutively shed exosomes with pleiotropic immunosuppressive effects [61, 62] that are protective and supportive of the tumor to facilitate escape from lymphocyte immunosurveillance [63]. Since released exosomes express molecules with biologic activity (such as Fas ligand, PD-1, MICA/B, mdr1, MMPs, CD44, and autoreactive antigens), the ability of exosomes to modulate lymphocyte and monocyte functions have been analyzed [64–66]. The immunological significance of exosomes, while far from clear, has been linked with their potential to modulate the host's immune system, which may be their major function. Supporting this view is that the topology of the macromolecules displayed on exosomes is analogous to that observed on the tumor's plasma membrane, making them well-positioned for interactions with target cell surface receptors. This positioning can mediate signal transduction without the need for direct cell–cell contact. Furthermore, these microvesicles can also fuse with the recipient cell, leading to the acquisition of novel molecules by the cells and the delivery of mRNA and miRNA through this route.

Tumor exosome release has been described as capable of modulating the evasion from antitumor immune responses [67]. It has been suggested that the antitumor immune response can be divided into three different phases and tumor-derived microvesicles/exosomes can exhibit roles in each phase [68]. The first step includes the recognition of tumor cells by innate immune cells. The progression and development of tumors is coordinated by biochemical and biophysical signals from the tumor microenvironment [69]. After growing to a certain threshold size, the solid tumors exceed their capacity to acquire oxygen and nutrients in this hypoxic environment [70, 71]. For tumor progression to occur, the "angiogenic switch" has been reported to be required [72]. The "angiogenic switch" promotes the transition of the tumor to a highly vascularized and progressive outgrowth [73]. This process further induces extracellular matrix remodeling and the production of a proinflammatory environment leading to the recruitment of innate immune cells, including NK, macrophages and dendritic cells and the mediation of T cells into the tumor microenvironment.

Recent studies have demonstrated that the MHC class I-related chain MICA and MICB ligands for the NK cell-activating receptor NKG2D is released by tumor cells as a component of exosomes [74]. This shedding of MICA/B by tumors not only prevents recognition of MICA/B expressing tumor cells, but also results in the downregulation of NKG2D expression on circulating CD8 T cells, NK cells, and $\gamma\delta$T cells, leading to impairment of the antitumor immune response [75]. Ashiru et al. [75] demonstrated that treatment of NK cells with MICA expressing exosomes suppressed NK expression of NKG2D on the cell surface and also suppressed NK cytotoxicity, which is independent of NKG2D ligand expression. Thus, their findings demonstrated exosomal MICA/B expression as a mechanism of NK suppression, facilitating immune escape and progression.

The second phase of an antitumor immune response involves the maturation and migration of tissue dendritic cells and priming of naïve T cells. Blood-derived exosomes from melanoma patients have been shown to promote the generation of myeloid-derived suppressor cells (MDSCs) from peripheral blood monocytes [76], which acts as one of the major mechanisms used by tumors to escape immune recognition [77]. MDSCs have potent immunosuppressive functions that can suppress T cell immune responses by a variety of mechanisms [78–80]. Furthermore, tumor exosomes have been shown to be involved in the regulation of the adaptive immune responses to cancer cells in animal models and cancer patients [81, 82] by impairing peripheral blood monocyte differentiation into dendritic cells [83]. As a result, the generation of tumor-specific T cells has been reported to be a very inefficient process. Most importantly, tumor exosomes have the ability to induce a series of functional defects in tumor-reactive effector T cells [84], through the expression of apoptosis-inducing ligands, such as FasL and TRAIL [85] or PDL-2, which directly stimulate their T cell targets to negatively regulate T cell activation [86]. An investigation of human prostate cancer exosomes added to activated T cells exposed a dose-dependent inhibition of $CD8^+$ T cell proliferation stimulated by Fas–FasL interaction [87]. Additionally, tumor-derived exosomes block innate immune effector cell function as seen in NK cells via production of exosome-associated MICA/B to downregulate NKG2D expression, thereby decreasing NKG2D-mediated killing [88]. Production of tumor exosome-associated MICA/B and FasL has been shown to decrease the effectiveness of not only the innate immune system, but also the adaptive immune system to reject the tumor [89]. It has been theorized that these released exosomes modulate lymphocyte functions by mimicking "activation induced cell death" (AICD) [90]. Lymphoid cells appear to release exosomes following activation and these appear to play an essential role in immunoregulation by preventing excessive immune responses and the development of autoimmunity [91]. It was postulated that exosome release by tumor cells is a reexpression of the fetal cell exosomes and that both constituted pathways to circumvent immunosurveillance.

The third and final phase involved in the antitumor response is the generation and homing of tumor-specific T cells. Tumor-reactive $CD4^+$ and $CD8^+$ T cells homing to the primary tumor site is an important step in eradicating the tumor [92]. Tumor exosomes express membrane-bound ICAM-1 that efficiently blocks the interaction between lymphocytes and endothelial cells [93] and therefore, decreases the recruitment of adaptive immune cells. Several studies have shown that tumor-infiltrating T cells are impaired by the tumor and display altered expression in intracellular signal transduction molecules such as CD3-zeta [88]. Tumor exosomes coincubation with

T cells leads to a decrease in CD3-zeta, which suggests that tumor exosomes may be an additional mechanism used by tumors to evade immune recognition [88]. The alteration of TCR-CD3-zeta has been observed in several types of tumors namely malignant melanoma [93], ovarian [94], and pancreatic [95]. Furthermore, studies have reported that cancer patients display a high frequency of suppressive peripheral blood regulatory T cells (T_{reg}) when compared to normal controls [96]. These cells have been shown to infiltrate the tumor and are involved in the induction of $CD8^+$ tumor-reactive cytotoxic T lymphocyte (CTL) apoptosis [97]. A study from Szajnik et al. shows that tumor-derived microvesicles expand and promote the suppressive activities of human T_{reg} by upregulating the expression of FasL, IL-10, TGF-β1, CTLA-4, granzyme B, and perforin [98]. Therefore, the immunoregulatory properties attributed to tumor-derived exosomes might be essential in regulating peripheral tolerance and promoting immune evasion of tumors. Collectively, these studies support a role for exosomes in adapting the host microenvironment to allow escape from immune surveillance via stimulation of angiogenesis and metastasis of tumors, which suggests that tumors may use exosomes to keep the host immune system under control without a direct interaction with host immune cells.

Critical components of the immune response, such as antigen-presenting cells, are significantly affected by interactions with tumor exosomes. These microvesicles not only impair the capacity of circulating monocytes to differentiate into functional DCs, but they also skew the differentiation of these cells toward altered $CD14^+$ monocytes expressing low or absent levels of HLA-DR [99]. These cells, which are present in relatively high numbers in PBMCs of melanoma patients, exert suppressive activity on lymphocyte proliferation and impair the expression of effector molecules (such as perforin and IFN-γ) in a TGF-β-mediated fashion. $CD14^+HLA\text{-}DR^-$/low cells behave as MSC, undergoing in vivo expansion upon administration of GM-CSF. These hallmark alterations induced by tumor microvesicles on target immune cells in vitro can also be detected on immune cells isolated from cancer patients supports the hypothesis that these suppressive pathways are present in vivo.

These exosomes activated a stronger proinflammatory response in the form of NF-κB activation and TNF-α release from untreated macrophages as compared to macrophages exposed to control exosomes. Parallel evaluations of the structural components of tumor-derived microvesicles demonstrated enhanced expression of tumor antigens on the vesicular surface [100]. In murine B16 melanoma cultures, the expression of surface glycoproteins on the isolated microvesicles represented a profile similar to that found in the melanoma cell membrane [100]. Continued characterization of tumor-derived microvesicles was achieved through biochemical analyses, which identified molecules with immunologic and biologic activity. Microvesicles released by late-stage tumor cells were found to exert a dose-dependent suppression of MHC II in monocytes/macrophages in comparison to early-stage tumor cells [101]. In addition, microvesicles suppressed lymphocyte activation induced by phytohemagglutinin, anti-CD3, concanavalin A [100], and interleukin-2 [102].

As evidence of the pleiotropic effect of tumor-secreted exosomes, tumor exosomes can interfere directly with T cell effector functions. Exosomal expression of bioactive FasL and TRAIL has been shown on exosomes derived from human tumors to induce apoptosis in activated tumor-specific T cells. This phenomenon highly resembles the one utilized under physiological conditions not only by T cells to downsize immune responses [103], but also by placenta cells that have recently been shown to promote a state of immune privilege by inducing FasL-mediated apoptosis and defects in the expression of crucial TCR signaling components (such as CD3-zeta and JAK3) in local T cells, which have been reported with microvesicles isolated from plasma of cancer patients and may help to explain the high frequency of apoptotic or CD3-zeta$^-$ lymphocytes that are often found in the peripheral circulation of these patients [104]. Natural killer cells lose their cytolytic potential, through the suppression perforin expression, upon encounter with tumor-secreted microvesicles.

A proinflammatory microenvironment is associated with the development and progression of cancer. Macrophages are prominent in the development of the proinflammatory environment, with IL-1β serving as a "master" cytokine regulator. Macrophages are critical for the resolution of inflammation by producing antiinflammatory cytokines and chemokines and by increasing phagocytic activity. Based on Th1/Th2 polarization, phenotypically polarized macrophages are termed proinflammatory M1 (classically activated) and antiinflammatory M2 (alternatively activated) [105]. In vitro, the macrophages can be polarized to the M1 state by treatment with IFN-γ and inducers of TNF-α, such as lipopolysaccharide (LPS) [106]. These M1 macrophages induce synthesis of proinflammatory cytokines and chemokines, including TNF-α, IL-12, IL-6, CCL2, and IL-1β, as well as increased production of reactive oxygen species [107]. Elevated levels of IL-1β are present in M1 polarized macrophages due to activation of the NF-κB and MAPK pathways [108], while no IL-1β protein is found in M2 polarized macrophages [109]. Using cytochalasin D, a known inhibitor of actin polymerization, while a dramatic suppression of exosome internalization was observed, this internalization was not essential for the induction of IL-1β mRNA and protein, following the exposure to exosomes.

IL-1β exhibits profound effects on immune cell function during inflammation. Components of the ECM have been demonstrated to be capable of stimulating the expression of IL-1β. Fibronectin is highly expressed in injured tissues and appears to be positioned to modulate the expression of IL-1β in diseased tissues [110]. In vitro, fibronectin stimulates the expression of IL-1β mRNA and its translation into the 31 kDa intracellular precursor protein, along with secretion of the 17 kDa active form in human mononuclear cells. This effect of fibronectin is mediated by specific cell surface α5β1 integrin receptor, which activate poorly understood intracellular signals to induce IL-1β expression [111]. Fibronectin contains a sequence, termed Arg-Gly-Asp (RGD), which promotes its attachment to integrin receptors [112]. Monocytes and macrophages have been shown to possess fibronectin receptors that recognize the RGD motif and mediate proinflammatory cytokine production. The effect of fibronectin has been shown to be dependent on binding of the RGD sequence of fibronectin to integrin receptors, as this effect could be inhibited by integrin receptor-blocking peptides (anti-RGD sequence mimics). The use of RGD antagonists demonstrates that exosome-induced IL-1β production by macrophages is mediated by exosome-associated fibronectin.

6.6 Conclusions

It has become increasingly clear, as new exosome studies are published, that these small bioactive vesicles are important in a number of biological functions. Exosomes exhibit important roles in intercellular communication and, under normal conditions, this communication mediates the activation of the immune response. However, in cancer, tumor exosomes can induce apoptosis of activated cytotoxic T cells, impairment of monocyte differentiation, induction of myeloid-suppressive cells and T regulatory cells, and suppression of lymphoid activation signaling molecules. Tumor-derived exosomes express molecules involved in angiogenesis promotion, stromal remodeling, signaling pathway activation through growth factor/receptor transfer, chemoresistance, and genetic intercellular exchange. Tumor exosomes induce a proinflammatory environment from macrophages due to expression of exosomal fibronectin. As a result of these exosomal effects, they can represent a central mediator of the tumor-supportive microenvironment. The removal of unwanted proteins from maturing reticulocytes, to their role in immune surveillance, the inventory of functions continues to grow.

References

1. Taylor DD, Doellgast GJ (1979) Quantitation of peroxidase-antibody binding to membrane fragments using column chromatography. Anal Biochem 98:53–59
2. Taylor DD, Gerçel-Taylor C (2011) Exosomes/microvesicles: mediators of cancer-associated immunosuppressive microenvironments. Seminars in Immunopathology, (J. Schifferli, ed), Springer, 33:441–454.
3. Trams EG, Lauter CJ, Salem Jr C, Heine U (1981) Exfoliation of membrane ecto-enzymes in the form of microvesicles. Biochim Biophys Acta 645:63–70
4. Pan BT, Blostein R, Johnstone RM (1983) Loss of the transferrin receptor during the maturation of sheep reticulocytes in vitro: an immunological approach. Biochem J 210:37–47
5. Harding C, Heuser J, Stahl P (1983) Receptor-mediated endocytosis of transferrin and recycling of the transferrin receptor in rat reticulocytes. J Cell Biol 97:329–339.
6. Johnstone RM, Adam M, Hammond JR, Orr L, Turbide C (1987) Vesicle formation during reticulocyte maturation. Association of plasma membrane activities with released vesicles (exosomes). J Biol Chem 262:9412–9420
7. Taylor DD, Homesley HD, Doellgast GJ (1980) Binding of specific peroxidase-labeled antibody to placental-type phosphatase on tumor-derived membrane fragments. Cancer Res 40:4064–4069
8. Taylor DD, Homesley HD, Doellgast GJ (1983) Membrane-associated immunoglobulins in cyst and ascites fluids of ovarian cancer patients. Am J Reprod Immunol 3:7–11
9. Taylor DD, Levy EM, Black PH (1985) Shed membrane vesicles: a mechanism for tumor induced immunosuppression. In: Mitchell MS, Reif AE (eds) Immunity to cancer. Academic, New York, pp. 369–373
10. Taylor DD, Black PH (1986) Shedding of plasma membrane fragments: neoplastic and develpomental importance. In: Steinberg M (ed) Developmental Biology vol. 3. Plenum New York pp. 33–57.
11. Taylor DD, Gerçel-Taylor C, Weese JL (1989) Expression and shedding of mdr-1 antigen by variants of the murine B16 melanoma. Surgical Forum 40:406–408
12. Manahan KJ, Taylor DD, Gercel-Taylor C (2001) Clonal heterogeneity of p53 mutations in ovarian cancer. Int J Oncol 19:387–394

13. Chinni SR, Gercel-Taylor C, Falchetto RA, Shabanowitz J, Hunt DF, Taylor DD (1997) Cathepsin D and glucose-regulated protein 78 recognized by the humoral response of ovarian cancer patients. Clin Cancer Res 3:1557–1564
14. Théry C, Ostrowski M, Segura E (2009) Membrane vesicles as conveyors of immune responses. Nat Rev Immunol 9:581–593
15. Atay S, Gercel-Taylor C, Kesimer M, Taylor DD (2011) Morphologic and proteomic characterization of exosomes released by cultured extravillous trophoblast cells. Exp Cell Res 317:1192–1202
16. Simons M, Raposo G (2009) Exosomes: vesicular carriers for intercellular communication. Curr Opin Cell Biol 21:575–581
17. Zumaquero E, Muñoz P, Cobo M, Lucena G, Pavón EJ, Martín A, Navarro P, García-Pérez A, Ariza-Veguillas A, Malavasi F, Sancho J, Zubiaur M (2010) Exosomes from human lymphoblastoid B cells express enzymatically active CD38 that is associated with signaling complexes containing CD81, Hsc-70, and Lyn. Exp Cell Res 316:2692–2706
18. Admyre C, Telemo E, Almqvist N, Lotvall J, Lahesmaa R, Scheynius A, Gabrielsson S (2008) Exosomes—nanovesicles with possible roles in allergy inflammation. Allergy 63:404–408
19. Montecalvo A, Shufesky WJ, Stolz DB, Sullivan MG, Wang Z, Divito SJ, Papworth GD, Watkins SC, Robbins PD, Larregina AT, Morelli AE (2008) Exosomes as a short-range mechanism to spread alloantigen between dendritic cells during T cell allorecognition. J Immunol 180:3081–3090
20. Rak J (2010) Microparticles in cancer. Semin Thromb Hemost, 36:888–906
21. Anand PK (2010) Exosomal membrane molecules are potent immune response modulators. Commun Integr Biol 3:405–408
22. Kesimer M, Scull M, Brighton B, DeMaria G, Burns K, O'Neal W, Pickles RJ, Sheehan JK (2009) Characterization of exosome-like vesicles released from human tracheobronchial ciliated epithelium: a possible role in innate defense. FASEB J 23:1858–1868
23. Ostrowski M, Carmo NB, Krumeich S, Fanget I, Raposo G, Savina A, Moita CF, Schauer K, Hume AN, Freitas RP, Goud B, Benaroch P, Hacohen N, Fukuda M, Desnos C, Seabra MC, Darchen F, Amigorena S, Moita LF, Thery C (2010) Rab27a and Rab27b control different steps of the exosome secretion pathway. Nat Cell Biol 12:19–30 (sup pp 1–13)
24. Valapala M, Vishwanatha JK (2011) Liid raft endocytosis and exosomal transport facilitate extracellular trafficking of annexin A2. J Biol Chem 286:30911–30925
25. Mathivanan S, Simpson RJ (2009) ExoCarta: a compendium of exosomalproteins and RNA. Proteomics 9:4997–5000
26. Yu X, Harris SL, Levine AJ (2006) The regulation of exosome secretion: a novel function of the p53 protein. Cancer Res 66:4795–4801
27. Graner MW, Alzate O, Dechkovskaia AM, Keene JD, Sampson JH, Mitchell DA, Bigner DD (2009) Proteomic and immunologic analyses of brain tumor exosomes. FASEB J 23:1541–1557
28. Taylor DD, Gercel-Taylor C (2005) Tumor-derived exosomes as mediates of T-cell signaling defects. Br J Cancer 92:305–311
29. Simons K, Gerl MJ (2010) Revitalizing membrane rafts: new tools and insights. Nat Rev Mol Cell Biol 11:688–699
30. Piper RC, Katzmann DJ (2007) Biogenesis and function of multivesicular bodies. Ann Rev Cell Dev Biol 23:519–547
31. Doring T, Gotthardt K, Stieler J, Prange R (2010) γ2-Adaptin is functioning in the late endosomal sorting pathway and interacts with ESCRT-I and -III subunits. Biochim Biophys Acta 1803:1252–1264
32. Park JE, Tan HS, Datta A, Lai RC, Zhang H, Meng W, Lim SK, Sze SK (2010) Hypoxic tumor cell modulates its microenvironment to enhance angiogenic and metastatic potential by secretion of proteins and exosomes. Mol Cell Proteomics 9:1085–1099
33. Marhaba R, Klingbeil P, Nuebel T, Nazarenko I, Buechler MW, Zoeller M (2008) CD44 and EpCAM: Cancer-initiating cell markers. Curr Mol Med 8:784–804
34. Xiang X, Poliakov A, Liu C, Liu Y, Deng ZB, Wang J, Cheng Z, Shah SV, Wang GJ, Zhang L, Grizzle WE, Mobley J, Zhang HG (2009) Induction of myeloid-derived suppressor cells by tumor exosomes. Int J Cancer 124:2621–2633.

35. Hong BS, Cho JH, Kim H, Choi EJ, Rho S, Kim J, Kim JH, Choi DS, Kim YK, Hwang D, Gho YS (2009) Colorectal cancer cell-derived microvesicles are enriched in cell cycle-related mRNAs that promote proliferation of endothelial cells. BMC Genomics 10:556
36. Nazarenko I, Rana S, Baumann A, McAlear J, Hellwig A, Trendelenburg M, Lochnit G, Preissner KT, Zöller M (2010) Cell surface tetraspanin Tspan8 contributes to molecular pathways of exosome-induced endothelial cell activation. Cancer Res 70:1668–1678
37. Keller S, König AK, Marmé F, Runz S, Wolterink S, Koensgen D, Mustea A, Sehouli J, Altevogt P (2009) Systemic presence and tumor-growth promoting effect of ovarian carcinoma released exosomes. Cancer Lett 278:73–81
38. Dolo V, D'Ascenzo S, Violini S, Pompucci L, Festuccia C, Ginestra A, Vittorelli ML, Canevari S, Pavan A (1999) Matrix-degrading proteinases are shed in membrane vesicles by ovarian cancer cells in vivo and in vitro. Clin Exp Metastasis 17:131–140
39. Dolo V, Ginestra A, Cassara D, Ghersi G, Nagase H, Vittorelli ML (1999) Shed membrane vesicles and selective localization of gelatinases and MMP-9/TIMP-1 complexes. Ann N Y Acad Sci 878:497–499
40. Graves LE, Ariztia EV, Navari JR, Matzel HJ, Stack MS, Fishman DA (2004) Proinvasive properties of ovarian cancer ascites-derived membrane vesicles. Cancer Res 64:7045–7049
41. Lewis CE, Pollard JW (2006) Distinct role of macrophages in different tumor microenvironments. Cancer Res 66:605–612
42. McLellan AD (2009) Exosome release by primary B cells. Crit Rev Immunol 29:203–217
43. Ratajczak J, Wysoczynski M, Hayek F, Janowska-Wieczorek A, Ratajczak MZ (2006) Membrane-derived microvesicles: important and underappreciated mediators of cell-to-cell communication. Leukemia 20:1487–1495
44. Ichim TE, Zhong Z, Kaushal S, Zheng X, Ren X, Hao X, Joyce JA, Hanley HH, Riordan NH, Koropatnick J, Bogin V, Minev BR, Min WP, Tullis RH (2008) Exosomes as a tumor immune escape mechanism: possible therapeutic implications. J Transl Med 6:37
45. Lenassi M, Cagney G, Liao M, Vaupotic T, Bartholomeeusen K, Cheng Y, Krogan NJ, Plemenitas A, Peterlin BM (2010) HIV Nef is secreted in exosomes and triggers apoptosis in bystander $CD4^+$ T cells. Traffic 11:110–122
46. Izquierdo-Useros N, Naranjo-Gómez M, Archer J, Hatch SC, Erkizia I, Blanco J, Borràs FE, Puertas MC, Connor JH, Fernández-Figueras MT, Moore L, Clotet B, Gummuluru S, Martinez-Picado J (2009) Capture and transfer if HIV-1 particles by mature dendritic cells converges with the exosome-dissemination pathway. Blood 113:2732–2741
47. Izquierdo-Useros N, Naranjo-Gómez M, Erkizia I, Puertas MC, Borràs FE, Blanco J, Martinez-Picado J (2010) HIV and mature dendritic cells: trojan exosomes riding the Trojan horse? PLoS Pathog 6:e1000740
48. Al-Nedawi K, Meehan B, Rak J (2009) Microvesicles: messengers and mediators of tumor progression. Cell Cycle 8:2014–2018
49. Al-Nedawi K, Meehan B, Micallef J, Lhotak V, May L, Guha A, Rak J (2008) Intercellular transfer of the oncogenic receptor EGFRvIII by microvesicles derived from tumour cells. Nat Cell Biol 10:619–624
50. Al-Nedawi K, Meehan B, Kerbel RS, Allison AC, Rak J (2009) Endothelial expression of autocrine VEGF upon the uptake of tumor-derived microvesicles containing oncogenic EGFR. Proc Natl Acad Sci USA 106:3794–3799
51. Viswanathan M, Sangiliyandi G, Vinod SS, Mohanprasad BK, Shanmugam G (2003) Genomic instability and tumor-specific alterations in oral squamous cell carcinomas assessed by inter-(simple sequence repeat) PCR. Clin Cancer Res 9:1057–1062
52. Camussi G, Deregibus MC, Bruno S, Cantaluppi V, Biancone L (2010) Exosomes/microvesicles as a mechanism of cell-to-cell communication. Kidney Int 78:838–848
53. Skog J, Wurdinger T, van Rijn S, Meijer DH, Gainche L, Sena-Esteves M, Curry Jr WT, Carter BS, Krichevsky AM, Breakefield XO (2008) Glioblastoma microvesicles transport RNA and proteins that promote tumour growth and provide diagnostic biomarkers. Nat Cell Biol 10:1470–1476

54. Janowska-Wieczorek A, Wysoczynski M, Kijowski J, Marquez-Curtis L, Machalinski B, Ratajczak J, Ratajczak MZ (2005) Microvesicles derived from activated platelets induce metastasis and angiogenesis in lung cancer. Int J Cancer 113:752–760
55. Koh W, Sheng CT, Tan B, Lee QY, Kuznetsov V, Kiang LS, Tanavde V (2010) Analysis of deep sequencing microRNA expression profile from human embryonic stem cells derived mesenchymal stem cells reveals possible role of let-7 microRNA family in downstream targeting of hepatic nuclear factor 4 alpha. BMC Genomics 11(Suppl 1):S6
56. Yuan XL, Chen L, Li MX, Dong P, Xue J, Wang J, Zhang TT, Wang XA, Zhang FM, Ge HL, Shen LS, Xu D (2010) Elevated expression of Foxp3 in tumor-infiltrating Treg cells suppresses T-cell proliferation and contributes to gastric cancer progression in a COX-2-dependent manner. Clin Immunol 134:277–288
57. Whiteside TL (2005) Tumour-derived exosomes or microvesicles: another mechanism of tumour escape from the host immune system? Br J Cancer 92:209–211
58. Valenti R, Huber V, Iero M, Filipazzi P, Parmiani G, Rivoltini L (2007) Tumor-released microvesicles as vehicles of immunosuppression. Cancer Res 67:2912–2915
59. Cocucci E, Racchetti G, Meldolesi J (2009) Shedding microvesicles: artefacts no more. Trends Cell Biol 19:43–51
60. Huber V, Fais S, Iero M, Lugini L, Canese P, Squarcina P, Zaccheddu A, Colone M, Arancia G, Gentile M, Seregni E, Valenti R, Ballabio G, Belli F, Leo E, Parmiani G, Rivoltini L (2005) Human colorectal cancer cells induce T-cell death through release of proapoptotic microvesicles: role in immune escape. Gastroenterol 128:1796–1804
61. Wieckowski EU, Visus C, Szajnik M, Szczepanski MJ, Storkus WJ, Whiteside TL (2009) Tumor-derived microvesicles promote regulatory T cell expansion and induce apoptosis in tumor-reactive activated CD8+ T lymphocytes. J Immunol 183:3720–3730
62. Huber V, Filipazzi P, Iero M, Fais S, Rivoltini L (2008) More insights into the immunosuppressive potential of tumor exosomes. J Transl Med 6:63
63. Taylor DD, Gercel-Taylor C, Lyons KS, Stanson J, Whiteside TL (2003) T-cell apoptosis and suppression of T-cell receptor/CD3-zeta by Fas ligand-containing membrane vesicles shed from ovarian tumors. Clin Cancer Res 9:5113–5119
64. Taylor DD, Gercel-Taylor C, Weese JL (1989) Expression and shedding of mdr-1 antigen by variants of the murine B16 melanoma. Surg Forum 40:406–408
65. Taylor DD, Gercel-Taylor C, Gall SA (1996) Expression and shedding of CD44 isoforms by gynecologic cancer patients. J Soc Gynecol Invest 3:289–294
66. Taylor DD, Lyons KS, Gercel-Taylor C (2002) Shed membrane fragment-associated markers for endometrial and ovarian cancers. Gynecol Oncol 84:443–448
67. Schorey JS, Bhatnagar S (2008) Exosome function: from tumor immunology to pathogen biology. Traffic 9:871–881
68. Zindl CL, Chaplin DD (2010) Immunology: tumor immune evasion. Science 328:697–698
69. Moserle L, Amadori A, Indraccolo S (2009) The angiogenic switch: implications in the regulation of tumor dormancy. Curr Mol Med 9:935–941
70. Gao D, Nolan D, McDonnell K, Vahdat L, Benezra R, Altorki N, Mittal V (2009) Bone marrow-derived endothelial progenitor cells contribute to the angiogenic switch in tumor growth and metastatic progression. Biochim Biophys Acta 1796:33–40
71. Zwirner NW, Croci DO, Domaica CI, Rabinovich GA (2010) Overcoming the hurdles of tumor immunity by targeting regulatory pathways in innate and adaptive immune cells. Curr Pharm Des 16:255–267
72. Soloski MJ (2001) Recognition of tumor cells by the innate immune system. Curr Opin Immunol 13:154–162
73. Park JE, Tan HS, Datta A, Lai RC, Zhang H, Meng W, Lim SK, Sze SK (2010) Hypoxic tumor cell modulates its microenvironment to enhance angiogenic and metastatic potential by secretion of proteins and exosomes. Mol Cell Proteomics 9:1085–1099
74. Ashiru O, Boutet P, Fernandez-Messina L, Aquera-Gonzalez S, Skepper JN, Vales-Gomez M, Reyburn HT (2010) Natural killer cell cytotoxicity is suppressed by exposure to the human NKG2D ligand MICA*008 that is shed by tumor cells in exosomes. Cancer Res 70:481–489

75. Groh V, Wu J, Yee C, Spies T (2002) Tumour-derived soluble MIC ligands impair expression of NKG2D and T-cell activation. Nature 419:734–738
76. Frey AB (2006) Myeloid suppressor cells regulate the adaptive immune response to cancer. J Clin Invest 116:2587–2590
77. Makarenkova VP, Bansal V, Matta BM, Perez LA, Ochoa JB (2006) CD11b+/Gr-1 + myeloid suppressor cells cause T cell dysfunction after traumatic stress. J Immunol 176:2085–2094
78. Yu S, Liu C, Su K, Wang J, Liu Y, Zhang L, Li C, Cong Y, Kimberly R, Grizzle WE, Falkson C, Zhang HG (2007) Tumor exosomes inhibit differentiation of bone marrow dendritic cells. J Immunol 178:6867–6875
79. Soderberg A, Barral AM, Soderstrom M, Sander B, Rosen A (2007) Redox-signaling transmitted in trans to neighboring cells by melanoma-derived TNF-containing exosomes. Free Radic Biol Med 43:90–99
80. Chalmin F, Ladoire S, Mignot G, Vincent J, Bruchard M, Remy-Martin JP, Boireau W, Rouleau A, Simon B, Lanneau D, De Thonel A, Multhoff G, Hamman A, Martin F, Chauffert B, Solary E, Zitvogel L, Garrido C, Ryffel B, Borg C, Apetoh L, Rébé C, Ghiringhelli F (2010) Membrane-associated Hsp72 from tumor-derived exosomes mediates STAT3-dependent immunosuppressive function of mouse and human myeloid-derived suppressor cells. J Clin Invest 120:457–471
81. Xie Y, Zhang H, Li W, Deng Y, Munegowda MA, Chibbar R, Qureshi M, Xiang J (2010) Dendritic cells recruit T cell exosomes via exosomal LFA-1 leading to inhibition of $CD8^+$ CTL responses through downregulation of peptide/MHC class I and Fas ligand-mediated cytotoxicity. J Immunol 185:5268–5278
82. Iero M, Valenti R, Huber V, Filipazzi P, Parmiani G, Fais S, Rivoltini L (2008) Tumour-released exosomes and their implications in cancer immunity. Cell Death Differ 15:80–88
83. Wieckowski E, Whiteside TL (2006) Human tumor-derived vs dendritic cell derived exosomes have distinct biologic roles and molecular profiles. Immunol Res 36:247–254
84. Tomihari M, Chung JS, Akiyoshi H, Cruz PD Jr, Ariizumi K (2010) DC-HIL/glycoprotein Nmb promotes growth of melanoma in mice by inhibiting the activation of tumor-reactive T cells. Cancer Res 70:5778–5787
85. Kim JW, Wieckowski E, Taylor DD, Reichert TE, Watkins S, Whiteside TL (2005) Fas ligand-positive membranous vesicles isolated from sera of patients with oral cancer induce apoptosis of activated T lymphocytes. Clin Cancer Res 11:1010–1020
86. Laulagnier K, Grand D, Dujardin A, Hamdi S, Vincent-Schneider H, Lankar D, Salles JP, Bonnerot C, Perret B, Record M (2004) PLD2 is enriched on exosomes and its activity is correlated to the release of exosomes. FEBS Lett 572:11–14
87. Abusamra AJ, Zhong Z, Zheng X, Li M, Ichim TE, Chin JL, Min WP (2005) Tumor exosomes expressing Fas ligand mediate $CD8^+$ T cell apoptosis. Blood Cells Mol Dis 35:169–173
88. Clayton A, Tabi Z (2005) Exosomes and the MICA-NKG2D system in cancer. Blood Cells Mol Dis 34:206–213
89. Clayton A, Mitchell JP, Court J, Linnane S, Mason MD, Tabi Z (2008) Human tumor-derived exosomes down-modulate NKG2D expression. J Immunol 180:7249–7258
90. Perone MJ, Larregina AT, Shufesky WJ, Papworth GD, Sullivan ML, Zahorchak AF, Stolz DB, Baum LG, Watkins SC, Thomson AW, Morelli AE (2006) Transgenic galectin-1 induces maturation of dendritic cells that elicit contrasting responses in naïve and activated T cells. J Immunol 176:7207–7220
91. Bamias A, Tsiatas ML, Kafantari E, Liakou C, Rodolakis A, Voulgaris Z, Vlahos G, Papageorgiou T, Tsitsilonis O, Bamia C, Papatheodoridis G, Politi E, Archimandritis A, Antsaklis A, Dimopoulos MA (2007) Significant differences of lymphocytes isolated from ascites of patients with ovarian cancer compared to blood and tumor lymphocytes. Association of $CD3^+CD56^+$ cells with platinum resistance. Gynecol Oncol 106:75–81
92. Kalinski P, Okada H (2010) Polarized dendritic cells as cancer vaccines: directing effector-type T cells to tumors. Semin Immunol 22(3):173–82
93. Maccalli C, Pisarra P, Vegetti C, Sensi M, Parmiani G, Anichini A (1999) Differential loss of T cell signaling molecules in metastatic melanoma patients' T lymphocyte subsets expressing distinct TCR variable regions. J Immunol 163:6912–6923

94. Nieuwland R, van der Post JA, Gemma CA, Kenter G, Sturk A (2010) Microparticles and exosomes in gynecologic neoplasias. Semin Thromb Hemost 36:925–929
95. Ristorcelli E, Beraud E, Verrando P, Villard C, Lafitte D, Sbarra V, Lombardo D, Verine A (2008) Human tumor nanoparticles induce apoptosis of pancreatic cancer cells. FASEB J 22:3358–3369
96. Wada J, Onishi H, Suzuki H, Yamasaki A, Nagai S, Morisaki T, Katano M (2010) Surface-bound TGF-beta1 on effusion-derived exosomes participates in maintenance of number and suppressive function of regulatory T cells in malignant effusions. Anticancer Res 30:3747–3757
97. Xie Y, Bai O, Yuan J, Chibbar R, Slattery K, Wei Y, Deng Y, Xiang J (2009) Tumor apoptotic bodies inhibit CTL responses and antitumor immunity via membrane-bound transforming growth factor-beta-1 inducing $CD8^{+}$ T-cell anergy and $CD4^{+}$ Tr1 cell responses. Cancer Res 69:7756–7766
98. Szajnik M, Czystowska M, Szczepanski MJ, Mandapathil M, Whiteside TL (2010) Tumor-derived microvesicles induce, expand and up-regulate biological activities of human regulatory T cells (Treg). PLoS One 5:e11469
99. Temme S, Eis-Hubinger AM, McLellan AD, Koch N (2010) The herpes simplex virus-1 encoded glycoprotein B diverts HLA-DR into the exosome pathway. J Immunol 184:236–243
100. Miguet L, Béchade G, Fornecker L, Zink E, Felden C, Gervais C, Herbrecht R, Van Dorsselaer A, Mauvieux L, Sanglier-Cianferani S (2009) Proteomic analysis of malignant B-cell derived microparticles reveals CD148 as a potentially useful antigenic biomarker for mantle cell lymphoma diagnosis. J Proteome Res 8:3346–3354
101. Taylor DD, Black PH (1985) Inhibition of macrophage Ia antigen expression by shed plasma membrane vesicles from metastatic murine melanoma lines. J Natl Cancer Inst 74:859–867
102. Pelton JJ, Taylor DD, Fowler WC, Gercel-Taylor C, Carp NZ, Weese JL (1991) Lymphokine-activated killer cell suppressor factor in malignant effusions. Arch Surg 126:476–480
103. Albanese J, Meterissian S, Kontogiannea M, Dubreuil C, Hand A, Sorba S, Dainiak N (1998) Biologically acive Fas antigen and its cognate ligand are expressed on plasma membrane-derived extracellular vesicles. Blood 91:3862–3874
104. Kim SH, Bianco NR, Shufesky WJ, Morelli AE, Robbins PD (2007) MHC class II+ exosomes in plasma suppress inflammation in an antigen-specific and Fas ligand/Fas-dependent manner. J Immunol 179(4):2235–41
105. Martinez FO, Sica A, Mantovani A, Locati M (2008) Macrophage activation and polarization. Front Biosci 13:453–461
106. Ehrt S, Schnappinger D, Bekiranov S, Drenkow J, Shi S, Gingeras TR, Gaasterland T, Schoolnik G, Nathan C (2001) Reprogramming of the macrophage transcriptome in response to interferon-gamma and Mycobacterium tuberculosis: signaling roles of nitric oxide synthase-2 and phagocyte oxidase. J Exp Med 194:1123–1140
107. Martinez FO, Gordon S, Locati M, Mantovani A (2006) Transcriptional profiling of the human monocyte-to-macrophage differentiation and polarization: new molecules and patterns of gene expression. J Immunol 177:7303–7311
108. Dinarello CA (1996) Biologic basis for interleukin-1 in disease. Blood 87:2095–2147
109. Scotton CJ, Martinez FO, Smelt MJ, Sironi M, Locati M, Mantovani A, Sozzani S (2005) Transcriptional profiling reveals complex regulation of the monocyte IL-1 beta system by IL-13. J Immunol 174:834–845
110. Roman J, Ritzenthaler JD, Fenton MJ, Roser S, Schuyler W (2000) Transcriptional regulation of the human interleukin 1beta gene by fibronectin: role of protein kinase C and activator protein 1 (AP-1). Cytokine 12:1581–1589
111. Pacifici R, Roman J, Kimble R, Civitelli R, Brownfield CM, Bizzarri C (1994) Ligand binding to monocyte alpha 5 beta 1 integrin activates the alpha 2 beta 1 receptor via the alpha 5 subunit cytoplasmic domain and protein kinase C. J Immunol 153:2222–2233
112. Ruoslahti E (1996) RGD and other recognition sequences for integrins. Ann Rev Cell Dev Biol 12:697–715

Chapter 7
Immune Modulation of T Cells and Natural Killer Cells by Tumor-Derived Exosomes

Theresa L. Whiteside

Abstract Tumor-derived exosomes (TEX) are present in body fluids of patients with cancer. By downregulating functions of immune cells, TEX promote tumor progression. In patients with cancer, apoptosis of activated tumor-specific T cells, depressed natural killer (NK) cell activity, and the increased frequency in the peripheral blood or tumor sites of $CD4^+CD25^{hi}FOXP3^+$ regulatory T cells (Treg) are common and consistent findings. Evidence suggests that TEX have immunoregulatory properties and contribute to $CD8^+$ effector T cell apoptosis, induction/expansion of Treg, and inhibition of NK cell activity. Membrane-associated molecules, such as TGF-β or Fas ligand, are carried by TEX and are responsible for the ability of TEX to regulate peripheral tolerance in patients with cancer.

7.1 Introduction

One of the major obstacles in cancer immunotherapy is tumor escape from the host immune system [1]. Human tumors develop the ability to downregulate functions of immune cells, and especially, functions of antitumor effector cells [1, 2]. As a consequence of tumor-driven suppression, immune cells, including $CD8^+$ and $CD4^+$T lymphocytes, natural killer (NK) cells, and dendritic cells (DC) are functionally disabled and susceptible to apoptosis. Several years ago, it has been observed that sera of patients with cancer, but not sera of normal donors (NC), can suppress functions of normal activated T cells following a brief incubation period [3–5]. This suppressive effect of cancer patients' sera was eventually attributed to a fraction containing small membranous vesicles ranging in size from 50 to 100 nm [3–5]. These vesicles, identified as exosomes by transmission electron microscopy, had a striking feature: they concentrated molecules that were characteristic components of tumor cell surface membranes [6].

The biogenesis of exosomes, including that of tumor-derived exosomes (TEX), is well known and begins with endosomes, which fuse forming multivesicular bodies (MVB) [7, 8]. Exosomes are generated through inward budding of the MVB membrane, and in the process of invagination, they enclose bits of cellular endoplasm,

T. L. Whiteside (✉)
University of Pittsburgh Cancer Institute, Pittsburgh, PA 15213, USA
e-mail: whitesidetl@upmc.edu

H.-G. Zhang (ed.), *Emerging Concepts of Tumor Exosome–Mediated Cell–Cell Communication,* DOI 10.1007/978-1-4614-3697-3_7,

including various endoplasmic components [7, 8]. Upon MVB fusion with the cell membrane, exosomes are released through an ATP-dependent process into extracellular space as double-membrane vesicles [9]. Exosomes are found in all body fluids, but those most extensively studied in man come from the peripheral circulation. Sera of patients with cancer are enriched in TEX, which exert adverse effects on cells of the immune system [3, 4].

This chapter will describe the role of TEX in mediating tumor-induced immune suppression. Today, this tumor-driven suppression of immune cells is recognized as a major problem in immunotherapy of cancer. Various types of abnormalities in functions of T cells and NK cells have been described in cancer patients and are especially evident in those with advanced malignancies. Current evidence suggests that TEX represent one of the mechanisms used by tumors for subversion of the host immune system, including antitumor activities of T and NK cells.

7.2 Molecular Composition of TEX

Exosomes are released by all cells. Consequently, sera of cancer patients contain exosomes derived from tumor cells as well as those derived from other, normal, cells. However, tumor cells release exosomes in a greater abundance as compared to noncancer cells, and the molecular profile of TEX is distinct from that of other exosomes [10]. For example, exosomes derived from antigen-presenting cells are enriched in the major histocompatibility complex (MHC) class I and II molecules , costimulatory molecules, as well as ICAM-1, while TEX concentrate death receptor ligands such as FasL or TRAIL [10]. While DC-derived exosomes (dexosomes) carry peptide-loaded MHC class I molecules and can stimulate $CD8^{+}$ T-cell responses [11, 12], TEX, which are enriched in FasL induce apoptosis of $CD95^{+}$ (Fas^{+}) T cells [4, 13]. Nevertheless, TEX also concentrate tumor antigens (TA) and are avidly taken up by DC, which process these TA and present TA-derived peptides complexed with MHC molecules to cognate T cells, thus contributing to the generation of TA-specific T cells [14, 15]. This capacity of TEX to both induce and inhibit T-cell responses is one of the most intriguing aspects of their biology. It appears that molecular content of exosomes is not random but rather it reflects a selective sorting process in the cell of origin [16]. In this respect, the possibility that sorting of TEX components in the parent tumor cells could define the target cell specificity is of great interest. Since the release of exosomes appears to be an important feature of lymphoid intercellular communication, the issue of whether and how TEX are targeted remains a fascinating puzzle.

To obtain TEX from sera or other body fluids of patients with cancer, we and others have used a two-step procedure consisting of size exclusion chromatography on a column of Sepharose 2B followed by ultracentrifugation of the exclusion fractions at 100,000 $\times$ g for 2 h to pellet exosomes [4, 5]. The protein and lipid content of TEX isolated from patients' sera or tumor cell supernatants can be determined in Western blots as previously described [4, 5] or by more sophisticated mass spectrometry

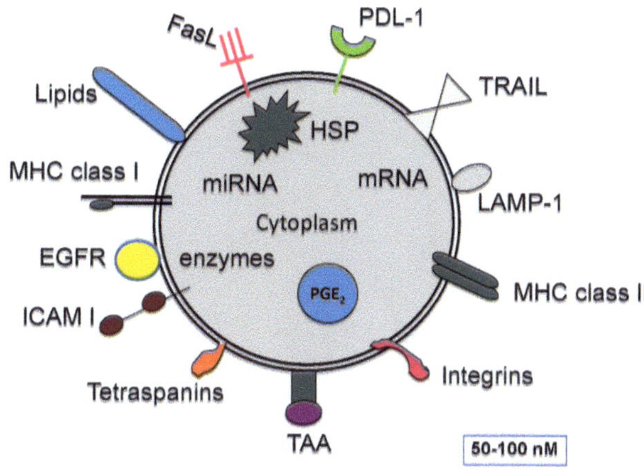

Fig. 7.1 A schematic view of tumor-derived exosome (TEX) showing the main classes of membrane-associated or cytoplasm–derived molecules carried by TEX. Note that TEX carry tumor-associated antigens (TAA), growth factor receptors (e.g., EGFR), ligands for death receptors (e.g., FasL), major histocompatibility complex molecules, as well as usual components of all exosomes such as tetraspanins

techniques. In fact, high throughput proteomic analysis greatly facilitates the definition of molecular profiles of purified exosomes [17, 18]. From surveying the protein/lipid composition of exosomes with different cellular origins, it can be concluded that in addition to a common profile of molecules characteristic of most exosomes, those derived from different cell types contain and concentrate a set of molecules that are unique or characteristic for each type of the parent cell. Consequently, TEX have a biochemical composition resembling that of tumor cells from which they derive. Figure 7.1 is a generalized schema depicting the molecular profile of TEX. It can be seen that in addition to the ubiquitous components of all exosomes, such as LAMP1, tetraspanins, annexins, and integrins or MHC class I and II molecules, TEX carry TA, growth factor receptors such as EGFR or HER-2, and membrane-associated enzymes such as, e.g., an ATP hydrolase CD39 and a 5′-ectonucleotidase CD73 [19]. Importantly, TEX also carry membrane forms of proapoptotic molecules such as FasL and TRAIL and suppression-inducing molecules such as PD-L1 and TGF-β.

7.3 Immunostimulatory Properties of TEX

The observation that TEX are enriched in TA and carry heat-shock proteins (HSP) as well as MHC-peptide complexes implies that TEX might serve as vehicles for delivery of TA to immune cells and activation of antitumor immune responses. Initially, it has been suggested that TEX represent a source of TA for cross-priming of cytolytic T lymphocytes (CTL) [15]. However, in vitro coincubation experiments, in which TEX purified from patients' sera or tumor supernatants were added to freshly purified human $CD4^+$ or $CD8^+$ T cells, showed that TEX did not promote proliferation of these cells, while exosomes derived from activated T cells or from in vitro-matured DC readily induced T cell proliferation (Fig. 7.2). In our hands, TEX did not activate immune cells [20]. There are literature reports of ex vivo studies with heat-stressed

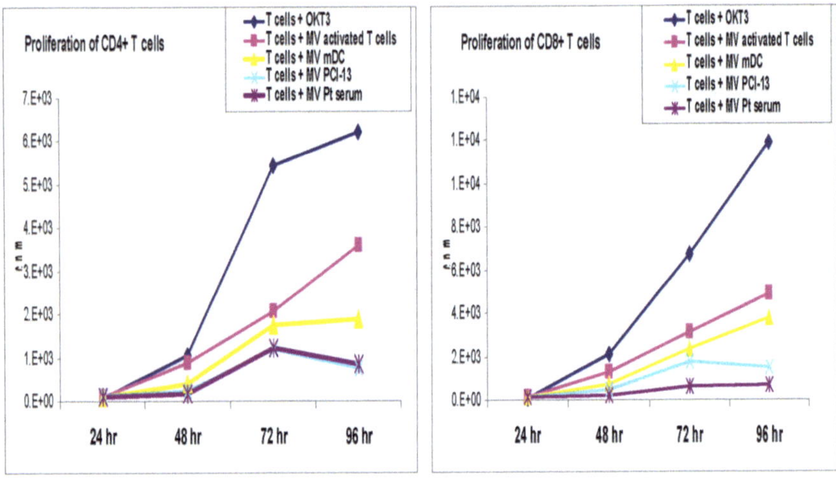

Fig. 7.2 TEX or microvesicles (MV) isolated from sera of patients with cancer or from supernatants of tumor cell lines (e.g., PCI-13, a human head and neck squamous cell carcinoma) do not or only weakly stimulate proliferation of OKT3-activated $CD4^+$ or $CD8^+$ T cell subsets isolated from peripheral blood of a normal donor using AutoMACS. In contrast, exosomes derived from supernatants of dendritic cell (DC) or activated T cells support proliferation of T cell subsets. Proliferation was measured in 3H-thymidine corporation assays

tumor cells, which released TEX enriched in HSP [21]. These TEX were found to induce NK cell activation [21]. Thus, a direct contact of TEX with immune cells may or may not result in cell activation, and this appears to be dependent on the molecular content of TEX and/or conditions existing in the tumor. On the other hand, TEX coincubated with DC were efficiently taken up by these cells, and TA carried by TEX were transferred to DC [22]. Furthermore, these DC were able to process the transferred TA and present the relevant peptides on their own MHC molecules to T cells, leading to the generation of tumor-reactive CTL [22]. In this case, TEX were immunogenic but only in an indirect way, upon a complete TA representation by DC. This strategy of TA transfer by TEX was used to produce exosome-based DC antitumor vaccines, which were extensively evaluated in murine models of tumor growth by Zitvogel and colleagues, and proved to be immunogenic, leading to tumor rejection [23, 24]. In subsequent human studies, the same strategy of vaccinating cancer patients with DC-derived exosomes has been less effective to date [25].

7.4 Immunosuppressive Properties of TEX

The ability to suppress antitumor immune responses appears to be a predominant biologic feature of TEX and it has received considerable attention recently as a potential mechanism responsible for tumor escape. In fact, there are several distinct mechanisms that TEX utilize to inhibit immune responses and these are reviewed below.

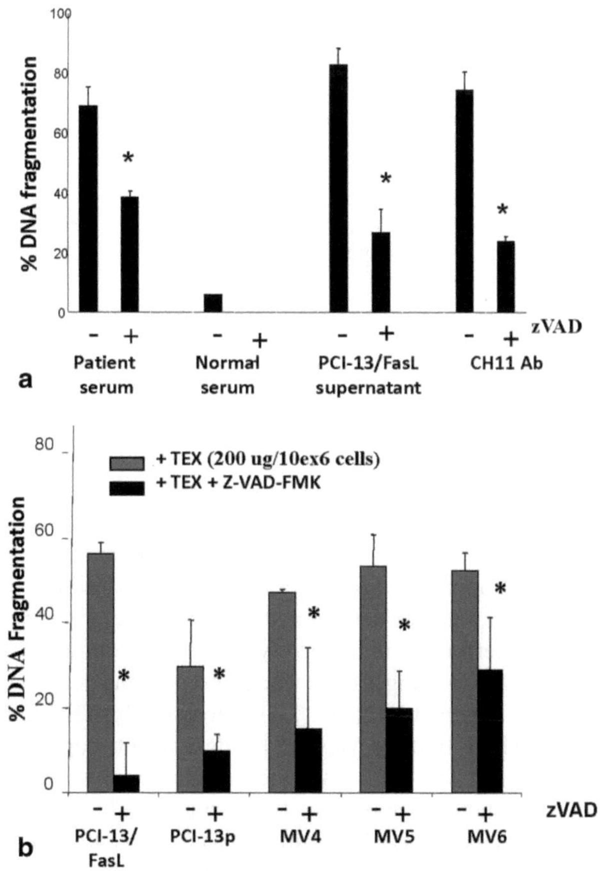

Fig. 7.3 (**a**) Sera of cancer patients or (**b**) TEX isolated from the cancer patients' sera induce DNA fragmentation in activated T lymphocytes as measured in 24 h JAM assays. In the presence of ZVAD, apoptosis is inhibited. *Asterisks* indicate significant differences at $p < 0.01$. (Adapted from Kim J-W et al., Clin Cancer Res 11:1010–1020, 2005)

TEX-induced apoptosis of activated T lymphocytes Evidence has accumulated indicating that sera of patients with cancer but not sera of healthy donors often inhibit various lymphocyte functions ex vivo or induce T cell apoptosis [3, 4]. The nature of these inhibitory serum factors remained a mystery for many years. Figure 7.3a shows that when OKT-3-activated T cells were incubated in the presence of serum from a cancer patient, a supernatant of cultured tumor cells or CH-11 antibody, which cross-links Fas, they underwent apoptosis as demonstrated by DNA fragmentation. Pretreatment of T cells with pan-caspase inhibitors (Z-VAD-FMK) partially blocked this effect [4]. More recently, we observed that exosome factions obtained from sera of cancer patients or tumor cell supernatants had the same inhibitory activity as unfractionated sera (Fig. 7.3b). The ability of TEX present in cancer patients' sera to induce death of activated $CD8^+$ T lymphocytes provided an explanation for spontaneous apoptosis seen by us in these patients' peripheral circulation [26–28]. Using Annexin V (ANXV), which detects a phosphatidyl serine flip in the T-cell membrane,

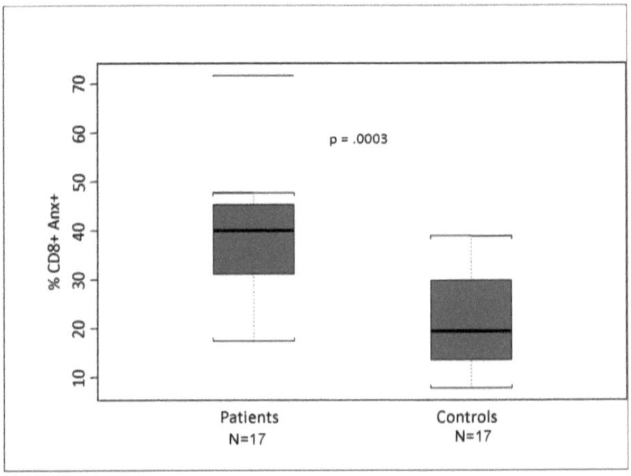

Fig. 7.4 Annexin V (ANXV) binding to the surface of $CD8^+$ T cells in freshly harvested peripheral blood of patients with cancer or normal age- and sex-matched donors. (The data presented as *box plots* were acquired by flow cytometry and adapted from Hoffmann et al., Clin Cancer Res 8:2553–2562, 2002)

thus serving as an early marker of apoptosis [29], we observed that over 50 % of circulating $CD8^+$ T cells in patients with cancer-bound ANXV were in the early stages of apoptosis [26]. T cells in very early stages of apoptosis can be identified and their frequency determined by flow cytometry using labeled ANXV, provided the assay is performed immediately after phlebotomy [29]. In contrast to cancer patients, few ANXV-binding T cells were present in the peripheral circulation of healthy age- and sex-matched NC (Fig. 7.4). Interestingly, TEX induced apoptosis of $CD8^+$ but not $CD4^+$ T lymphocytes [28]. These results suggested that tumor-infiltrating or -circulating $CD8^+$ T lymphocytes in patients with cancer might be especially sensitive to TEX-mediated apoptosis. Furthermore, only TEX and not exosomes derived from normal cells, such as DC or fibroblasts, induced death of $CD8^+$ T cells [10]. The high frequency of ANXV-binding circulating $CD8^+$ T cells was observed in patients with head and neck squamous cell cancer (HNSCC), melanoma, and ovarian carcinoma [30–32].

The ability of TEX to induce death of $CD8^+$ T cells in patients with HNSCC was linked to the enrichment of TEX in membrane-associated form (42 kDa) of FasL and in MHC class I molecules [20, 32, 33]. TEX with the highest content of these molecules were most active in inducing T cell apoptosis, which could be partially blocked by anti-Fas or anti-MHC class I antibodies and completely blocked in the presence of both antibodies [4]. In the circulation of HNSCC patients, nearly all $CD8^+$ lymphocytes express surface CD95 [28], and because TEX present in these patients' sera carry FasL [20], the Fas–FasL pathway is, in a large part, responsible for the demise of $CD8^+$ T cells observed in vivo [26]. The role of MHC class I molecules expressed on TEX in inducing $CD8^+$ T cell apoptosis remains unclear, although it has been suggested that the direct engagement of MHC class I molecules with the CD8 receptor activates the endogenous Fas/FasL pathway, leading to apoptosis [34, 35]. The ex vivo studies with human T cells and purified TEX as well as the phenomenon of spontaneous apoptosis of circulating $CD8^+$ T cells observed in patients with cancer clearly implicate TEX in $CD8^+$ T cell demise. However, evidence demonstrating that a transfer of TEX from tumor-bearing mice to naïve mice prior to or at the time of tumor injection induces T cell death, eliminates T-cell responses, and promotes tumor growth is still lacking.

Molecular signals delivered by TEX to T cells In contrast to exosomes derived from normal cells, TEX predominantly function as harbingers of tolerogenic signals. TEX-induced unresponsiveness of T cells is associated with: (a) inhibition of T cell receptor (TcR) and IL-2R signaling, (b) signals inhibiting T cell proliferation, (c) interference with cytokine production, and (d) signals for apoptosis of activated T cells. T cells receive and process activation stimuli via the TcR/CD3 complex. The CD3ζ chain is responsible for transfer of activating signals to the cell nucleus. A loss of ζ chain alters functions of TcR and interferes with T cell functions [36]. We and others have observed that sera of patients with cancer as well as exosomes isolated from these sera mediate dose- and time-dependent suppression of CD3-ζ chain expression in T cells [4, 6]. TEX were shown to be able to downregulate mRNA for ζ and induce a loss of ζ protein (Fig. 7.5a). It has been previously shown that a loss of ζ is a common feature of T cells in cancer patients and may function as a biomarker of poor prognosis, because it is associated with suppression of antitumor immune functions [36, 37]. Furthermore, as shown in Fig. 7.5b, a loss or downregulation of ζ correlated with the level of FasL expression on TEX and thus their ability to induce T cell apoptosis [4, 5]. Caspase activation was observed in Jurkat cells incubated in the presence of TEX but not exosomes obtained from sera of NC (Fig. 7.5c).

The ζ chain is not the only signaling element downregulated upon T cell coincubation with TEX. As shown in Fig. 7.6a, b, the coincubation of T cells with TEX also resulted in reduced expression of JAK3 in these cells [20]. As integrity of the JAK pathway is critical for function of receptors for cytokines, which express the common γ chain, including IL-2, IL-7, and IL-15, suppression of this pathway results in a failure of T cells to proliferate and produce cytokines. Studies by Taylor and colleagues indicated that placenta-derived exosomes suppressed IL-2 production in T cells [38, 39], and we and others observed that TEX impair human lymphocyte responses to IL-2 [4, 40]. The addition of TEX to activated $CD8^+$ T cells also dramatically reduced expression of phosphorylated STAT5 in our hands. In contrast, TMV increased phosphorylated STAT5 expression in activated $CD4^+$T cells (Fig. 7.6c). Importantly, while TEX added to preactivated $CD8^+$T cells failed to promote and tended to inhibit proliferation, activated $CD4^+$T cells expanded in the presence of TEX (Fig. 7.7). In aggregate, these data are consistent with the conclusion that activated $CD8^+$T cells are selectively targeted by TEX, which interfere with TcR- and IL-2R-mediated signaling, thus inhibiting cellular proliferation. Thus, TEX can regulate T cell expansion, and $CD8^+$ vs. $CD4^+$T cells are differentially regulated by TEX [20]. TEX can also interfere with CTL generation by blocking differentiation of human monocytes into DC, as demonstrated by Rivoltini and colleagues [41]. Human DC generated in the presence of TEX had low expression of costimulatory molecules, produced inhibitory cytokines, and induced dose-dependent suppression of T cell proliferation and antitumor cytotoxicity [41].

The TEX-mediated signals leading to apoptosis of T cells induce early membrane changes (ANXV binding), caspase-3 cleavage, cytochrome C release from mitochondria, loss of the mitochondrial membrane potential (MMP), and, ultimately,

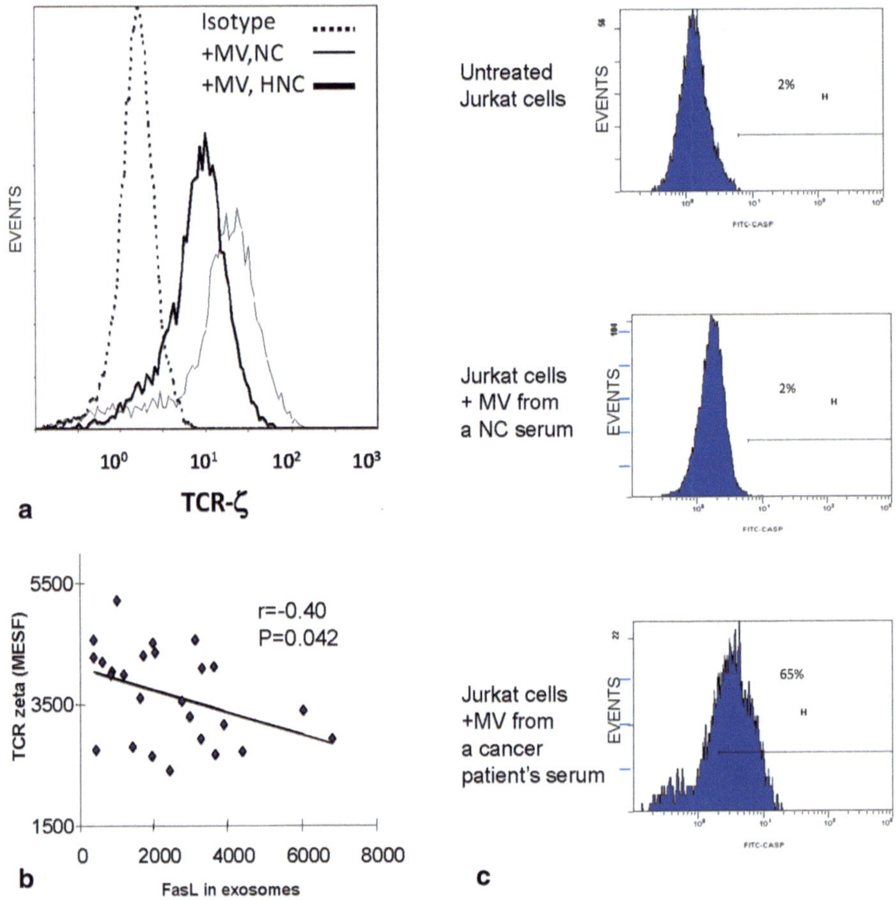

Fig. 7.5 Signals delivered by TEX to T cells result in the TcR-associated CD3ζ chain downregulation and caspase activation. **a** In downregulation of the CD3ζ chain expression in T cells coincubated with TEX obtained from serum of a patient with head and neck cancer (HNC) but not with MV obtained from serum of a normal donor. **b** In a reverse correlation between the FasL content of MV obtained from patients with HNC and the CD3ζ chain expression in T cells coincubated with these MV fractions for 24 h. **c** In Jurkat cells coincubated with TEX from a cancer patients' serum upregulated caspase expression. (Adapted from Kim J-W et al., Clin Cancer Res 11:1010–1020, 2005)

DNA fragmentation [28, 42]. We have demonstrated that TEX induce T cell death by engaging both the extrinsic and intrinsic apoptotic pathways [42, 43]. It appears that PI3K/Akt pathway is a central TEX target in activated T cells, because their exposure to TEX causes a dramatic time-dependent Akt dephosphorylation and its inactivation, which is accompanied by the downregulation of antiapoptotic Bcl-2 family member (i. e., Bcl-2, Bcl-XL, and Mcl-1) expression and upregulation of the proapoptotic protein, Bax [42]. Overall, our studies indicated that in vitro exposure

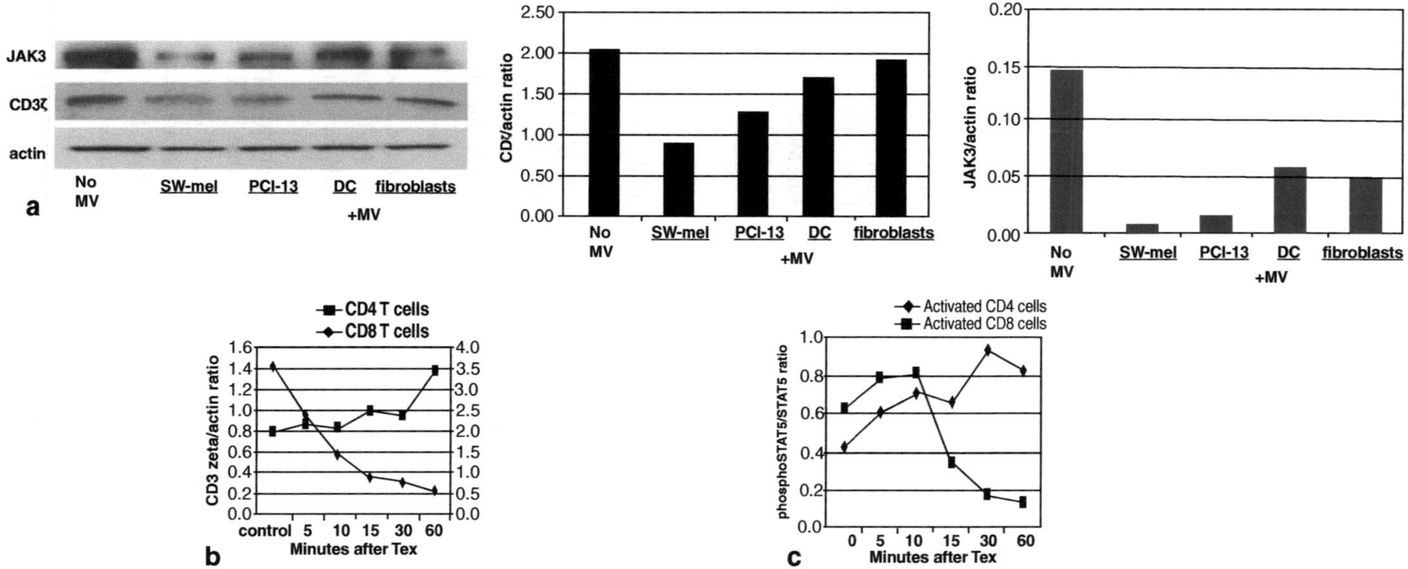

Fig. 7.6 TEX added to primary activated T cells induced downregulation of CD3ζ chain and JAK3 expression. **a** In Western blots showing low expression of CD3ζ and JAK3 expression in activated T cells after coincubation with TEX but not after coincubation with MV obtained from normal cell supernatants. On the right, the JAK3/actin ratios for the Western blots and on the left CD3ζ/actin ratios for the Western blots are shown. **b** In expression of CD3ζ decreases in $CD8^+$ primary T cells within minutes after TEX addition. In contrast, no such decrease in CD3ζ expression occurs in $CD4^+$T cells incubated with TEX. **c** In phosphorylated STAT5/total STAT5 ratios in activated $CD8^+$ and $CD4^+$ T cell subsets incubated $\pm$TEX. Phosphorylated STAT5 expression decreases after incubation of $CD8^+$ but not $CD4^+$ T cells with TEX. (Adapted from Wieckowski et al., J. Immunol. 183:3720–3730, 2009)

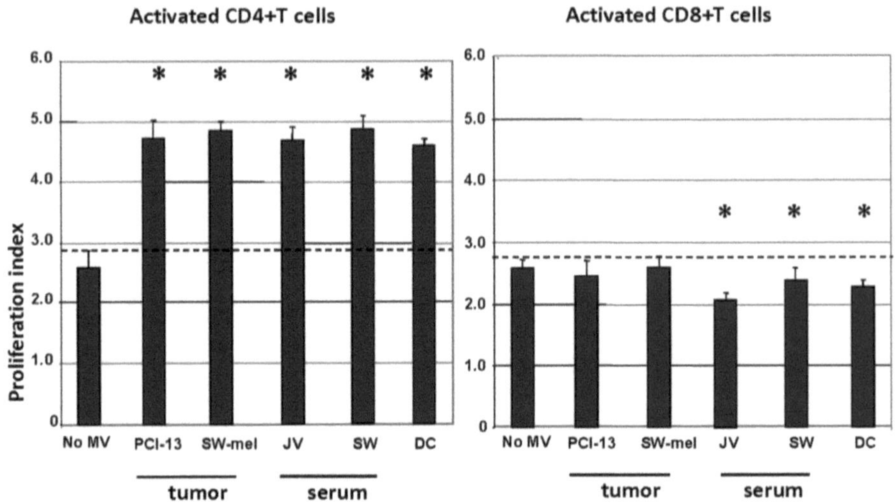

Fig. 7.7 Effects of TEX obtained from sera of patients with cancer (JV and SW) or from supernatants of tumor cell lines on proliferation of activated $CD4^+$ and $CD8^+$ T cell subsets. TEX of various origins were coincubated with CFSE-labeled, OKT3-activated T cell subsets, and proliferation was assessed by flow cytometry. The proliferation index was calculated using Modfit software. The data are mean values ± SD. *Asterisks* indicate significant differences at $p < 0.001$ for $CD4^+$ T cells and at $p < 0.01$ for $CD8^+$ T cells. (Adapted from Wieckowski et al., J. Immunol. 183:3720–3730, 2009)

of T cells to TEX, but not to exosomes produced by normal cells, leads to a loss of balance between pro- and antiapoptotic Bcl-2 family proteins and dysfunction of signaling via the TcR, IL-2R, and PI3K pathways. Ultimately, it results in T cell death. These in vitro studies are consistent with reports by us and others of similar changes in Bcl-2, Bcl-XL, or Bax in circulating T cells of patients with different malignancies [28, 44, 45] potentially implicating TEX in inducing these events in vivo [46].

TEX-mediated expansion of regulatory T cells (Treg) We have observed that the frequency of $CD4^+CD25^{high}FOXP3^+$ Treg was elevated in body fluids of cancer patients relative to NC ($p < 0.0001$) and that these body fluids were highly enriched in exosomes.

This observation has motivated us to more closely examine a potential relationship between TEX and Treg and to test the hypothesis that TEX contribute to induction/expansion of human $CD4^+CD25^{high}FOXP3^+$ Treg. To test this hypothesis, we sorted $CD4^+CD25^{high}$ Treg from the peripheral blood and incubated them with TEX or exosomes (MV) derived from normal cells. Only TEX isolated from supernatants of tumor cells but not MV from normal cells induced the Treg generation and enhanced their expansion (Fig. 7.8). TEX also mediated conversion of conventional $CD4^+CD25^{neg}$ T cells into $CD4^+CD25^{high}FOXP3^+$ Treg [47]. Upon coincubation with TEX, Treg had increased expression of FasL, IL-10, TGF-β1,

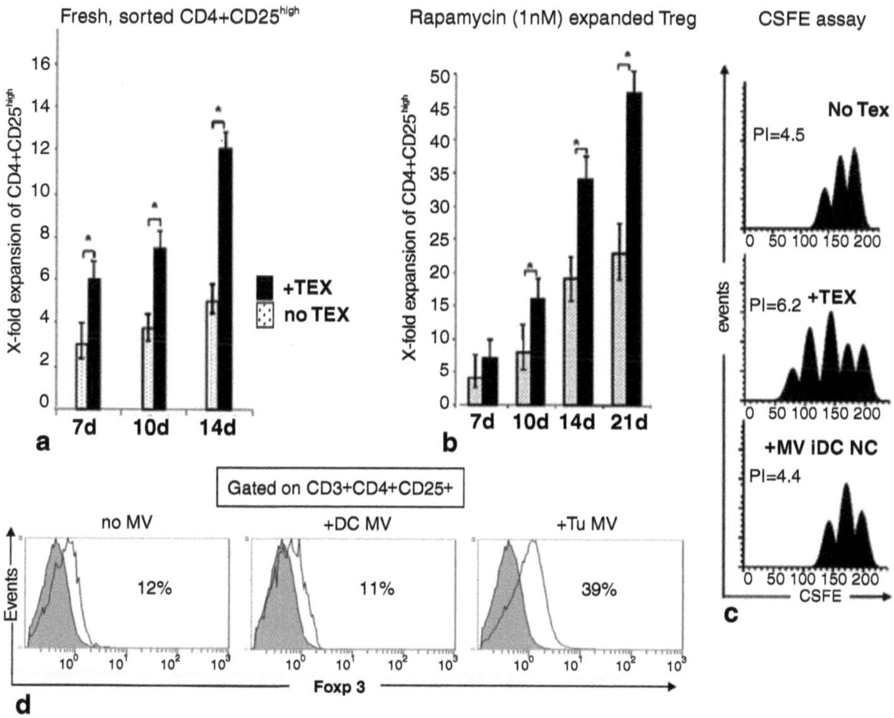

Fig. 7.8 TEX but not MV from normal cells promote proliferation of human Treg freshly sorted from the peripheral blood (**a**) or Treg previously induced to proliferate in the presence of 1 nm of rapamycin (**b**). Proliferation of Treg in the presence or absence of TEX was measured in CFSE-based assays. *Asterisks* indicate significant differences in cell expansion at $p < 0.05$. The data are means ± SD from six independent cocultures. In (**c**) proliferation indices for T cells cultured ± TEX or + MV obtained from supernatants of immature DC generated from normal donors and used as controls are shown (adapted from Szajnik et al., PLoS ONE 5:e11469, 2011). In (**d**), induction of Treg from $CD4^+$ conventional T cells cultured in the presence or absence of TEX or MV derived from control normal DC. (Adapted from Wieckowski et al., J. Immunol., 183:3720–3730, 2009)

CTLA-4, granzyme B, and perforin expression ($p < 0.05$ for all) and mediated higher suppression of autologous responder cell proliferation ($p < 0.01$). Treg were resistant to TEX-mediated apoptosis, while conventional $CD8^+$ T cells were sensitive. TEX were enriched in membrane-form TGF-β1 and they increased expression of phosphorylated SMAD-2/3 and phosphorylated STAT-3 in Treg. These effects of TEX on T cells were dependent on TGF-β1 and to a lesser degree on IL-10, as neutralizing antibodies specific for these cytokines blocked the ability of TEX to expand Treg. In aggregate, these results indicated that TEX have immunoregulatory properties. Interactions of TEX with Treg represent a newly defined mechanism that might be involved in regulating peripheral tolerance by tumors and in supporting immune evasion of human cancers.

TEX-mediated dysfunction of NK cells $CD8^+$ T cells are not the only lymphocytes, which are targeted by TEX and are sensitive to their inhibitory effects. We and others have reported earlier that spontaneous apoptosis often accompanies tumor progression in man and that NK cells in cancer patients have low expression levels of activating receptors, NKp30, NKp46, NKG2C, and NKG2D [48, 49]. More recently, TEX were shown to inhibit cytolysis mediated by NK cells ex vivo, and preliminary experiments suggest that the treatment of mice with TEX decreased the percent of NK cells in the spleen and lungs (our unpublished data). TEX also decreased the protective effects of syngeneic DC-based vaccines in mice bearing TS/A or 4.1 mammary tumors by blocking NK cell activity [50]. This study clearly implicates TEX in blocking NK cell-mediated antitumor responses in vivo. Zhang and colleagues reported that TEX suppresses NK cell functions by the mechanisms targeting perforin but not granzymes and downregulating expression of JAK3 in murine NK cells [50]. In addition to suppressing cytolytic functions of NK cells, TEX also inhibited IL-2-stimulated expansion of mouse NK cells by downregulating expression of cyclin D3 and decreasing Rb phosphorylation, the events, which play a critical role in the G1 to S transition in lymphocytes [50]. A different mechanism of TEX-mediated inhibition of NK cell activity in man has been described by Clayton's group [51]. TEX enriched in NKG2D ligands, MICA and MICB, as well as TGF-β1 were shown to downregulate the NKG2D receptor expression and to inhibit NKG2D-dependent functions upon incubation with peripheral blood-derived NK cells [52]. It was initially thought that MICA and MICB ligands were responsible for blocking of this activating receptor. However, MICA-specific Abs only partially attenuated this effect, while neutralizing antibodies to TGF-β1 strongly abrogated NKG2D downregulation. In fact, it has been previously reported that impaired NK cell cytotoxicity in patients with cancer was related to downmodulated NKG2D expression induced by elevated TGF-β1 levels in sera of these patients [53]. This finding introduced a possibility that TGF-β1 carried by TEX was the primary mechanism responsible not only for low expression of NKG2D but also for concomitantly impaired NKG2D-dependent killing activity [51]. Using TEX isolated from sera of patients with acute myelogenous leukemia (AML) at the time of diagnosis, we demonstrated similar effects on the NK cell phenotype and functions. Sera of AML patients were enriched in blast-derived exosomes carrying CD34, CD33, and CD117 as well as TGF-β1 [54]. These TEX decreased cytolytic activity of NK cells isolated from normal donors, downregulated NKG2D receptor expression and induced SMAD phosphorylation in NK cells. Neutralization of TGF-β1 carried by TEX derived from leukemic blasts significantly but not completely abrogated these TEX-mediated effects (Fig. 7.9). In addition, we reported that while sera of AML patients contained elevated levels of soluble TGF-β1, considerably more TGF-β1 resided in TEX and could be released and measured upon their disruption by detergents [54]. These data suggest that TEX-associated TGF-β1 is indeed the principal mechanism responsible for NK cell dysfunction in patients with AML and perhaps other malignancies as well.

TEX and the adenosinergic pathway Adenosine is a well-known immunosuppressive factor [55]. Operating via its receptors (A_1, A_{2A}, A_{2B}, A_3) present on various cell types, it upregulates cAMP levels thereby reducing cellular functions [55]. Recently,

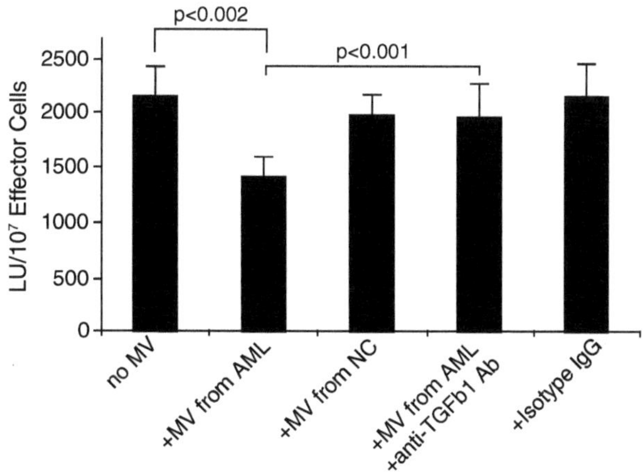

Fig. 7.9 Inhibition of NK cell activity by TEX obtained from sera of patients with AML but not MV from normal donors' sera. Neutralizing antibodies to TGF-β1 but not isotype control IgG restored normal NK activity. Exogenous recombinant TGFβ1 also inhibited NK cell activity, as expected. (Adapted from Szczepanski et al., Haematologica 96:1302–1309, 2011)

the ability of Treg to produce adenosine has been described, and the presence of ectonucleotidases CD39 (ATP hydrolase) and CD73 (5′-nucleotidase) on the surface murine Treg was reported [56, 57]. In humans, few natural (n) Treg coexpress these enzymes, although inducible (i) Treg generated in vitro and those found in tumor tissues of patients with cancer can express both and can produce adenosine (Fig. 7.10a) [58]. Although human nTreg mainly express CD39, they can produce adenosine in vivo, as TEX ubiquitously present in body fluids of cancer patients carry CD73 and have 5′-nuclotidase activity [19]. Exosomes obtained from sera of normal donors are also enriched in CD39 and CD73 and can perform ATP hydrolysis to adenosine (unpublished data, P. Schuler and T. Whiteside). It is possible that B lymphocytes might be a source of $CD39^+$ and $CD73^+$ exosomes in these sera (unpublished data, P. Schuler and T. Whiteside). Thus, exosomes deliver enzymatically active CD73 to nTreg (or other cells), complementing their ability to produce adenosine (Fig. 7.10b). As tumor cells are often enriched in CD73 [59], TEX are especially well-equipped to deliver CD73 to sites of ATP hydrolysis, enabling adenosine production and thus negatively modulating T cells in the tumor microenvironment.

7.5 Clinical Importance of TEX-Mediated Effects on Lymphocytes

The presence of TEX in body fluids of patients with cancer and TEX-mediated effects on T cells are clinically significant. TMV isolated from sera of most HNSCC patients induced caspase activation in Jurkat cells (used as a surrogate for primary T cells in cellular apoptosis assays). In contrast, exosomes from sera of normal donors did not induce caspase activation in Jurkat cells [60]. While ANXV binding to $CD8^+$ T cells readily discriminated between normal donors and patients with HNSCC ($p < 0.001$), it did not distinguish patients with active disease from those with no evident disease

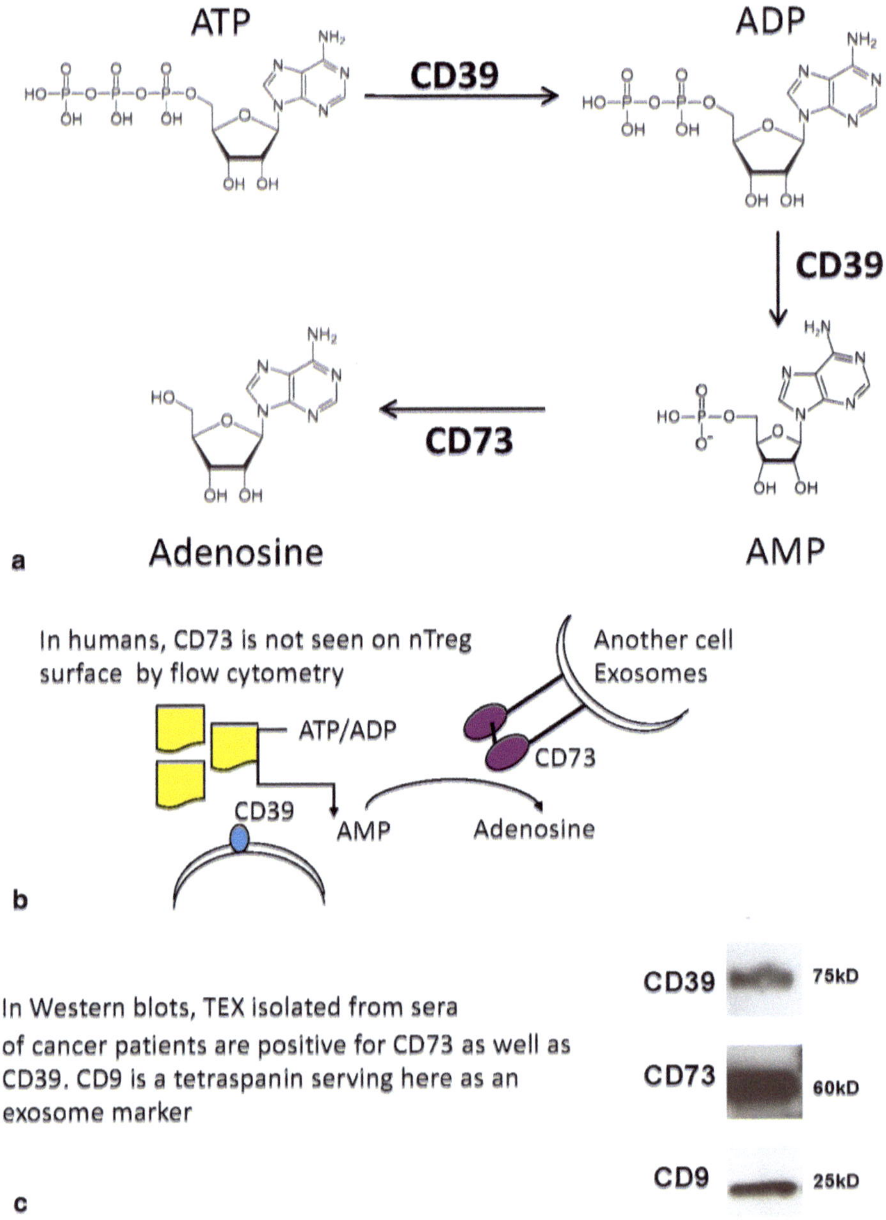

Fig. 7.10 The presence of ectonucleotidases CD39 and CD73 on human TEX. **a** In the pathway of exogenous adenosine production involving hydrolysis of exogenous ATP to ADP and AMP by an ATPase, CD39, and then of AMP to adenosine by 5′nucleotidase, CD73. **b** In a schema showing how exosomes or nonregulatory T cells carrying or expressing these enzymes, respectively, can help human Treg, which express low levels of or no CD73, to produce immunoinhibitory adenosine. **c** In Western blots showing that TEX isolated from sera of cancer patients or normal donors carry CD39, CD73, and CD9

after therapy [27, 28]. Not surprisingly, the presence in sera of TEX carrying high levels of FasL discriminated HNSCC patients with advanced (stage T3/T4) tumors ($p < 0.009$) from those with stage T1/T2 tumors whose TEX had low levels of FasL [4]. TEX from sera of patients with nodal metastases and poor prognosis were enriched in FasL and induced apoptosis in Jurkat cells discriminating this cohort of HNSCC patients ($n = 32$) from those with no nodal involvement and good prognosis ($n = 28$) at $p < 0.02$ [60]. TEX isolated from sera of AML patients at diagnosis varied widely in the levels of membrane-associated TGF-β [54], and our preliminary results suggest that these levels correlate with the blast counts and thus could serve as prognostic biomarkers. Based on these preliminary results, it is possible to predict that TEX will become increasingly important as biomarkers of disease, especially after technologies for separation of TEX from other exosomes present in body fluids become available in the near future.

The clinical importance of TEX is also emphasized by their ability to sequester tumor-reactive therapeutic antibodies delivered to patients with cancer [61]. For example, HER2 is present on TEX derived from HER-2-expressing cancer cells. Trastuzumab, a humanized antibody used for the adjuvant treatment of breast cancer, binds to HER2 receptor amplified on tumor cells, and it also mediates ADCC; both these mechanisms considered to be largely responsible for its therapeutic effects. However, HER2 is also carried on TEX, and such TEX interfere with the antibody binding to tumor cells and with ADCC [61]. In this context, TEX may be responsible for resistance of some patients to trastuzumab therapy.

7.6 Protection of Immune Cells from TEX

Since TEX contribute to tumor escape by inducing dysfunction and death of immune effector cells in cancer patients, elimination of TEX or blocking of TEX-mediated effects should be beneficial [62]. Several strategies for TEX removal have been considered. Ichim and colleagues proposed a physical approach based on the extra-corporeal removal of exosomes from plasma of cancer patients [63]. Huber et al. suggest interventions, which intervene with TEX secretion by inhibiting upstream pathways, e.g., drugs interfering with the microtubule stability or proton pump inhibitors (PPI) to modify secretory metabolism [64]. Yet another strategy aims to protect antitumor effector cells from functional impairments and death by using cytokines [65]. We have demonstrated that the survival cytokines, IL-7 and IL-15, as well as a biologic, IRX-2, produced by ex vivo-stimulated human PBMC and containing natural cytokines and growth factors, provide strong protection to T cells from apoptosis mediated by TEX (Fig. 7.11). Our studies indicated that the pre-treatment of T cells with these cytokines restored the balance between the pro- and antiapoptotic Bcl-2 family members and normalized JAK3 and CD3ζ expression in these cells, using the PI3 K/Akt pathway as the key regulatory mechanism [42, 43]. These results are in agreement with previously reported clinical and experimental data showing that the survival cytokines using the common receptor γ-chain are able to protect activated T cells from tumor-induced death and enhance their antitumor activity [65].

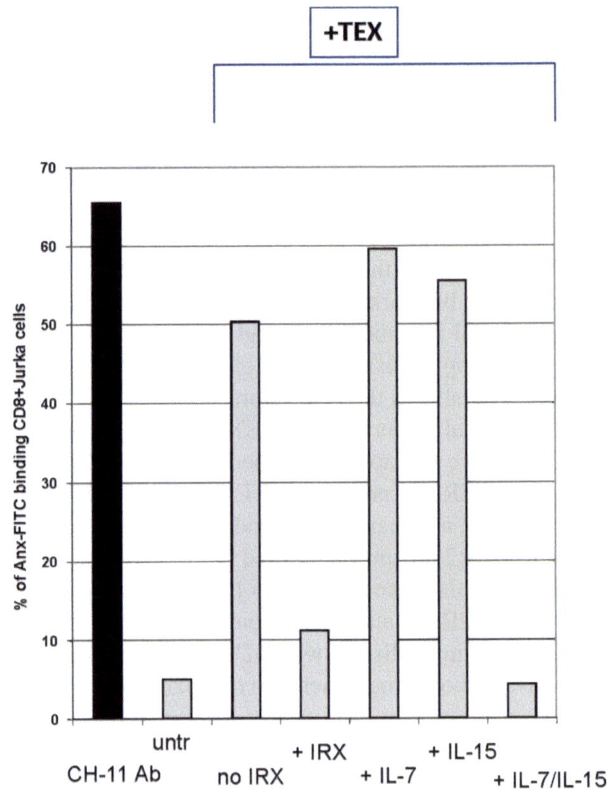

Fig. 7.11 Protective effects of cytokines on Jurkat cells coincubated in the presence of TEX. Jurkat cells were incubated in the presence or absence of survival cytokines, IL-7 and/or IL-15, for 24 h and then were treated with PCI-13-derived TEX for 3 h. Percentages of ANXV$^+$T cells were measured by flow cytometry (adapted from Czystowska et al., Cell Death Diff 16:708–718, 2009)

7.7 Conclusions

The ability of TEX to modulate functions of T cells and NK cells, which are responsible for antitumor activity in patients with cancer, represents an ingenious mechanism of tumor escape from the host immune system. Although evidence for suppression of immune cells by tumors has been known for many years, the role of TEX in this process has emerged relatively recently and is gaining attention. This may be a consequence of disappointing results seen with therapeutic antitumor vaccines in man [66]. These vaccines have not been as effective as originally expected in controlling tumor progression, despite the fact that in many cases, vaccination increases the frequency of TA-specific immune effector cells [67]. Among many tumor-orchestrated suppression mechanisms, the release by the tumor of nanosized TEX that distribute throughout the body and are able to suppress or eliminate T and NK cells far away from the tumor site, might explain, at least in part, the lack of clinical success with antitumor vaccines. From a clinical viewpoint, the recognition of TEX as immunoregulatory elements is important in charting novel strategies for therapeutic interventions aimed at the protection of immune effector from detrimental signaling delivered by TEX. An expanding understanding of the immunobiology associated with TEX signaling represents an opportunity for the development of novel cancer therapies able to counterbalance suppressive effects of TMV on the host immune system.

References

1. Whiteside TL, Mandapathil M, Szczepanski M, Szajnik M (2011) Mechanisms of tumor escape from the immune system: adenosine-producing Treg, exosomes and tumor-associated TLRs. Bull Cancer 98:E25–E31
2. Whiteside TL (2006) Immune suppression in cancer: effects on immune cells, mechanisms and future therapeutic intervention. Sem Cancer Biol 16:3–15
3. Huber V, Fais S, Iero M, Lugini L, Canese P, Squarcina P, Zaccheddu A, Colone M, Arancia G, Gentile M, Seregni E, Valenti R, Ballabio G, Belli F, Leo E, Parmiani G, Rivoltini L (2005) Human colorectal cancer cells induce T-cell death through release of proapoptotic microvesicles: role in immune escape. Gastroenterology 128:1796–1804
4. Kim J-W, Wieckowski E, Taylor DD, Reichert TE, Watkins S, Whiteside TL (2005) FasL+ membraneous vesicles isolated from sera of patients with oral cancer induce apoptosis of activated T lymphocytes. Clin Cancer Res 11:1010–1020
5. Taylor DD, Gercel-Taylor C (2005) Tumour-derived exosomes and their role in cancer-associated T-cell signaling defects. Br J Cancer 92:305–311
6. Taylor DD, Gercel-Taylor C, Lyons KS, Stanson J, Whiteside TL (2003) T-cell apoptosis and suppression of T-cell receptor/CD3-zeta by Fas ligand-containing membrane vesicles shed from ovarian tumors. Clin Cancer Res 9:5113–5119
7. Chaput N, Thery C (2011) Exosomes: immune properties and potential clinical implementations. Semin Immunopathol 33:419–440
8. Bobrie A, Colombo M, Raposo G, Thery C (2011) Exosome secretion: molecular mechanisms and roles in immune responses. Traffic. doi:10.1111/j.1600–0854.2011.01225
9. Thery C, Zitvogel L, Amigorena S (2002) Exosomes: composition, biogenesis and function. Nat Rev 2:569–579
10. Wieckowski E, Whiteside TL (2006) Human tumor-derived vs. dendritic cell-derived exosomes have distinct biologic roles and molecular profiles. Immunol Res 36:247–254
11. Clayton A, Court J, Navabi H, Adams M, Mason MD, Hobot JA, Newman GR, Jasani B (2001) Analysis of antigen presenting cell derived exosomes, based on immuno-magnetic isolation and flow cytometry. J Immunol Meth 247:163–174
12. Thery C, Regnault A, Garin J, Wolfers J, Zitvogel L, Ricciardi-Castagnoli P, Raposo G, Amigorena S (1999) Molecular characterization of dendritic cell-derived exosomes. Selective accumulation of the heat shock protein hsc73. J cell Biol 147:599–610
13. Andreola G, Rivoltini L, Castelli C, Huber V, Perego P, Deho P, Squarcina P, Accornero P, Lozupone F, Lugini L, Stringaro A, Molinari A, Arancia G, Gentile M, Parmiani G, Fais S (2002) Induction of lymphocyte apoptosis by tumor cell secretion of FasL-bearing microvesicles. J Exp Med 195:1303–1316
14. Andre F, Schartz NE, Chaput N, Flament C, Raposo G, Amigorena S, Angevin E, Zitvogel L (2002) Tumor-derived exosomes: a new source of tumor rejection antigens. Vaccine 20 Suppl 4:A28–31
15. Wolfers J, Lozier A, Raposo G, Regnault A, Thery C, Masurier C, Flament C, Pouzieux S, Faure F, Tursez T, Angevin E, Amigorena S, Zitvogel L (2001) Tumor-derived exosomes are a source of shared tumor rejection antigens for CTL cross-priming. Nat Med 7:297–303
16. Thery C, Ostrowski M, Segura E (2009) Membrane vesicles as conveyors of immune responses. Nat Rev Immunol 9:581–593
17. Mathivanan S, Lim JW, Tauro BJ, Ji H, Moritz RL, Simpson RJ (2010) Proteomics analysis of A33 immunoaffinity-purified exosomes released from the human colon tumor cell line LIM1215 reveals a tissue-specific protein signature. Mol Cell Proteomics 9:197–208

18. Hegmans JP, Bard MP, Hemmes A, Luider TM, Kleijmeer MJ, Prins JB, Zitvogel L, Burgers SA, Hoogsteden HC, Lambrecht BN (2004) Proteomic analysis of exosomes secreted by human mesothelioma cells. Am J Path 164:1807–1815
19. Clayton A, Al-Taei S, Webber J, Mason MD, Tabi Z (2011) Cancer exosomes express CD39 and CD73, which suppress T cells through adenosine production. J Immunol 187:676–683
20. Wieckowski E, Visus C, Szajnik M, Szczepanski MJ, Storkus WJ, Whiteside TL (2009) Tumor-derived microvesicles promote regulatory T-cell expansion and induce apoptosis in tumor-reactive activated CD8 + T lymphoctyes. J Immunol 183:3720–30
21. Gastpar R, Gehemann M, Bausero MA, Asea A, Gross C, Schroeder JA, Multhoff G (2005) Heat shock protein 70 surface-positive exosomes stimulate migratory and cytolytic activity of natural killer cells. Cancer Res 65:5238–5247
22. Andre F, Chaput N, Schartz NE, Flament C, Aubert N, Bernard J, Lemonnier F, Raposo G, Escudier B, Hsu DH, Tursz T, Amigorena S, Angevin E, Zitvogel L (2004) Exosomes as potent cell-free peptide-based vaccine. I. Dendritic cell-derived exosomes transfer functional MHC class I/peptide complexes to dendritic cells. J Immunol 172:2126–2136
23. Zitvogel L, Regnault A, Lozier A, Wolfers J, Flament C, Tenza D, Ricciardi-Castagnoli P, Raposo G, Amigorena S (1998) Eradication of established murine tumors using a novel cell-free vaccine: dendritic cell-derived exosomes. Nat Med 4:594–600
24. Chaput N, Schartz NE, Andre F, Zitvogel L (2003) Exosomes for immunotherapy of cancer. Adv Exp Med Biol 532:215–221
25. Viaud S, Thery C, Ploix S, Tursz T, Lapierre V, Lanz O, Zitrogel L, Chaput N (2010) Dendritic-cell derived exosomes for cancer immunotherapy: what's next? Cancer Res 70:1281–1285
26. Hoffmann TK, Dworacki G, Meidenbauer N, Gooding W, Johnson JT, Whiteside TL (2002) Spontaneous apoptosis of circulating T lymphocytes in patients with head and neck cancer and its clinical importance. Clin Cancer Res 8:2553–2562
27. Reichert TE, Strauss L, Wagner EM, Gooding W, Whiteside TL (2002) Signaling abnormalities and reduced proliferation of circulating and tumor-infiltrating lymphocytes in patients with oral carcinoma. Clin Cancer Res 8:3137–3145
28. Kim J-W, Tsukishiro T, Johnson JT, Whiteside TL (2004) Expression of pro- and anti-apoptotic proteins in circulating CD8 + T cells of patients with squamous cell carcinoma of the head and neck (SCCHN). Clin Cancer Res 10:5101–5110
29. Saito T, Kuss I, Dworacki G, Gooding W, Johnson JT, Whiteside TL (1999) Spontaneous ex vivo apoptosis of peripheral blood mononuclear cells in patients with head and neck cancer. Clin Cancer Res 5:1263–1273
30. Whiteside TL (2002) Tumor-induced death of immune cells: its mechanisms and consequences. Sem Cancer Biol 12:43–50
31. Saito T, Dworacki G, Gooding W, Lotze MT, Whiteside TL (2000) Spontaneous apoptosis of CD8 + T lymphocytes in peripheral blood of patients with advanced melanoma. Clin Cancer Res 6:1351–1364
32. Martinez-Lorenzo MJ, Anel A, Alava MA, Pineiro A, Naval J, Lasierra P, Larrad L (2004) The human melanoma cell line MelJuSo secretes bioactive FasL and APO2 L/TRAIL on the surface of microvesicles. Possible contribution to tumor counterattack. Exp Cell Res 295:315–329
33. Abrahams VM, Straszewski SL, Kamsteeg M, Hanczaruk B, Schwartz PE, Rutherford TJ, Mor G (2003) Epithelial ovarian cancer cells secrete functional Fas ligand. Cancer Res 63:5573–5581
34. Contini P, Ghio M, Poggi A, Filaci G, Indiveri F, Ferrone S, Puppo F (2003) Soluble HLA-A,-B,-C and -G molecules induce apoptosis in T and NK CD8 + cells and inhibit cytotoxic T cell activity through CD8 ligation. Eur J Immunol 33:125–134
35. Contini P, Ghio M, Merlo A, Poggi A, Indiveri F, Puppo F (2005) Apoptosis of antigen-specific T lymphocytes upon the engagement of CD8 by soluble HLA class I molecules is Fas ligand/Fas mediated: evidence for the involvement of p56lck, calcium calmodulin kinase II, and Calcium-independent protein kinase C signaling pathways and for NF-kappaB and NF-AT nuclear translocation. J Immunol 175:7244–7254

36. Whiteside TL (2004) Down-regulation of zeta-chain expression in T cells: a biomarker of prognosis in cancer? Cancer Immunol Immunother 53:865–878
37. Reichert TE, Day R, Wagner EM, Whiteside TL (1998) Absent or low expression of the z chain in T cells at the tumor site correlates with poor survival in patients with oral carcinoma. Cancer Res 58:5344–5347
38. Taylor DD, Akyol S, Gercel-Taylor C (2006) Pregnancy-associated exosomes and their modulation of T cell signaling. J Immunol 176:1534–1542
39. Sabapatha A, Gercel-Taylor C, Taylor DD (2006) Specific isolation of placenta-derived exosomes from the circulation of pregnant women and their immunoregulatory consequences. Am J Reprod Immunol 56:345–355
40. Clayton A, Mitchell JP, Court J, Mason MD, Tabi Z (2007) Human tumor-derived exosomes selectively impair lymphocyte responses to interleukin-2. Cancer Res 67:7458–7466
41. Valenti R, Huber V, Filipazzi P, Pilla L, Sovena G, Villa A, Corbelli A, Fais S, Parmiani G, Rivoltini L (2006) Human tumor-released microvesicles promote the differentiation of myeloid cells with transforming growth factor beta-mediated suppressive activity on T lymphocytes. Cancer Res 66:9290–9298
42. Czystowska M, Han J, Szczepanski MJ, Szajnik M, Quadrini K, Brandwein H, Hadden JW, Signorelli K, Whiteside TL (2009) IRX-2, a novel immunotherapeutic, protects human T cells from tumor-induced cell death. Cell Death Diff 16:708–718
43. Czystowska M, Szczepanski MJ, Szajnik M, Quadrini K, Brandwein H, Hadden JW, Whiteside TL (2011) Mechanisms of T-cell protection from death by IRX-2, a new immunotherapeutic. Cancer Immunol Immunother 60:495–506
44. Groneberg C, Pickartz T, Binder A, Ringel R, Srock S, Sieber T, Schoeler D, Schriever F (2003) Clinical relevance of CD95 (Fas/Apo-1) on T cells of patients with B-cell chronic lymphocytic leukemia. Exp Hematol 31:682–685
45. Massaia M, Borrione P, Attisano C, Barral P, Beggiato E, Montacchini L, Bianchi A, Boccadoro M, Pileri A (1995) Dysregulated Fas and Bcl-2 expression leading to enhanced apoptosis in T cells of multiple myeloma patients. Blood 85:3679–3687
46. Ghosh AK, Secreto CR, Knox TR, Ding W, Mukhopadhyay D, Kay NE (2010) Circulating microvesicles in B-cell chronic lymphocytic leukemia can stimulate marrow stromal cells: implications for disease progression. Blood 115:1755–1764
47. Szajnik M, Czystowska M, Szczepanski MJ, Mandapathil M, Whiteside TL (2010) Tumor-derived microvesicles induce, expand and up-regulate biological activities of human regulatory T cells (Treg). PLoS ONE 5:e11469
48. Bauernhofer T, Kuss I, Henderson B, Baum AS, Whiteside TL (2003) Preferential apoptosis of CD56dim natural killer cell subset in patients with cancer. Eur J Immunol 33:19–124
49. Szczepanski MJ, Szajnik M, Welsh A, Foon KA, Whiteside TL, Boyiadzis M (2010) Interleukin-15 enhances NK cell cytotoxicity in patients with acute myeloid leukemia by upregulating the activating NK cell receptors. Cancer Immunol Immunother 59:73–79
50. Liu C, Yu S, Zinn K, Wang J, Zhang L, Jia Y, Kappes JC, Barnes S, Kimberly RP, Grizzle WE, Zhang HG (2006) Murine mammary carcinoma exosomes promote tumor growth by suppression of NK cell function. J Immunol 176:1375–1385
51. Clayton A, Mitchell JP, Court J, Linnane S, Mason MD, Tabi Z (2008) Human tumor-derived exosomes down-modulate NKG2D expression. J Immunol 180:7249–7258
52. Clayton A, Tabi Z (2005) Exosomes and the MICA-NKG2D system in cancer. Blood Cells Mol Dis 34:206–213
53. Lee J-C, Lee K-M, Kim W-D, Heo DS (2004) Elevated TGF- β1 secretion and down-modulation of NK62D underlies impaired NK cytotoxicity in cancer patients. J Immunol 172:7335–7340
54. Szczepanski MJ, Szajnik M, Welsh A, Whiteside TL, Boyiadzis M (2011) Blast-derived microvesicles in sera of acute myeloid leukemia patients suppresses NK cell function via membrane-associated TGF-β1. Haematologica 96:1302–1309
55. Zhang H, Conrad DM, Butler JJ, Zhao C, Blay J, Hoskin DW (2004) Adenosine acts through A2 receptors to inhibit IL-2-induced tyrosine phosphorylation of STAT5 in T lymphocytes: role of cyclic adenosine 3′5′-monophosphate and phosphatases. J Immunol 173:932–944

56. Deaglio S, Dwyer KM, Gao W, Friedman D, Usheva A, Erat A, Chen JF, Enjyoji K, Linden J, Oukka M, Kuchroo VK, Strom TB, Robson SC (2007) Adenosine generation catalyzed by CD39 and CD73 expressed on regulatory T cells mediates immune suppression. J Exp Med 204:1257–1265
57. Borsellino G, Kleinewietfeld M, Di Mitri D, Sternjak A, Diamantini A, Giometto R, Hopner S, Centonze D, Bernardi G, Dell'Acqua ML, Rossini PM, Battistini L, Rotzschke O, Falk K (2007) Expression of ectonucleotidase CD39 by Foxp3 + Treg cells: hydrolysis of extracellular ATP and immune suppression. Blood 110:1225–1232
58. Mandapathil M, Szczepanski MJ, Szajnik M, Ren J, Jackson EK, Johnson JT, Gorelik E, Lang S, Whiteside TL (2010) Adenosine and prostaglandin E2 cooperate in the suppression of immune responses mediated by adaptive regulatory T cells. J Biol Chem 285:27571–27580
59. Zhang B (2010) CD73: A novel target for cancer immunotherapy. Cancer Res 70:6407–6411
60. Bergman C, Strauss L, Wieckowski E, Czystowska M, Albers A, Wang Y, Zeidler R, Lang S, Whiteside TL (2008) Tumor-derived microvesicles in sera of patients with head and neck cancer and their role in tumor progression. Head Neck 31:371–380
61. Battke C, Ruiss R, Welsch U, Wimberger P, Lang S, Jochum S, Zeidler R (2011) Tumour exosomes inhibit binding of tumour-reactive antibodies to tumour cells and reduce ADCC. Cancer Immunol Immunother 60:639–648
62. Iero M, Valenti R, Huber V, Filipazzi P, Parmiani G, Fais S, Rivoltini L (2008) Tumour-released exosomes and their implications in cancer immunity. Cell Death Diff 15:80–88
63. Ichim TE, Zhong Z, Kaushal S, Zheng X, Ren X, Hao X, Joyce JA, Hanley HH, Riordan NH, Koropatnick J, Bogin V, Minev BR, Min WP, Tullis RH (2008) Exosomes as a tumor immune escape mechanism: possible therapeutic implications. J Transl Med 6:37
64. Huber V, Filipazzi P, Iero M, Fais S, Rivoltini L (2008) More insights into the immunosuppressive potential of tumor exosomes. J Transl Med 6:63–66
65. Smyth MJ, Cretney E, Kerskaw MH, Hakayawa Y (2004) Cytokines in cancer immunity and immunotherapy. Immunol Rev 202:275–293
66. Alpizar YA, Chain B, Collins MK, Greenwood J, Katz D, Strauss HJ, Mitchison NA (2011) Ten years of progress in vaccination against cancer: the need to conunteract cancer evasion by dual targeting in future therapies. Cancer Immunol Immunother 60:1127–1135
67. Baumgaertner P, Jandus C, Rivals JP, Derre L, Lovgren T, Baitsch L, Guillaum P, Luescher IF, Berthod G, Matter M, Rufer N, Michielin O, Speiser DE (2011) Vaccination-induced functional competence of circulating human tumor-specific CD8 T-cells. Int J Cancer doi:10.1002/ijc.26297

Chapter 8
The Role of Tumor Exosomes in Tumorigenicity

Johan K. O. Skog

Abstract Tumor growth and metastasis is dependent on a complex interplay between the tumor cells and the tumor stroma as well as tumor-tumor interaction. This communication occurs through direct interaction, secreted factors as well as microvesicles carrying proteins and nucleic acids. The tumor microvesicles carry a wide range of cargo, including oncogenic proteins, fragmented DNA, mRNA and non-coding RNA with regulatory functions that can be horizontally transferred between different cells. In addition, these microvesicles are enriched in retro transposable elements that have been implicated with genetic instability and transcriptional dysregulation.

8.1 Tumor–Stroma Interactions

The tumor environment comprises tumor cells as well as stromal and vascular cells participating in the events necessary for tumor growth. There is a complex interaction between the cell types in the tumor shown in Fig. 8.1 to facilitate tumor growth and metastasis. The tumor cell needs to hijack the normal endothelial cells to form new blood vessels and the tumor also needs to escape the immune response, which otherwise would limit its growth capacity. Tumor cells and stromal cells communicate with each other through secreted factors [9] as well as microvesicles. Microvesicles can transfer molecules that are not secreted through the classical secretory pathway and the effects of these vesicles include inducing angiogenesis, tumor proliferation, and blunting the immune response [24, 43, 54].

Metalloproteinases (MMP) are produced by both tumor cells and stromal fibroblasts and they play a critical role in the metastatic spread of tumors. MMP2 and MMP9 are particularly important for tumor invasion and metastasis. The primary substrate for these enzymes is type IV collagen, a major component of the basement membrane and a barrier for tumor invasion. Tumor microvesicles are rich in MMPs, including MMP2 and MMP9. The proteolytic activity of tumor microvesicles shed from human cancer cells has been shown to correlate to their invasiveness [21]. In many tumors, MMPs are produced by the stromal fibroblasts rather than

J. K. O. Skog (✉)
Exosome Diagnostics Inc., The Lasker Biomedical Research Building,
3960 Broadway, Suite 540, New York, NY 10032, USA

H.-G. Zhang (ed.), *Emerging Concepts of Tumor Exosome–Mediated Cell–Cell Communication,* DOI 10.1007/978-1-4614-3697-3_8,

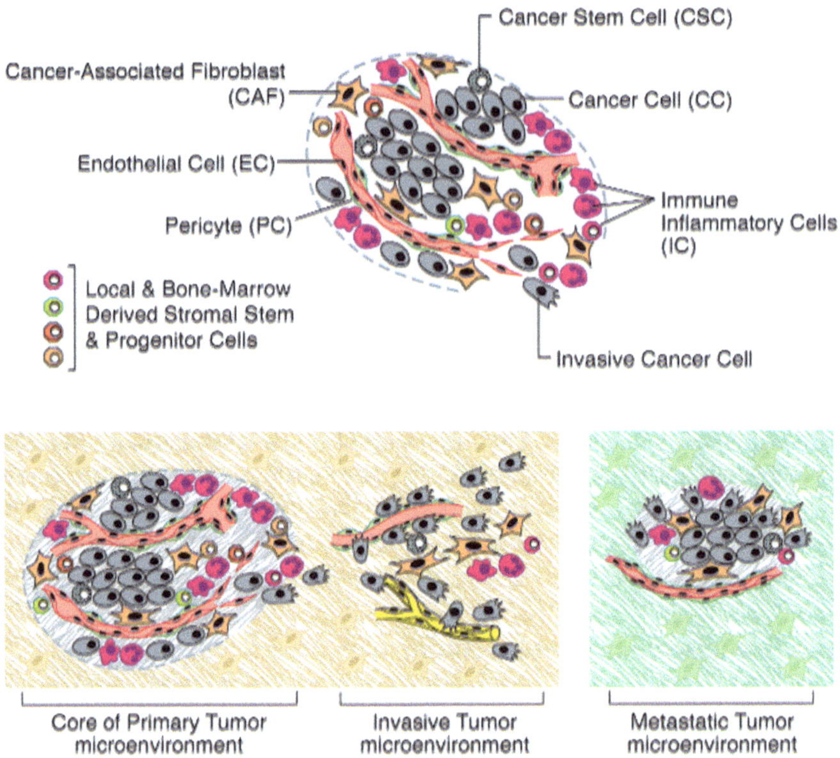

Fig. 8.1 The complex interplay between different cell types in the tumor. There are several cell types present in a tumor, all with important roles in the tumor progression. Intercellular signaling occurs with the release of secreted molecules like VEGF and bFGF, but also through release of microvesicles containing different proteins, lipids, and nucleic acids. Tumor cells release an abundance of microvesicles that can bombard the neighboring cells with proteins and nucleic acids over long periods of time. (Adapted from Hanahan D, Weinberg RA (2011) Hallmarks of cancer: the next generation. Cell 144(5):646–674)

the tumor cells and this expression is in part regulated by the extracellular matrix metalloproteinase inducer (EMMPRIN or CD147).

EMMPRIN is a type 1 transmembrane glycoprotein often enriched on the surface of tumor cells and can induce the protease production in fibroblasts [46]. It has long been suggested that microvesicles are involved in cell movement and tissue remodeling, in particular, during embryogenesis [45]. In normal cells, the release of microvesicles is under temporal, physiological, or hormonal control, whereas cancer cells seem to have a general activation of the pathways leading to shedding of the vesicles [45].Tumor microvesicles containing EMMPRIN can stimulate MMP production in fibroblasts [42] as well as endothelial cells and induce proangiogenic responses in endothelial cells, including increased invasiveness, increased MMP, and increased tubular formation in vitro [34].

Microvesicles are capable of inducing anchorage-independent growth and enhanced survival capability, characteristics of transformed cells, on to normal fibroblasts and epithelial cells by transfer of tissue transglutaminase and fibronectin that cooperatively activated the mitogenic signals in the recipient cells [5]. Tumor microvesicles can also stimulate growth of tumor cells as well as endothelial cells pointing to their role as autocrine and paracrine mediators of growth [43]. They can carry growth factors [44], but also receptors for growth factors such as EGFR or the mutated constitutively active EGFRv3 to stimulate EGFR-dependent responses such as the MAPK and Akt pathways and VEGF production in endothelial cells [2, 3].

8.1.1 Regulation of Tumor Microvesicle Release

It has been shown that phorbol myristate acetate (PMA) that induces the protein kinase C, calcium mobilization, and the mitogen-activated protein kinase (MAP kinase) pathway also stimulate the release of microvesicles [42]. This pathway is frequently upregulated in cancer, and is a common target for new chemotherapeutic drugs. The anticancer MEK1/2 inhibitors currently under development [47] could potentially also affect the microvesicle release in vivo. MEK1/2 activates the MAPK and it has been shown that MEK1/2 inhibitor scan reverse the upregulated shedding of vesicles [31, 42]. Many tumors have activation of the MAPK/ERK pathway, leading to uncontrolled growth, and may also explain the abundant microvesicle release from tumor cells. Other oncogenic events, such as expression of the constitutively active EGFRv3 [3], activating mutations in K-RAS, or loss of P53 also lead to an increased MV release [56]. Several members of the RAS superfamily seem to be involved in the microvesicle trafficking and release from cells, including the Rab and Arf subfamily previously shown to be involved in vesicle trafficking [35, 37, 52]. In cells with functional P53, it has been shown that P53 can upregulate the transmembrane protein TSAP6 and that this protein later promotes responses ranging from apoptosis, cell cycle regulation, to vesicular trafficking. Cells with wild-type P53 exposed to stress had an increased shedding of microvesicles even in the absence of apoptosis whereas cells with mutant P53 were not affected [4, 57].

8.2 Microvesicle-Mediated Horizontal Gene Transfer

Horizontal gene transfer is a phenomenon where nucleic acid is transferred from one cell (donor) to another cell (recipient). The nucleic acid can be DNA or RNA and can sometimes lead to expression/translation in the recipient cell. Horizontal gene transfer is a common mechanism in prokaryotes to rapidly adapt to a changing environment and is a driver for prokaryotic evolution, but it has previously been thought to be of limited or no importance in eukaryotic organisms. It was not clear how nucleic acids could shuttle between eukaryotic cells, especially for RNA that would be very susceptible for degradation if released unprotected in the extracellular milieu. It is now becoming evident that nucleic acids and other molecules that are not secreted directly into the extracellular space can still be transferred by these microvesicles and mediate cell–cell communication and gene transfer also in human cells [6, 7, 43, 48]. This finding added a new layer of complexity to intercellular communication, where different types of nucleic acids can play a role in tissue development and homeostasis as well as tumor cell manipulation of its environment. Although most of the tumor transcriptome can be found in the microvesicles, there is an enrichment of certain RNA species in the microvesicles, suggesting that there is a specific packaging mechanism of the genetic material, similar to viruses [43]. A large fraction of the transcriptome can be detected in a batch of purified microvesicles, including mRNA, miRNA and other non-coding RNA, but not all RNA transcripts will be in the same vesicle. One limitation of the packaging capacity of a microvesicle would be its size. Microvesicles are very heterogeneous both in size and content based on transmission electron microscopy, but on average the size of a microvesicle is similar to a retrovirus particle. A retrovirus particle can package two single-stranded RNA molecules of around 7–10 kb in length and a few selected host RNAs [13, 36], although it is unclear how pure the retrovirus preparations are from microvesiclepreparation. When looking at the RNA profile of a microvesicle preparation, the RNA levels are reflected as the average of these heterogeneous particles.

Microvesicles, including apoptotic bodies have been shown to transfer RNA and DNA between cells. Apoptotic bodies can transfer oncogene sequences horizontally and induce a tumorigenic phenotype in nontransformed cells. It was shown that the protein P53 had a role in protecting the recipient cell against incorporation of the foreign DNA since recipient cells deficient of P53 (a common trait in cancer) was more susceptible to incorporation of the horizontally transferred oncogenes [10]. Tumor microvesicles also carry amplified oncogenes such as c-myc [7]. C-myc is a versatile oncogene involved in several steps of the oncogenic process. It can regulate tumor growth, induce genomic instability and has also been shown to be involved in modulation of the tumor stroma, inflammatory response activation and induction of angiogenesis [53].

The effect of the microvesicle RNA transfer has mostly been studied in vitro and the effects in vivo are still very preliminary. We have not yet understood all mechanisms involved in microvesicle gene transfer including factors involved in cell tropism and determinants of the fate of the nucleic acids once they reach the recipient cell. The effect of the transfer to recipient cells will be the net effect of all the different lipids, proteins, sugar moieties, and nucleic acids present in and on the microvesicles (Fig. 8.2). The content of microvesicles depends on the type of vesicle and cell of origin. Tumor microvesicles have a different content than, for example, the dendritic cell microvesicles [54]. Transfer of vesicles will always involve a wide array of different lipids, proteins, sugar moieties, and nucleic acids. This can make the vesicle a more potent messenger when there is a need for simultaneous presentation of several effector molecules to have the desired effect (co-stimulatory molecules), but makes it more difficult to study the effects of individual proteins and nucleic acids carried by the vesicles. There are preliminary reports on specific microvesicle RNAs

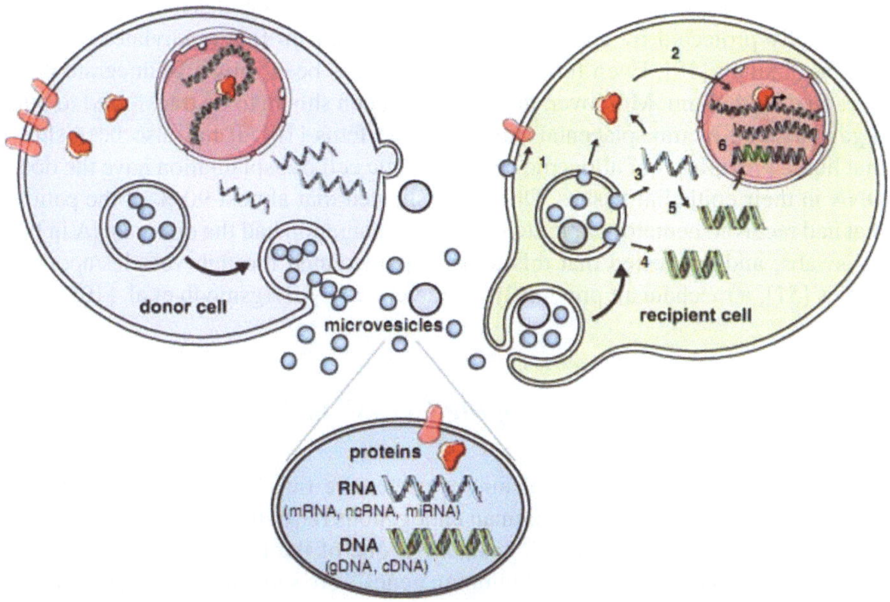

Fig. 8.2 Microvesicle-mediated communication. The donor cell (e.g., tumor cell) incorporates proteins (such as growth factors, growth factor receptors, MMPs), RNAs (including mRNA, retrotransposons, miRNA, and other ncRNA), and DNA. *1* Transmembrane proteins can be transferred to the plasma membrane and activate signaling. *2* Transcriptional regulators can potentially be transferred to the nucleus and regulate transcriptional activity. *3* RNA can be transferred to influence translational activity. *4* Donor-derived DNA, including oncogenes such as c-myc, can be delivered to the recipient cell. *5* Reverse transcriptase found in some microvesicles may generate single stranded DNA. *6* Aid genomic integration of transposable elements. (Adapted from Van der Vos K. et al. [49])

having effects in vivo, and a paper by Zhang et al. indicated that microvesicle miRNA in food sources can affect target genes in mammals [58], however, the significance of this finding needs to be further studied.

Microvesicles have also been shown to function as a "release hatch" for unwanted material from the cell, the typical example being release of the transferrin receptor during reticulocyte maturation. We still don't know how much the microvesicle-mediated transfer is "accidental transfer" where the primary reason of vesicle shedding is removal of certain proteins and nucleic acids rather than sending a message to surrounding cells. However, the net effect of the tumor microvesicle shedding seem to be promoting tumor growth through stimulation of angiogenesis, promoting tumor cell proliferation, blunting the immune system, and promoting metastasis.

8.2.1 Susceptibility of Horizontal Gene Transfer

It is becoming increasingly clear that the human genome is more dynamic than previously thought. The human genome is constantly under attack by foreign DNA; viral, transposons, food-ingested DNA, etc. When new DNA integrates in the genome, we are usually protected by an ancient defense mechanism that methylates the DNA making it silent [18]. Even food-ingested DNA has been shown to integrate in the intestinal epithelium. Moreover, this DNA has been shown to be transferred to other organs as well as transplacental transfer to the fetus [18]. It has also been shown that human recipients of allogenic hematopoietic cell transplantation have the donor DNA in their epithelial tissues. One study showed that almost 90 % of the patients that had received hematopoietic stem cell transplantation had the donor DNA in buccal swabs, and suggested that this transfer is mediated through vesicles/apoptotic bodies [51], a mechanism previously characterized by Bergsmedh et al. [10].

8.2.2 Transposons or "Jumping Genes" in Microvesicles

It was recently shown that tumor microvesicles are rich in transposable elements such as LINE-1 (L1), ALU, and human endogenous retroviruses (HERV) [7]. These elements are mediators of genetic instability, one of the hallmarks of cancer [25]. Retrotransposons, also known as "jumping genes," are sequences that can transpose themselves in the genome through an RNA intermediate. The retrotransposon RNA is reverse transcribed back into DNA and integrated into the genome by a reverse transcriptase and integrase, thus for each retrotransposition event it generates an additional copy of itself elsewhere in the genome. This is in contrast to the DNA transposons that excise themselves from the genome and move as a "cut and paste" mechanism. DNA transpositions can still leave scars in the genome since the target sequences are duplicated during integration and will leave traces in the form of short repeats. However, the DNA transposons are believed to be silenced in the human genome [38]. These genes have played an important role in driving evolution and increasing the plasticity of our genomes [15], changing our genetic makeup, and affecting gene regulation [39]. Around 45 % of our genome comprises retrotransposon sequences. These include about one million ALU sequences, half a million L1 elements, of which around 100 are transcriptionally active and encode the proteins necessary for retrotransposition, and thousands of HERV sequences [22]. Transposable elements need to invade the germline of a new host to persist in a population, and will persist until it, by something called vertical inactivation, becomes inactive and remains only as fossils. To ensure the survival of a transposon, it must be passed on to a new genome by horizontal gene transfer, something that has been described to happen even between different phyla [33]. Since transposable elements themselves are not infectious, it is not known how these elements are able to invade new genomes, but the use of parasites such as viruses has been proposed [27]. However, since these genetic elements are found in the microvesicles [7], microvesicle-mediated transfer could be an alternative mechanism.

Activation of transposable elements can have both good and bad effects on our genome. Syncytin, the gene that is critical for formation of the placental syncytium, is derived from the envelope protein gene of HERV-W [12]. Transposable elements are very prevalent in conserved nongenic sequences that are involved in gene expression regulation. In fact, genomic comparative studies of mice and rat as well as human and chimpanzee showed that these regions have contributed significantly to the evolution of gene expression between these species [39, 40].

8.2.3 Human Endogenous Retroviruses

HERVs entered the human host by infecting the germ-line cells. Most of these have since then acquired mutations and deletions, but some still retain coding capacity for functional proteins. Retroviruses have long been suspected to play a role in human cancers. They have been shown to be the causative agent of several types of cancers in mice and chicken. We still don't have the large picture of how these mobile elements are regulated and learned to coexist in our genomes. The HERVs are still very similar to the retroviruses from which they are derived. The human HERVs have acquired mutations rendering them noninfectious, although certain cancers that reactivate HERV transcription produce HERV particles with envelope proteins and viral RNA much like a normal retrovirus [11, 14]. Some retroviruses are transforming cells by capturing oncogenes during the course of virus replication or insertion mutagenesis, whereas other retroviruses in humans are less well understood [26].

Interestingly, microvesicles and retroviruses such as HIV-1 released from T cells share mechanisms of biogenesis and content so retroviruses and microvesicles and have been proposed to share a common origin [19, 20, 23, 29]. HERV RNA sequences seem to be specifically packaged into microvesicles and can be horizontally transferred from one cell to another [7, 43]. Some HERV sequences are upregulated in tumors and there seems to be a tissue-specific upregulation indicating that they are differentially regulated in tissues and may have different functions [1]. HERV-K has, for example, been implicated in the malignant transformation of melanomas [41], and HERV-H that is highly expressed in fetal brain [1] was interestingly the most enriched HERV in glioblastoma microvesicles [7].

8.2.4 Long Interspersed Nuclear Elements 1 (LINE1)

LINE1 are the most active autonomous retrotransposons in mammalian genomes. They possess reverse transcriptase activity that can retrotranspose not only themselves but also other mRNA (generating pseudogenes) and ALU sequences. Cancer cells have a variable but typically a higher level of endogenous LINE1 mRNA expression [8]. Tumor cells are often hypomethylated in the LINE1 genomic regions, unlocking their transcriptional activity. LINE1 can often affect transcription levels of other genes, and insertional mutagenesis has been shown in cancers such as colon and breast cancers [8]. LINE1 elements are packaged into microvesicles and in some tumor types there seem to be a specific enrichment of these elements [7]. The LINE1 endonuclease have been implicated in generation of gene fusions in prostate cancer, such as the TMPRSS2 fused to ERG or ETV1 [32].

8.2.5 *Short Interspersed Elements (SINEs)*

ALUs are the most abundant group SINEs. They are typically 90–300 bp in length and are nonautonomous in the sense that they depend on LINE1 for their transposition. There is on average one ALU insertion every 3 kb in the genome [15, 30]. Most of them are derived from various tRNA genes or the 7SL RNA gene. ALU elements have been associated with gene dysregulation as well as insertional mutagenesis in cancers such as retinoblastoma, breast cancer, and neurofibromatosis [8]. These elements were recently shown to be very abundant in the microvesicles and are interestingly coinciding in size with an abundant RNA population in microvesicles [7].

8.2.6 *Regulation of Transposable Elements*

The retrotransposon activity in the cell needs to be tightly controlled in normal tissue to avoid deleterious effects on the genome. In most cells, they are regulated by mechanisms such as methylation on the genomic level, expression of piwi-interacting RNAs (piRNAs) [28] or expression of the cytidine deaminase family APOBEC. However, the retrotransposon activity seems to have a crucial role during certain stages of development. Embryonic cells as well as cancer cells have an increased expression of these elements as compared to differentiated cells. Induced pluripotent stem cells (iPSCs) express higher levels of LINE1 sequences, and retrotransposition of a human LINE1 element was tenfold more efficient in the iPSCs than in the parental human dermal fibroblasts [55]. Hypomethylation of Line1 and ALU sequences in cancers have been shown to correlate to genomic instability [17]. This upregulated activity of transposable elements may impact the genomic integrity and plasticity of these progenitor cells and LINE-1 transposition has been shown to be involved in regulation of the developing brain [16, 50]. We are just starting to understand the role of these transposable elements and other repetitive sequences in our genomes, and the potential of horizontal gene transfer through microvesicles is adding to the complexity of gene regulation in multicellular organisms.

References

1. Ahn K, Kim HS (2009) Structural and quantitative expression analyses of HERV gene family in human tissues. Mol Cells 28(2):99–103
2. Al-Nedawi K, Meehan B et al (2009) Endothelial expression of autocrine VEGF upon the uptake of tumor-derived microvesicles containing oncogenic EGFR. Proc Natl Acad Sci U S A 106(10):3794–3799

3. Al-Nedawi K, Meehan B et al (2008) Intercellular transfer of the oncogenic receptor EGFRvIII by microvesicles derived from tumour cells. Nat Cell Biol 10(5):619–624
4. Amzallag N, Passer BJ et al (2004) TSAP6 facilitates the secretion of translationally controlled tumor protein/histamine-releasing factor via a nonclassical pathway. J Biol Chem 279(44):46104–46112
5. Antonyak MA, Li B et al (2011) Cancer cell-derived microvesicles induce transformation by transferring tissue transglutaminase and fibronectin to recipient cells. Proc Natl Acad Sci U S A 108(12):4852–4857
6. Baj-Krzyworzeka M, Szatanek R et al (2006) Tumour-derived microvesicles carry several surface determinants and mRNA of tumour cells and transfer some of these determinants to monocytes. Cancer Immunol Immunother 55(7):808–818
7. Balaj L, Lessard R et al (2010) Tumour microvesicles contain retrotransposon elements and amplified oncogene sequences. Nat Commun 2:180
8. Belancio VP, Roy-Engel AM et al (2010) All y'all need to know' bout retroelements in cancer. Semin Cancer Biol 20(4):200–210
9. Bergers G, Benjamin LE (2003) Tumorigenesis and the angiogenic switch. Nat Rev Cancer 3(6):401–410
10. Bergsmedh A, Szeles A et al (2001) Horizontal transfer of oncogenes by uptake of apoptotic bodies. Proc Natl Acad Sci U S A 98(11):6407–6411
11. Bieda K, Hoffmann A et al (2001) Phenotypic heterogeneity of human endogenous retrovirus particles produced by teratocarcinoma cell lines. J Gen Virol 82(Pt 3):591–596
12. Blond JL, Beseme F et al (1999) Molecular characterization and placental expression of HERV-W, a new human endogenous retrovirus family. J Virol 73(2):1175–1185
13. Coffin JM (1992) Structure and classification of retroviruses. Plenum, New York
14. Contreras-Galindo R, Kaplan MH et al (2008). Human endogenous retrovirus K (HML-2) elements in the plasma of people with lymphoma and breast cancer. J Virol 82(19):9329–9336
15. Cordaux R, Batzer MA (2009) The impact of retrotransposons on human genome evolution. Nat Rev Genet 10(10):691–703
16. Coufal NG, Garcia-Perez JL et al (2009) L1 retrotransposition in human neural progenitor cells. Nature 460(7259):1127–1131
17. Daskalos A, Nikolaidis G et al (2009) Hypomethylation of retrotransposable elements correlates with genomic instability in non-small cell lung cancer. Int J Cancer 124(1):81–87
18. Doerfler W, Hohlweg U et al (2001) Foreign DNA integration–perturbations of the genome–oncogenesis. Ann N Y Acad Sci 945:276–288
19. Fang Y, Wu N et al (2007) Higher-order oligomerization targets plasma membrane proteins and HIV gag to exosomes. PLoS Biol 5(6):e158
20. Gan X, Gould SJ (2011) Identification of an inhibitory budding signal that blocks the release of HIV particles and exosome/microvesicle proteins. Mol Biol Cell 22(6):817–830
21. Ginestra A, La Placa MD et al (1998) The amount and proteolytic content of vesicles shed by human cancer cell lines correlates with their in vitro invasiveness. Anticancer Res 18(5A):3433–3437
22. Goodier JL, Kazazian HH Jr (2008) Retrotransposons revisited: the restraint and rehabilitation of parasites. Cell 135(1):23–35
23. Gould SJ, Booth AM et al (2003) The Trojan exosome hypothesis. Proc Natl Acad Sci U S A 100(19):10592–10597
24. Grange C, Tapparo M et al (2011) Microvesicles released from human renal cancer stem cells stimulate angiogenesis and formation of lung premetastatic niche. Cancer Res 71(15):5346–5356
25. Hanahan D, Weinberg RA (2011) Hallmarks of cancer: the next generation. Cell 144(5):646–674
26. Kannian P, Green PL (2010) Human T lymphotropic virus type 1 (HTLV-1): molecular biology and oncogenesis. Viruses 2(9):2037–2077
27. Kidwell MG (1992) Horizontal transfer. Curr Opin Genet Dev 2(6):868–873

28. Klattenhoff C, Theurkauf W (2008) Biogenesis and germline functions of piRNAs. Development 135(1):3–9
29. Krishnamoorthy L, Bess JW Jr et al (2009) HIV-1 and microvesicles from T cells share a common glycome, arguing for a common origin. Nat Chem Biol 5(4):244–250
30. Lander ES, Linton LM et al (2001) Initial sequencing and analysis of the human genome. Nature 409(6822):860–921
31. Li M, Yu D et al (2010) Tobacco smoke induces the generation of procoagulant microvesicles from human monocytes/macrophages. Arterioscler Thromb Vasc Biol 30(9):1818–1824
32. Lin C, Yang L et al (2009) Nuclear receptor-induced chromosomal proximity and DNA breaks underlie specific translocations in cancer. Cell 139(6):1069–1083
33. Lohe AR, Moriyama EN et al (1995) Horizontal transmission, vertical inactivation, and stochastic loss of mariner-like transposable elements. Mol Biol Evol 12(1):62–72
34. Millimaggi D, Mari M et al (2007) Tumor vesicle-associated CD147 modulates the angiogenic capability of endothelial cells. Neoplasia 9(4):349–357
35. Muralidharan-Chari V, Clancy J et al (2009) ARF6-regulated shedding of tumor cell-derived plasma membrane microvesicles. Curr Biol 19(22):1875–1885
36. Onafuwa-Nuga AA, King SR et al (2005) Nonrandom packaging of host RNAs in moloney murine leukemia virus. J Virol 79(21):13528–13537
37. Ostrowski M, Carmo NB et al (2010) Rab27a and Rab27b control different steps of the exosome secretion pathway. Nat Cell Biol 12(1):19–30 sup pp 1–13
38. Pace JK 2nd, Feschotte C (2007) The evolutionary history of human DNA transposons: evidence for intense activity in the primate lineage. Genome Res 17(4):422–432
39. Pereira V, Enard D et al (2009) The effect of transposable element insertions on gene expression evolution in rodents. PLoS One 4(2):e4321
40. Polavarapu N, Arora G et al (2011) Characterization and potential functional significance of human-chimpanzee large INDEL variation. Mob DNA 2:13
41. Serafino A, Balestrieri E et al (2009) The activation of human endogenous retrovirus K (HERV-K) is implicated in melanoma cell malignant transformation. Exp Cell Res 315(5):849–862
42. Sidhu SS, Mengistab AT et al (2004) The microvesicle as a vehicle for EMMPRIN in tumor-stromal interactions. Oncogene 23(4):956–963
43. Skog J, Wurdinger T et al (2008) Glioblastoma microvesicles transport RNA and proteins that promote tumour growth and provide diagnostic biomarkers. Nat Cell Biol 10(12):1470–1476
44. Taraboletti G, D'Ascenzo S et al (2006) Bioavailability of VEGF in tumor-shed vesicles depends on vesicle burst induced by acidic pH. Neoplasia 8(2):96–103
45. Taylor DD, Black PH (1986) Shedding of plasma membrane fragments. Neoplastic and developmental importance. Dev Biol (N Y 1985) 3:33–57
46. Taylor PM, Woodfield RJ et al (2002) Breast cancer cell-derived EMMPRIN stimulates fibroblast MMP2 release through a phospholipase A(2) and 5-lipoxygenase catalyzed pathway. Oncogene 21(37):5765–5772
47. Trujillo JI (2011) MEK inhibitors: a patent review 2008–2010. Expert Opin Ther Pat 21(7):1045–1069
48. Valadi H, Ekstrom K et al (2007) Exosome-mediated transfer of mRNAs and microRNAs is a novel mechanism of genetic exchange between cells. Nat Cell Biol 9(6):654–659
49. van der Vos KE, Balaj L, Skog J, Breakefield XO (2011) Brain tumor microvesicles: insights into intercellular communication in the nervous system. Mol Cell Neurobiol 31(6):949–959 (Epub 2011 May 8)
50. Vogel G (2011) Retrotransposons. Do jumping genes spawn diversity? Science 332(6027):300–301
51. Waterhouse M, Themeli M et al (2011) Horizontal DNA transfer from donor to host cells as an alternative mechanism of epithelial chimerism after allogeneic hematopoietic cell transplantation. Biol Blood Marrow Transplant 17(3):319–329
52. Wennerberg K, Rossman KL et al (2005) The Ras superfamily at a glance. J Cell Sci 118 (Pt 5):843–846

53. Whitfield JR, Soucek L (2011) Tumor microenvironment: becoming sick of Myc. Cell Mol Life Sci 69(6):931–934
54. Wieckowski E, Whiteside TL (2006) Human tumor-derived vs dendritic cell-derived exosomes have distinct biologic roles and molecular profiles. Immunol Res 36(1–3):247–254
55. Wissing S, Munoz-Lopez M et al (2011) Reprogramming somatic cells into iPS cells activates LINE-1 retroelement mobility. Hum Mol Genet 21:208–218
56. Yu JL, May L et al (2005) Oncogenic events regulate tissue factor expression in colorectal cancer cells: implications for tumor progression and angiogenesis. Blood 105(4):1734–1741
57. Yu X, Harris SL et al (2006) The regulation of exosome secretion: a novel function of the p53 protein. Cancer Res 66(9):4795–4801
58. Zhang L, Hou D et al (2011) Exogenous plant MIR168a specifically targets mammalian LDLRAP1: evidence of cross-kingdom regulation by microRNA. Cell Res 22:107–126

Chapter 9
Microenvironmental Regulation of Metastasis by Exosomes

Haiying Zhang, Guillermo Garcia-Santos, Héctor Peinado and David C. Lyden

Abstract Exosomes and microvesicles are small spherical membrane fragments shed by several cell types during physiological processes. Increasing attention has been focused on the role of exosomes in malignant progression and metastasis. This chapter highlights the functional roles of tumor exosomes in influencing pre-metastatic niche formation and metastatic behavior. We discuss their interaction with cells within the tumor microenvironment (bone marrow-, endothelial-, and mesenchymal-derived cells) and attempt to shed light on the mechanisms underlying these interactions. We address the emerging role of tumor-secreted exosomes as new players that change the tumor microenvironment during tumor progression. Furthermore, we comment on the contribution of exosomes secreted by non-tumor cells such as endothelial, bone marrow, and mesenchymal cells to metastatic progression. Data are presented to highlight the role of exosomes in influencing the metastatic behavior of tumors. We conclude that exosome research has the potential to uncover new targets and lead to new therapies against tumor and metastatic progression.

Haiying Zhang and Guillermo Garcia-Santos contributed equally to this chapter.

H. Peinado (✉)
Department of Pediatrics, Cell and Developmental Biology, Weill Cornell Medical College, 515 E 71st Street, S-725, Box 284, New York, NY 10021, USA
e-mail: hps2002@med.cornell.edu

D. C. Lyden (✉)
Department of Pediatrics, Cell and Developmental Biology, Weill Cornell Medical College, 515 E 71st Street, S-726, Box 284, New York, NY 10021, USA
e-mail: dcl2001@med.cornell.edu

H. Zhang · G. Garcia-Santos
Department of Pediatrics, Cell and Developmental Biology, Weill Cornell Medical College, 515 E 71st Street, New York, NY 10021, USA

H. Peinado · D. C. Lyden · H. Zhang · G. Garcia-Santos
Champalimaud Metastasis Programme, Lisbon, Portugal

H.-G. Zhang (ed.), *Emerging Concepts of Tumor Exosome–Mediated Cell–Cell Communication,* DOI 10.1007/978-1-4614-3697-3_9,

9.1 Tumor Exosomes in Tumor Malignancy and Metastasis

Intercellular communication mediated by exosomes is an emerging concept and an increasing body of evidence has implicated the potential roles of exosomes not only in normal physiological processes, such as the immune response and communication within the nervous system [1–4], but also in pathogenic events, and cancer progression and metastasis [5–8]. Exosomes are vesicles of endocytic origin secreted by most cell types, and recent studies have revealed selective recruitment of proteins such as membrane-bound tyrosine kinase receptors and angiogenic factors into these vesicles [9–11]. Furthermore, genetic material such as mRNAs and miRNAs is also packaged into exosomes in a cell type-specific pattern [12, 13]. Importantly, exosomes can mediate the horizontal transfer of these functional molecules into different cell types, resulting in epigenetic reprogramming in these recipient cells [14–16].

Many types of tumor cells secrete large amounts of exosomes, which can be detected in the serum and ascitic fluid of cancer patients [17–20]. The level of circulating tumor exosomes in the serum has been associated with cancer. For example, patients diagnosed with ovarian cancer exhibited significantly elevated levels of Ep-CAM-positive tumor-derived exosomes (as assayed for exosomal proteins and compared with benign disease or controls) in the peripheral circulation, with the highest levels observed in those with the most advanced stages of disease [19]. Similarly, it has been demonstrated that the number of circulating microvesicles (MVs) of nonplatelet origin ($CD61^-$) is higher in gastric cancer patients [18]. Recent animal studies have demonstrated the role of exosomes in cancer progression and metastasis. Due to the intrinsic genetic instability of tumor cells, heterogeneity among tumor cells with distinct metastatic potential has been well recognized [21]. However, such metastatic capacity could be propagated among tumor cells via exosomes. For example, uptake of exosomes and MVs released from highly metastatic melanoma B16-F10 cells enables the less metastatic melanoma B16-F1 cells to colonize and thrive in lung [22, 23]. However, the molecular mechanism underlying these differences has not been described. Tumor exosomes have also been associated to tumor growth and metastases in murine mammary carcinoma and melanoma models [24, 25].

In the rat pancreatic adenocarcinoma model, exosomes in combination with the soluble matrix assembled by CD44v wild-type (WT) tumor cells can influence the metastatic organ site and enhance tumor cell seeding and growth in lung tissue and lymph nodes [26]. Finally, melanoma exosomes have been found to home in sentinel lymph nodes after intrafootpad injection and induce expression of metastatic factors resulting in altered melanoma cell distribution [27]. There is increasing interest in investigating the molecular mechanisms underlying these phenomena and in uncovering the contributions of tumor exosomes to various aspects of tumor metastasis, such as angiogenesis/vasculogenesis, tumor invasion, and pre-metastatic niche formation.

9.1.1 Contribution of Tumor Exosomes to the Establishment of Metastatic Niches

In the past decade, several publications have dramatically changed our understanding of metastasis biology. These studies showed that both cellular and molecular changes occur in metastatic organs before initial tumor cell dissemination from the primary tumor [28–30]. Factors derived from primary tumors can prepare distant sites for future metastasis by generating a suitable microenvironment called the "pre-metastatic niche" before the arrival of the cancer cells [28–32]. Soluble factors such as the angiogenic cytokines (vascular endothelial growth factor A [VEGF-A] and placental growth factor [PlGF] [30]), inflammatory chemokines (transforming growth factor β [TGF-β] and tumor necrosis factor α [TNF-α]) [28], and the hypoxia -regulated extracellular protein lysyl oxidase (LOX) [31, 32] are several molecules that have been well characterized at pre-metastatic sites.

One of the main cell types attracted to pre-metastatic sites during tumor progression are bone marrow (BM)-derived hematopoietic progenitor cells (HPCs). HPCs are pluripotent cells with the capacity to differentiate into specific hematopoietic cell types belonging to the erythroid, megakaryocytic, lymphoid, and myeloid lineages [33]. In a seminal study, Kaplan et al. [30] demonstrated that vascular endothelial growth factor receptor 1 positive ($VEGFR1^+$) HPCs migrate to pre-metastatic sites before the arrival of the first tumor cell. These cells express the fibronectin receptor VLA-4 (integrin α4β1), which allows them to anchor onto fibronectin, an extracellular matrix component that is overexpressed at the future metastatic site. The arrival of HPCs at the future metastatic site triggers interactions with other cell types such as stromal cells and endothelial cells leading to the formation of a microenvironment which favors the engraftment of tumor cells [34]. Newly disseminated tumor cells trigger angiogenic signals that recruit $VEGFR2^+$ BM-derived endothelial progenitor cells (EPCs) to the metastatic site, allowing the maturation of the metastatic lesion [30, 35, 36].

In addition to growth factors, cytokines and soluble factors, tumor-derived exosomes and MVs may also play a critical role during pre-metastatic niche formation (Fig. 9.1). Tumor exosomes may not only impact on the constituents of the local microenvironment, but also travel long distances through the peripheral circulation and ascitic fluids to target distal organs. When protected within the membrane vesicles, a large variety of molecules can be delivered by exosomes to the metastatic organs and cause more profound responses at these sites. For example, it has been shown that $CD105^+$ human renal carcinoma cell-derived MVs enhanced lung metastasis compared with control vesicles 5 weeks after intravenous injection of renal tumor cells [5]. When the lung microenvironment was evaluated, the authors observed elevated VEGF, VEGFR1, and matrix metalloproteinase 2 (MMP-2) expression in lung endothelial cells and increased MMP-9 in total lung tissue after the intravenous administration of MVs. Interestingly, these factors have been previously implicated in the formation of the lung pre-metastatic niche [37]. Similarly, Hood et al. [27] showed that when melanoma exosomes were administered to mice by intrafootpad

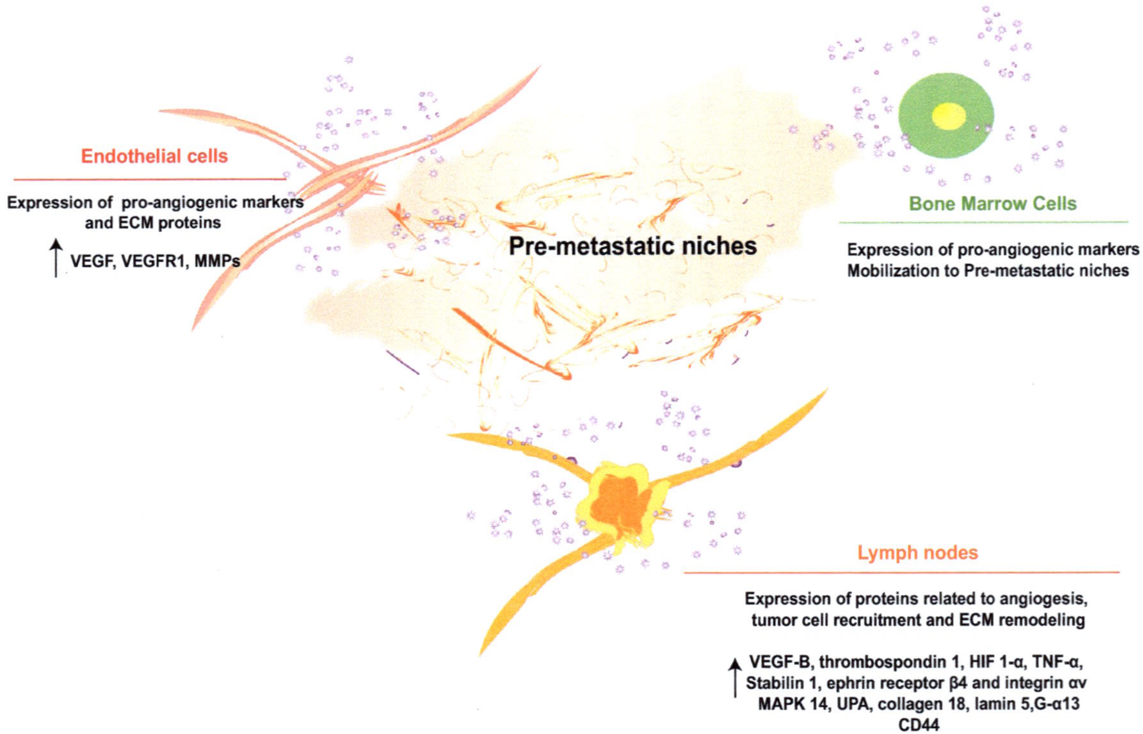

Fig. 9.1 Contribution of tumor exosomes to the establishment of pre-metastatic niches. Tumor-secreted exosomes have been found to influence the behavior of different cell types such as endothelial cells [5] and bone marrow cells [43] in pre-metastatic niches (i.e. lungs) a well as prepare a pre-metastatic microenvironments in sentinel lymph nodes [27]. The upregulation of pro-angiogenic molecules (i.e., VEGF, VEGF-B, and thrombospondin) and ECM remodeling proteins (i.e., MMPs, collagen 18, UPA, and lamin 5) in these areas by tumor exosomes promote the generation of suitable microenvironments for the metastasis to occur. (pre-metastatic niches)

injection, they specifically homed to sentinel lymph nodes and influenced the lymph node distribution pattern of free melanoma cells by recruiting more melanoma cells to the exosome-enriched sites in sentinel lymph nodes. When the lymph nodes preconditioned by melanoma exosomes were evaluated for changes in metastatic factors, the authors observed elevated expression of several genes implicated in angiogenesis (VEGF-B, thrombospondin 1, hypoxia inducible factor 1α [HIF 1-α] and TNF-α), tumor cell recruitment (Stabilin 1, ephrin receptor β4, and integrin α_v), and extracellular matrix modification (MAPK 14, urokinase plasminogen activator protease, collagen 18, lamin 5, and G protein subunit G-α13). It will be of further interest to identify the cell types targeted by these exosomes in the lymph nodes responsible of the upregulation of these pro-metastatic factors.

These studies support a role for tumor-secreted exosomes in preparing metastatic tissues such as lymph nodes and lungs for metastasis. Tumor exosomes and soluble factors act in conjunction to promote the expression of proangiogenic molecules and extracellular matrix (ECM) remodeling in pre-metastatic tissues. These changes favor the formation of a pre-metastatic microenvironment that would help tumor cells to metastasize (Fig. 9.1).

It is worthy to point out that the soluble fraction and exosomes secreted by tumor cells work in a cooperative manner to establish metastatic niches. One illustration in support of this process is the study carried out by Jung et al. [26]. In this study, they used the rat pancreatic adenocarcinoma as the model system and showed that intrafootpad injections of conditioned media (CM) collected from CD44v WT-, but not CD44v knock down (KD)- tumor cells promoted lymph node and lung metastasis. When the impact exerted by the soluble fraction versus exosomes was evaluated, the authors found that although exosomes were the driving force of metastasis, they still required the assistance of the soluble fraction since the CD44v6-WT tumor cell-secreted soluble fraction could complement the CD44v6 KD tumor cell-derived exosomes, but not vice versa.

Recently, using vesicles derived from lung tissue, Aliota et al. [40] demonstrated the ability of MVs to transfer specific characteristics from lung cancer cells to BM cells and reprogram BMDC behavior. Specifically, MV transfer induced the expression of lung epithelial cell mRNAs in BM cells and increased their ability to engraft in the lung. Interestingly, the expression of lung-specific mRNAs in the BM cells persists for as long as 3 weeks after the exposure to MVs, suggesting a semipermanent phenotypic change. Similar effects have also been observed in the case of human prostate cancer cells [41]. In this study, the incubation of BM cells with MVs derived from primary human prostate cancer cultures induced the expression of prostate-specific antigen mRNA in human BM cells.

As mentioned, BM cells are also involved in pre-metastatic niche formation. Exosomes and MVs also have long-distance effects on BM cells and their mobilization. Some studies have shown that tumor-derived microvesicles can directly influence BM stromal cells [38]. MVs derived from the plasma of patients with B cell chronic lymphoid leukemia (CLL) can induce the activation of the AKT/p70S6K/HIF-1α-signaling axis and increase the production of VEGF-A by BM stromal cells [39].

Despite the limited data available regarding the effect of tumor-derived exosomes/MVs on BMDCs, it is not unreasonable to assume that tumor-derived particles can also drive BM cells to a acquire pro-metastatic phenotype, promoting their migration to the pre-metastatic site. In fact, our laboratory has recently shown that tumor-derived exosomes can reprogram BM progenitor cells [42]. In a series of *in vivo* and *in vitro* experiments, we have found that exosomes from highly metastatic melanoma cells increased the metastatic behavior of primary tumors by "educating" BM progenitor cells, while exosomes from low metastatic melanoma cells did not alter the incidence of metastases. Melanoma-derived exosomes induced vascular leakiness at pre-metastatic sites and promoted BM progenitor mobilization and cell "education" toward a provasculogenic phenotype in BM through MET oncogene. These studies define a novel pathway in metastatic progression involving the cross-talk between tumor-derived exosomes and BM progenitor cells leading to BM "education," mobilization, and participation in metastatic progression (Figs. 9.1 and 9.2; [42]).

The contribution of tumor exosomes to pre-metastatic niche formation has just begun to be explored, and many questions remain to be addressed. For example, the access of tumor exosomes to different organs and tissues, the targeting mechanism that mediates the interaction between tumor exosomes and recipient cells, the identity of target cells and their response to exosome uptake, and last but not least, the molecular mechanisms underlying these responses are all areas of great interest for future studies.

9.1.2 Contribution of Tumor Exosomes to Metastatic Behavior

Metastasis is a multistage process in which malignant cells disseminate from the primary tumor, survive and travel in the circulation, and eventually colonize in distant organs [43]. Each of these steps is fulfilled by cooperative interaction between cancer cells and non-malignant cells in the tumor microenvironment [44]. Invasion is a hallmark of metastatic cells and is required for tumor cells to escape from the primary tumor, engraft, and outgrow at metastatic organs [45]. Over the past decade, significant progress has been made towards our understanding of the mechanisms underlying tumor invasion. The epithelial-mesenchymal transition (EMT) program and ECM remodeling have been implicated in the regulation of multiple aspects of tumor invasion and metastasis [45–47]. Secreted proteins can also play a role at metastasis; for example, breast cancer cells support their own metastasis-initiating ability by expressing tenascin C (TNC) at metastatic sites, a protein associated with the aggressiveness of pulmonary metastasis [48]. Angiogenesis is another fundamental property of tumor cells that accounts for their metastatic behavior. Angiogenesis is crucial for providing nutrient and oxygen supply and evacuating metabolic waste and carbon dioxide from the growing tumor. Beyond that, it promotes tumor metastasis by providing cancer cells access to circulation and for the persistent growth of tumor cells at the metastatic sites. Recent studies have revealed that tumors

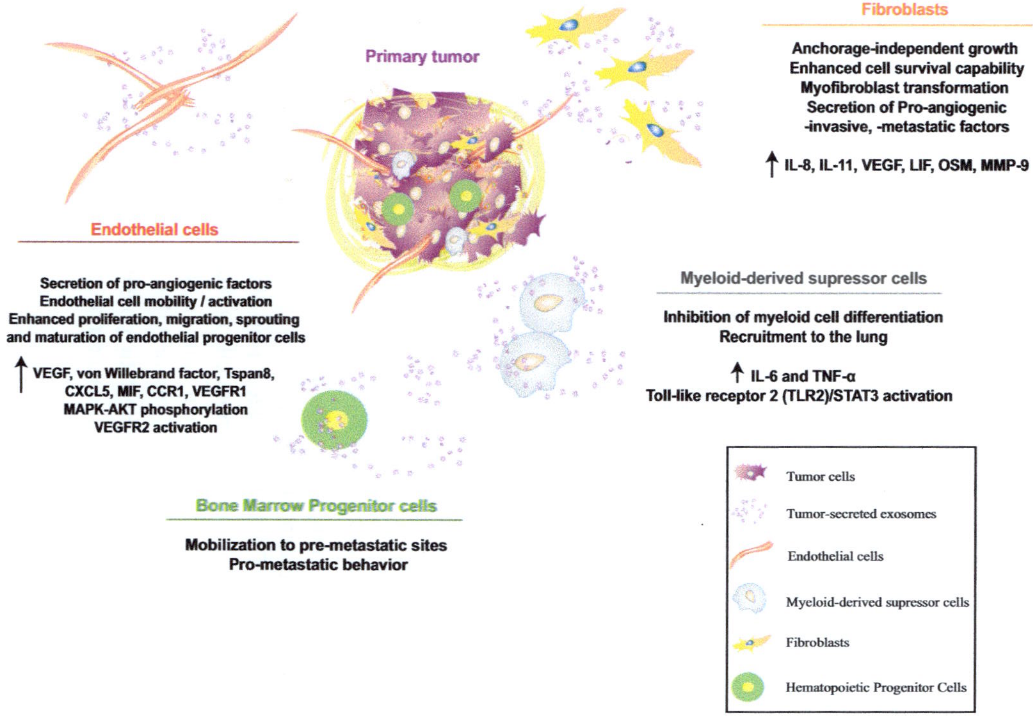

Fig. 9.2 Role of tumor-derived exosomes in metastasis. As the primary tumor is developing, secreted exosomes and microvesicles influence specific populations of fibroblast [52–56], endothelial [5, 10, 11, 63], and bone marrow (BM) cells [42, 67–70, 25] collaborating to support metastasis. Tumor exosomes and microvesicles promote the secretion of proangiogenic factors in fibroblast cells (i.e., VEGF and LIF) and endothelial cells (i.e., VEGF, von Willebrand factor, Tspan8, CXCL5, CCR1, VEGFR1, and MIF) triggering their pro-metastatic behavior. Tumor-secreted vesicles also promote the activation of endothelial cells, bolstering their survival, proliferation, and activation. Similarly, tumor vesicles exosomes enhance fibroblast survival and promote their myofibroblast differentiation as well as anchorage-independent growth. Exosomes secreted by the primary tumor could also influence BMDCs such as BM progenitor cells and myeloid-derived suppressor cells. Tumor exosomes have been found to inhibit myeloid cell differentiation by increasing IL-6 and TNF-α expression in these populations as well as promote pro-metastatic behavior in BM progenitor cells recruiting them to pre-metastatic sites therefore increasing the formation of metastatic foci

can achieve angiogenesis/vasculogenesis by adopting multiple strategies, including activation of normal quiescent endothelial cells, as well as recruitment and differentiation of BM-derived and tissue-resident progenitor cells [5, 49, 50].

In the following section, we will explore existing data supporting the hypothesis that tumor-derived exosomes contribute to metastasis by mediating the reciprocal-signaling cross-talk between tumor cells and their associated microenvironmental constituents, such as mesenchymal/fibroblast cells, endothelial cells, and BMDCs (Fig. 9.2). Recent studies in the field which highlight the cross-talk between tumor cells and individual types of nonmalignant cells in the microenvironment via exosomes will be discussed.

9.1.2.1 Role of Tumor Exosomes in Regulating Fibroblast Behavior During Metastasis

Fibroblasts are the major constituent of the tumor stroma and it has been suggested that activated fibroblasts play a central role in promoting the migration and invasion of tumor cells [44, 51]. Little is known about the role of tumor exosomes in mediating the reciprocal signaling between tumor cells and associated fibroblasts (Fig. 9.2). Tumor MVs can confer transformed characteristics to normal fibroblast and endothelial cells, such as anchorage-independent growth and enhanced survival capability, or trigger fibroblast to myofibroblast transformation (Fig. 9.2) [52, 53].

The role of tumor exosomes in mediating cross-talk between cancer cells and stromal fibroblasts to enhance tumor invasion and metastasis has been directly illustrated by the study performed by Wysoczynski et al. [54]. In this study, the authors showed that lung cancer cell-derived MVs can stimulate stromal fibroblasts to express proangiogenic factors such as interleukins 8 and 11 (IL-8/IL-11) , VEGF, leukemia inhibitory factor (LIF), oncostatin (OSM), and MMP-9. Fibroblasts stimulated by tumor MVs secrete factors that, in turn, lead to the phosphorylation, and thus activation of the signal transducer and activator of transcription 3 (STAT3), AKT, and MAPKp42/44 in lung cancer cells. The CM harvested from fibroblasts stimulated by tumor MVs enhanced the mobility of treated tumor cells and the adhesion of tumor cells to endothelial cells *in vitro*. Importantly, when tumor cells pretreated with CM harvested from the fibroblasts and stimulated by tumor MVs were administered to mice intravenously, they demonstrated higher metastatic potential *in vivo*. However, the identity of the effector molecules in this reciprocal communication have not been uncovered and need to be tackled in future studies.

Recent studies revealed that CD147 released from lung carcinoma via membrane vesicles, which are typically larger than exosomes, stimulates MMP expression in fibroblasts and hence facilitates tumor invasion and metastasis [55]. In addition, membrane vesicles derived from prostate carcinoma cells can activate fibroblasts (increaseing migration/mobility and resistance to apoptosis) and promote MV shedding from activated fibroblasts, which increase (at least partially) the migration and invasion of tumor cells through the Chemokine (C-X3-C motif) ligand/receptor 1 axis

(CX3CL1-CX3CR1) [56]. Whether exosomes function in a similar fashion to regulate the availability of these particular molecules and other MMP inducers/MMPs, chemokines/cytokines and their corresponding receptors remains to be determined.

9.1.2.2 Role of Tumor Exosomes in Regulating Endothelial Cell Behavior During Metastasis

Previous studies have documented that soluble angiogenic factors secreted by tumor cells and other stromal cells can induce angiogenesis [57]. However, accumulating evidence indicates that molecules delivered via exosomes can also promote a proangiogenic response in endothelial cells; the stimulatory effector molecules range from lipids to membrane-bound and soluble factors, as well as to genetic material (Fig. 9.2).

When compared with the plasma membrane, exosomes contain a relatively high level of certain types of lipids, such as sphingomyelin, cholesterol, and phosphatidylserine [58]. Kim et al. [59] reported that sphingomyelin is the major active component in human fibrosarcoma- and prostate carcinoma-derived MVs and induces endothelial cell migration, morphogenesis, and *in vivo* neovascularization. Interestingly, sphingomyelin levels are found to be significantly elevated in highly metastatic cancer cells compared with less metastatic variants [60]. It would be interesting to determine whether exosomal sphingomyelin-mediated angiogenesis contributes to the high metastatic potential of these cells.

Tumor-derived exosomes may encapsulate angiogenic factors and release them into the tumor microenvironment to induce an angiogenic response. For example, elevated amounts of VEGF were detected in human ovarian carcinoma-derived MVs and released upon induction of vesicle burst by acid pH, which is characteristic of the tumor microenvironment [10]. The availability of VEGF released from vesicles significantly stimulated endothelial cell mobility. Tumor exosomes harbor a variety of membrane-bound proteins that play critical roles in promoting angiogenesis. It has been reported that active oncogenic endothelial growth factor receptor (EGFR) can be transferred to endothelial cells in a human squamous cell carcinoma model *in vitro* and *in vivo*, and the uptake of EGFR activates not only MAPK- and AKT-signaling pathways but also autocrine VEGF production and VEGFR2 activation in endothelial cells [11].

Tetraspanins are constitutive components of exosomes [9, 61]. Interestingly, expression of tetraspanin 8 (Tspan8) correlates with metastatic phenotype in gastrointestinal cancer models [62]. By using a rat pancreatic adenocarcinoma line overexpressing Tspan8 as a model, Nazarenko et al. [63] explored the specific effect of tumor-derived Tspan8-containing exosomes on endothelial cells. Their study demonstrated that Tspan8 can specifically recruit integrin CD49d and vascular cell adhesion molecule (VCAM/CD106) into exosomes, allowing the exosomes to bind to and be internalized by endothelial cells. In addition to recruiting receptors such as CD49d and CD106 into exosomes, Tspan8 also contributes to selective packaging of mRNAs and cytokines into exosomes; some of these molecules may have roles in angiogenesis, such as fibroblast growth factor-binding protein 1 (FGFBP1) and growth and differentiation factor 3 (GDF3). Uptake of Tspan8-containing exosomes induced angiogenesis in aortic endothelial cells, including enhanced proliferation, migration, sprouting, and maturation of endothelial progenitor cells. This phenomenon was accompanied by upregulation of several angiogenesis-related genes, including von Willebrand factor, Tspan8, C-X-C motif chemokine 5 (CXCL5), macrophage migration inhibitory factor (MIF), chemokine receptor 1 (CCR1), and VEGFR1, in a VEGF-independent manner.

It has been demonstrated that endothelial progenitor cell-derived MVs activate an angiogenic program in endothelial cells via horizontal transfer of mRNAs [14]. Tumor exosomes may function in a similar manner by selectively recruiting and delivering a panel of angiogenesis-related mRNA and miRNAs to target cells [63]. CD105+ human renal carcinoma cells release MV containing VEGF, fibroblast growth factor 2 (FGF-2), angiopoietin1, ephrin A3, and MMP-2 and MMP-9 mRNAs [5]. However, the functional contribution of these exosome-packaged mRNAs and miRNAs to angiogenesis needs to be further validated. These studies demonstrate that tumor exosomes are driving a proangiogenic response in endothelial cells promoting their activation, proliferation, and maturation during metastasis. Surface molecules such as exosomal tetraspanins and endothelial membrane receptors seem to be the main molecules involved in the regulation of these mechanisms. Future research will uncover these mechanisms and facilitate the identification of new targets to block the collaboration between tumor cells and endothelial cells by exosomes during metastasis.

9.1.2.3 Role of Tumor Exosomes in Regulating BMDC Behavior During Metastasis

Among the BMDC subsets that are affected by tumor-derived exosomes during tumor progression and metastasis, the myeloid-derived suppressor cell (MDSC) population is the most extensively studied (Fig. 9.2). Under normal physiological conditions, MDSCs are a heterogeneous population of immature BMDCs that differentiates into mature granulocytes, macrophages, and dendritic cells [64]. Under pathological conditions such as cancer, BM cell differentiation is partially blocked resulting in the expansion of the immature MDSC population. These immature MDSCs have been found to negatively regulate immune responses by suppressing T cell responses in the setting of cancer [65]. The mechanisms underlying MDSC expansion during tumor progression are thought to include the secretion of soluble factors such as VEGF, colony-stimulating factor-1 (SDF-1), IL-6, prostaglandin E2 (PGE2), and granulocyte-macrophage colony-stimulating factor (GM-CSF) by tumor cells [66].

Recent studies suggest that tumor exosomes can play a critical role in MDSC phenotype. The laboratory of Dr. Zhang has demonstrated that tumor exosomes derived from breast cancer cells can be taken up by BM myeloid cells, which can impair the ability of the immature myeloid cells to differentiate into functional antitumorigenic dendritic cells, and promote tumor growth [67, 68]. Tumor-derived exosomes also play a role in the activation of MDSCs, triggering their immunosuppressive function through autocrine production of IL-6 and TNF-α [67, 69]. Although it has been suggested that the effect exerted by tumor-derived exosomes on MDSCs is dependent on Toll-like receptor 2 (TLR2)/STAT3 pathway activation [68], a recent publication contradicts this suggestion arguing that exosomes released from tumor cells that are not within a tumor microenvironment may not realistically represent the role of tumor exosomes *in vivo* [70]. This work points out that *in vitro* models of exosome function may not always accurately reflect *in vivo* conditions, a fact that should be considered when interpreting the role of exosomes in tumor biology.

It is well known that MDSCs are also important for the promotion of tumor angiogenesis, cell invasion, and metastasis [71–73]. Infiltration of MDSCs at the tumor invasive front promotes tumor angiogenesis and facilitates tumor cell invasion and metastasis through the production of various MMPs and TGF-β1 [72]. Liu et al. [25] demonstrated that melanoma and breast cancer exosomes significantly increase the recruitment of MDSCs to the lung. This mobilization of MDSCs to the metastatic organ correlated with a significant acceleration of tumor metastasis. Interestingly, these effects were blocked in MyD88 KO mice (a mouse model with impaired inflammatory responses) highlighting the role of inflammation in general and MyD88 in particular in MDSC recruitment by tumor exosomes.

It is also well known that specific subpopulations of BMDCs significantly contribute to *de novo* vessel formation during tumor growth and metastasis [50]. Asahara et al. [74] were the first to demonstrate that EPCs can be mobilized from the BM niches to the peripheral circulation and differentiate into endothelial cells that then contribute to neovessel formation. These cells can be mobilized from the BM not only in response to various stress/traumatic signals related to tissue injury (ischemic conditions, trauma, and chemotherapy) but also by tumor growth [75]. The critical role of these cells in cancer progression was demonstrated by different studies using animal models of cancer, which showed that blocking EPC mobilization from BM led to a significant decrease in tumor growth and vasculature development [35, 36, 76, 77]. Interestingly, it has been suggested that EPC arrival at metastatic sites and their differentiation to mature endothelial cells is a key determinant of progression from micrometastases to macrometastatic lesions [36].

While there are several studies investigating microvesicles and exosomes secreted by nontumor cells and their ability to reprogram EPCs toward a proangiogenic phenotype (see below), none of them addresses the direct effects of tumor-derived exosomes on EPCs. However, it is reasonable to speculate that tumor exosomes could promote the recruitment of EPCs to tumor and metastatic sites by direct interaction and may act as paracrine mediators in the metastatic environment. Much work is needed, however, to clarify the mechanisms driving this critical interaction during the metastatic process.

Despite recent advances in our understanding of the role played by tumor exosomes in metastasis and in the cross-talk between tumor and BM cells, the roles of tumor exosomes in regulating tumor invasion and angiogenesis/vasculogenesis remain largely unknown. For example, the EMT regulatory program is an important mediator of invasion and metastasis [45]. Whether exosomes contribute to the EMT transition has yet to be determined, however. Interestingly, proteomic analysis of the secretome (a combination of MVs and soluble factors) of MDCK cells in which EMT is induced by Ras transformation revealed that multiple proteases and factors that promote cell migration were upregulated during EMT [78]. Whether exosomes function as carriers of some of these molecules and therefore modulate the EMT process will be of great interest for future studies.

Furthermore, it is worthy to point out that to date, all studies on exosomes utilized exosomes that had been harvested from cultured tumor cells. However, the composition and function of tumor exosomes, especially their contribution to metastasis, can be influenced by many factors that tumor cells encounter *in vivo*. For example, hypoxia can stimulate glioblastoma cells to express significantly higher levels of tissue factor (TF), which is released via exosomes to trigger angiogenic phenotypes in endothelial cells [79]. Tumors develop through the reciprocal signaling between tumor cells and the microenvironment and this process may lead to differential recruitment of molecules into tumor exosomes.

This notion is supported by our data demonstrating that, in exosomes harvested from the plasma of melanoma patients, the level of exosomal proteins per particle increased as disease proceeds to more advanced stages [42]. In addition, when co-cultured with tumor-infiltrating leukocytes, breast carcinoma cells secreted exosomes that carried high levels of fibronectin and enhanced tumor cell invasion [80]. Therefore, it seems that malignant cell behavior is correlated with increassed packaging of proteins and other molecules such as RNAs, small RNAs and DNA on exosomes [42].

Finally, the heterogeneity of tumor cells *in vivo* may produce exosomes carrying different cargo and exerting different functions. Indeed, Grange et al. [5] showed that exosomes released by CD105^{+}, but not CD105-negative human renal tumor-initiating cells isolated from the same tumor mass, triggered angiogenesis and pre-metastatic niche formation (see discussion below). Therefore, to fully understand the functional roles of tumor exosomes at each stage of metastasis, it will be important to dissect the interaction between tumor cells and their associated environments and to understand how such interactions influence exosome behavior.

9.1.3 Role of Non-tumor Exosomes in Metastasis

It is likely that in addition to tumor exosomes, exosome secretion by stromal cells within the tumor microenvironment could also be affected by tumor progression and subsequently modify tumor cell behavior. To date, only physical contact and cytokines or chemokines have been described as regulators of cell–cell communication during tumor progression [81, 82]. However, exosomes shed within the tumor microenvironment by many cell types (i.e., endothelial and BMDCs) may influence tumor cell behavior, as well as the behavior of other cell types within [6, 8]. The main challenge in addressing these interactions is that the mix of exosomes and MVs from multiple cell types does not allow for the identification of the specific origin and role of these exosomes.

Although the number of reports regarding the influence of exosomes secreted by other cell types on tumor cell behavior is limited, in the past few years we have observed an increasing number of publications in this field. It has been shown that embryonic stem cell-derived MVs can enhance the survival of murine HPCs and increase their expansion by reinforcing the expression of stem cell markers. Stem cell-derived MVs upregulate the expression of pluripotent and hematopoietic stem cell markers (Oct-4, Nanog, Rex-Scl, HoxB4, and GATA 21) through a process that involves the horizontal transfer of mRNA and protein [83]. Initial work on platelet-derived microvesicles first established a role for MVs in stimulating the mRNA expression of angiogenic factors such as MMP-9, VEGF, IL-8, and hepatocyte growth factor (HGF) during tumor progression/metastasis and angiogenesis in lung cancer [84]. More recently, IL-4-activated macrophages have been found to secrete more invasion-potentiating cytokines, enzymes, and miRNAs, including miR-223, which, in turn, regulate breast cancer cell phenotype and behavior [85]. Furthermore, stromal BM-secreted exosomes have been proposed to regulate tumor cell dormancy *via* miRNA transfer. Lim et al. [86] have reported that stromal-derived exosomes containing miRNAs contributed to breast cancer cell quiescence, and that miRNAs were transferred via gap junctions. These studies demonstrated that extracellular miRNAs circulating systemically either freely or packaged in exosomes may serve as biomarkers for breast cancer patient monitoring [87].

Zernecke et al. [88] have recently shown that miR-126 can be transferred by apoptotic particles at sites of tissue damage, inducing paracrine stimulation in surrounding cells, which, in turn, increases the production of chemokine stromal-derived factor-1α (SDF-1α also known as CXCL12), through the chemokine (C-X-C motif) receptor 4 (CXCR4) pathway. This upregulation of SDF-1 promotes the mobilization of Sca-1^+ progenitor cells from the BM and enhances their contribution to angiogenesis [88]. Interestingly, miR-126 controls angiogenic signals, acting as a regulator of several components of the VEGF pathway [89], and has been found in circulating exosomes from ovarian and bladder cancer patients [19, 90, 91]. Future research analyzing the role of miRNA-126 in the cross-talk between endothelial and tumor cells by exosomes will define the role of this molecule in metastasis.

Regarding the role of exosomes in the regulation of EPC behavior, it has been reported that circulating plasma microvesicles can reprogram EPCs toward a pro-angiogenic phenotype [92]. The incubation of BM cells with circulating plasma microvesicles increased the expression of pro-angiogenic factors (VEGF, SDF-1, and angiopoietin-1) and the differentiation of the EPCs toward the endothelial lineage as demonstrated by the expression of mature endothelial cells markers (VEGFR2, CD144, CD31, and CD54), through the activation of the PPARα/Akt/NF-κB pathway. Recently, Mause et al. [93] have shown that platelet-derived microvesicles are able to restore endothelial integrity after vascular injury. These microvesicles induced expression of endothelial markers (CD31, CD143, and CXCR4) and modulated the composition of the secretome of endothelial cells by increasing levels of pro-angiogenic factors (VEGF, EGF, HGF, IL-1β, and TNF-α) [93]. However, analysis of the content of exosomes in these studies did not permit the identification of specific molecules in the vesicles responsible for pro-angiogenic effects. Moreover, several reports showed that microvesicles from endothelial cells can also affect EPC function and angiogenesis. Microvesicles from apoptotic endothelial cells are able to activate a pro-angiogenic program *in vitro*, in cultured human EPCs isolated from peripheral blood, inducing their proliferation and differentiation toward mature endothelial cells [94]. Similar results have been reported using BM-derived mononuclear cells [95]. In this study, the microvesicles from apoptotic endothelial cells promoted *in vitro* BM-derived mononuclear cell differentiation and *in vivo* postnatal vasculogenesis through a mechanism involving reactive oxygen species production [95]. It has also been shown that the proangiogenic behavior of exosomes depends not only on their cargo but also on the quantity of specific molecules packaged in exosomes. Lacroix et al. [96] demonstrated that endothelial-derived microvesicles can favor or inhibit tube formation by EPCs in a dose-dependent manner: low amounts of endothelial-derived MVs increased tube formation and were associated with matrix metalloproteinase activation, whereas higher concentrations inhibited both processes.

Interestingly, human-circulating $CD34^+$ BM cells can secrete exosomes with angiogenic activity and this phenomenon significantly contributes to the paracrine effects of progenitor cells [97]. Although this study proposes the use of these exosomes and progenitor cells for therapeutic purposes after BM transplantation, it also suggests that exosomes secreted by specific populations of BM progenitor cells could be playing an essential role in regulating the angiogenic behavior of tumors and metastasis. Similarly, exosomes derived from cardiac progenitor cells influence the migratory capacity of endothelial cells [98]. Endothelial-derived microvesicles and exosomes are known to play important roles in coagulation, inflammation, endothelial function, and angiogenesis and thus may disturb vascular homeostasis, contributing to the progression of vascular diseases. In fact, elevated levels of these vesicles have been found in plasma from patients with vascular diseases and endothelial-derived MVs and exosomes have been found to promote cell survival, anti-inflammatory effects, or endothelial regeneration [99].

Therefore, although the role of endothelial exosomes in tumorigenesis has yet to be described, it is likely that they could reinforce specific pathways promoting primary tumor growth and metastasis. This field of study deserves further attention and will likely lead to the identification of essential roles for endothelial released exosomes in metastasis. Thus far, HSP70 shed by endothelial cell exosomes has been described as a potential protein activating monocytes and resulting in monocyte adhesion to endothelial cells [100]. These results may indicate that exosome-mediated release of HSP70 from endothelial cells provides a novel paracrine mechanism to regulate vascular endothelial functional integrity, which should be studied during cancer progression.

Finally, monocyte MVs have been found to increase endothelial thrombogenicity and apoptosis, while having a pro-angiogenic effect [101]. These findings suggest a role for monocyte MVs in endothelial dysfunction associated with inflammatory and hypercoagulatory disease [101]. Similarly, mast cell-derived exosomes influence endothelial cell behavior increasing PAI-1 levels and procoagulant states [102].

Extensive experimental evidence shows that platelets are involved in tumor metastasis. Platelet activation and the coagulation system have a crucial role in cancer progression [103]. Interestingly, procoagulant MVs released by BM-derived cells contribute to cancer thrombogenicity and vascular cell dysfunction and angiogenesis [101, 104]. Future studies analyzing the role of exosome-mediated cross-talk between BM-derived and endothelial cells in procoagulatory pathologies during cancer metastasis will shed light on the mechanisms regulating these processes.

9.2 Closing Remarks

As discussed above, membrane microvesicles and exosomes shed by cancer cells contribute to the horizontal transfer of proteins and genetic material (mRNAs and small RNAs). Exosomes can alter the tumor microenvironment by inducing short-term and long-term effects in other cell types most likely by inducing epigenetic and genetic changes, which alter the gene expression profile of cells that uptake these exosomes.

It is worth to menction that in this blooming field of research, there is still controversy about vesicle nomenclature and composition. Exosomes, microvesicles and other secreted vesicles need to be better defined and characterized. In this chapter, we focused our attention on the role of tumor-secreted exosomes in influencing the main cell types that have been found to contribute to pre-metastatic niche formation and metastasis (i.e., BM-derived, endothelial, and stromal cells). All of these studies indicate that tumor-secreted exosomes play an important and active role in tumor progression by shaping the tumor microenvironment. Therefore, nontumor cells could be permanently altered by tumor exosomes, which prime them to support metastatic behavior. In fact, our current work demonstrates that tumor exosomes can alter the BM in a "permanent" manner by promoting BM progenitor cell "education" toward a prometastatic phenotype. Determining the mechanisms through which tumor exosomes promote BM-derived progenitor education will lead to the identification of new biomarkers that could be utilized to monitor metastasis before and after therapeutic intervention.

Similarly, markers present on circulating exosomes may be utilized as potential metastatic indicators. The specific protein composition of circulating plasma exosomes could be used as a molecular signature in the clinical setting, not only as an indicator of widespread metastatic disease, but also as a predictive factor for metastatic potential. Several studies have found specific exosome markers that correlate with specific pathologies; however, there is no information available on the use of these exosome markers to predict the outcome of metastatic disease. Future work will address the utility of exosome signatures not only for diagnostic purposes but also for monitoring prognosis and response to therapy. It is also likely that exosomes released into the peripheral circulation by different types of cancer cells have distinct exosomal signatures that will allow categorization of tumor types, particularly in those patients who present with disease of unknown origin.

Since exosomes represent a new mode of interaction between the tumor and its microenvironment that contributes to tumor progression, it is tempting to speculate that therapeutic blocking of exosome secretion by tumor cells could help improve current anti-metastatic treatments. Novel therapies could be designed to target the unique proteins expressed on exosomes, thus preventing exosomal material transfer to other cell types. Alternatively, specific drugs targeting exosome secretion pathways could reduce exosome secretion by tumor cells. Determining the specific pathways governing exosome secretion and identifying exosome markers specific to each cancer type will define new molecular targets for combinational therapies. These data will lead to the development of "personalized medicine" approaches, which match the treatment to the specific disease of each patient in a timely fashion, thereby increasing patient survival by hindering tumor metastasis.

Acknowledgments We thank Irina Matei, Ayuko Nitadori-Hoshino, and Bruno Costa Da Silva for their critical reading of this chapter and their suggestions. We also thank the members of the laboratories of Dr. Lyden for their helpful discussions. Our work is supported by grants from the Children's Cancer and Blood Foundation, Stavros S. Niarchos Foundation, Champalimaud Foundation, Nancy and Daniel Paduano Foundation, The Hartwell Foundation, AHEPA 5th District, Malcolm Hewitt Wiener Foundation, NCI, National Foundation for Cancer Research, Susan G. Komen for the Cure, Mary Kay Foundation. The Manning Foundation, The George Best Costacos Foundation, National Foundation for Cancer Research, Fundación para el Fomento en Asturias de la Investigación Científica Aplicada y la Tecnología, Fundación Universidad de Oviedo and the Beth E. Tortolani Foundation.

References

1. Kramer-Albers EM, Bretz N, Tenzer S, Winterstein C, Mobius W, Berger H, Nave KA, Schild H et al (2007) Oligodendrocytes secrete exosomes containing major myelin and stress-protective proteins: trophic support for axons? Proteomics Clin Appl 1(11):1446–1461
2. Lachenal G, Pernet-Gallay K, Chivet M, Hemming FJ, Belly A, Bodon G, Blot B, Haase G et al (2011) Release of exosomes from differentiated neurons and its regulation by synaptic glutamatergic activity. Mol Cell Neurosci 46(2):409–418
3. Potolicchio I, Carven GJ, Xu X, Stipp C, Riese RJ, Stern LJ, Santambrogio L (2005) Proteomic analysis of microglia-derived exosomes: metabolic role of the aminopeptidase CD13 in neuropeptide catabolism. J Immunol 175(4):2237–2243
4. Raposo G, Nijman HW, Stoorvogel W, Liejendekker R, Harding CV, Melief CJ, Geuze HJ (1996) B lymphocytes secrete antigen-presenting vesicles. J Exp Med 183(3):1161–1172
5. Grange C, Tapparo M, Collino F, Vitillo L, Damasco C, Deregibus MC, Tetta C, Bussolati B et al (2011) Microvesicles released from human renal cancer stem cells stimulate angiogenesis and formation of lung premetastatic niche. Cancer Res 71(15):5346–56
6. Castellana D, Toti F, Freyssinet JM (2010) Membrane microvesicles: macromessengers in cancer disease and progression. Thromb Res 125(Suppl 2):S84–88
7. Muralidharan-Chari V, Clancy JW, Sedgwick A, D'Souza-Schorey C (2010) Microvesicles: mediators of extracellular communication during cancer progression. J Cell Sci 123(Pt 10):1603–1611
8. Peinado H, Lavotshkin S, Lyden D (2011) The secreted factors responsible for pre-metastatic niche formation: old sayings and new thoughts. Semin Cancer Biol 21(2):139–146
9. Mathivanan S, Ji H, Simpson RJ (2010) Exosomes: extracellular organelles important in intercellular communication. J Proteomics 73(10):1907–1920

10. Taraboletti G, D'Ascenzo S, Giusti I, Marchetti D, Borsotti P, Millimaggi D, Giavazzi R, Pavan A et al (2006) Bioavailability of VEGF in tumor-shed vesicles depends on vesicle burst induced by acidic pH. Neoplasia 8(2):96–103
11. Al-Nedawi K, Meehan B, Kerbel RS, Allison AC, Rak J (2009) Endothelial expression of autocrine VEGF upon the uptake of tumor-derived microvesicles containing oncogenic EGFR. Proc Natl Acad Sci U S A 106(10):3794–3799
12. Baj-Krzyworzeka M, Szatanek R, Weglarczyk K, Baran J, Urbanowicz B, Branski P, Ratajczak MZ, Zembala M (2006) Tumour-derived microvesicles carry several surface determinants and mRNA of tumour cells and transfer some of these determinants to monocytes. Cancer Immunol Immunother 55(7):808–818
13. Collino F, Deregibus MC, Bruno S, Sterpone L, Aghemo G, Viltono L, Tetta C, Camussi G (2010) Microvesicles derived from adult human bone marrow and tissue specific mesenchymal stem cells shuttle selected pattern of miRNAs. PLoS ONE 5(7):e11803
14. Deregibus MC, Cantaluppi V, Calogero R, Lo Iacono M, Tetta C, Biancone L, Bruno S, Bussolati B et al (2007) Endothelial progenitor cell derived microvesicles activate an angiogenic program in endothelial cells by a horizontal transfer of mRNA. Blood 110 B(7):2440–2448
15. Skog J, Wurdinger T, van Rijn S, Meijer DH, Gainche L, Sena-Esteves M, Curry WT Jr, Carter BS et al (2008) Glioblastoma microvesicles transport RNA and proteins that promote tumour growth and provide diagnostic biomarkers. Nat Cell Biol 10(12):1470–1476
16. Valadi H, Ekstrom K, Bossios A, Sjostrand M, Lee JJ, Lotvall JO (2007) Exosome-mediated transfer of mRNAs and microRNAs is a novel mechanism of genetic exchange between cells. Nat Cell Biol 9(6):654–659
17. Andre F, Schartz NE, Movassagh M, Flament C, Pautier P, Morice P, Pomel C, Lhomme C et al (2002) Malignant effusions and immunogenic tumour-derived exosomes. Lancet 360(9329):295–305
18. Baran J, Baj-Krzyworzeka M, Weglarczyk K, Szatanek R, Zembala M, Barbasz J, Czupryna A, Szczepanik A et al (2010) Circulating tumour-derived microvesicles in plasma of gastric cancer patients. Cancer Immunol Immunother 59(6):841–850
19. Taylor DD and Gercel-Taylor C (2008) MicroRNA signatures of tumor-derived exosomes as diagnostic biomarkers of ovarian cancer. Gynecol Oncol 110(1):13–21
20. Valenti R, Huber V, Filipazzi P, Pilla L, Sovena G, Villa A, Corbelli A, Fais S et al (2006) Human tumor-released microvesicles promote the differentiation of myeloid cells with transforming growth factor-beta-mediated suppressive activity on T lymphocytes. Cancer Res 66(18):9290–9298
21. Fidler IJ and Kripke ML (1977) Metastasis results from preexisting variant cells within a malignant tumor. Science 197(4306):893–895
22. Hao S, Ye Z, Li F, Meng Q, Qureshi M, Yang J, Xiang J (2006) Epigenetic transfer of metastatic activity by uptake of highly metastatic B16 melanoma cell-released exosomes. Exp Oncol 28(2):126–131
23. Poste G and Nicolson GL (1980) Arrest and metastasis of blood-borne tumor cells are modified by fusion of plasma membrane vesicles from highly metastatic cells. Proc Natl Acad Sci U S A 77(1):399–403
24. Liu C, Yu S, Zinn K, Wang J, Zhang L, Jia Y, Kappes JC, Barnes S et al (2006) Murine mammary carcinoma exosomes promote tumor growth by suppression of NK cell function. J Immunol 176(3):1375–1385
25. Liu Y, Xiang X, Zhuang X, Zhang S, Liu C, Cheng Z, Michalek S, Grizzle W et al (2010) Contribution of MyD88 to the tumor exosome-mediated induction of myeloid derived suppressor cells. Am J Pathol 176 (5):2490–2499
26. Jung T, Castellana D, Klingbeil P, Cuesta Hernandez I, Vitacolonna M, Orlicky DJ, Roffler SR, Brodt P et al (2009) CD44v6 dependence of premetastatic niche preparation by exosomes. Neoplasia 11(10):1093–1105
27. Hood JL, San RS, Wickline SA (2011) Exosomes released by melanoma cells prepare sentinel lymph nodes for tumor metastasis. Cancer Res 71(11):3792–3801.

28. Hiratsuka S, Watanabe A, Aburatani H, Maru Y (2006) Tumour-mediated upregulation of chemoattractants and recruitment of myeloid cells predetermines lung metastasis. Nat Cell Biol 8(12):1369–1375
29. Kang SY, Halvorsen OJ, Gravdal K, Bhattacharya N, Lee JM, Liu NW, Johnston BT, Johnston AB et al (2009) Prosaposin inhibits tumor metastasis via paracrine and endocrine stimulation of stromal p53 and Tsp-1. Proc Natl Acad Sci U S A 106(29):12115–12120
30. Kaplan RN, Riba RD, Zacharoulis S, Bramley AH, Vincent L, Costa C, MacDonald DD, Jin DK et al (2005) VEGFR1-positive haematopoietic bone marrow progenitors initiate the pre-metastatic niche. Nature 438(7069):820–827
31. Erler JT, Bennewith KL, Cox TR, Lang G, Bird D, Koong A, Le QT, Giaccia AJ (2009) Hypoxia-induced lysyl oxidase is a critical mediator of bone marrow cell recruitment to form the premetastatic niche. Cancer Cell 15(1):35–44
32. Wong CC, Gilkes DM, Zhang H, Chen J, Wei H, Chaturvedi P, Fraley SI, Wong CM et al (2011) Hypoxia-inducible factor 1 is a master regulator of breast cancer metastatic niche formation. Proc Natl Acad Sci U S A 108(39):16369–16374
33. Rafii S, Lyden D, Benezra R, Hattori K, Heissig B (2002) Vascular and haematopoietic stem cells: novel targets for anti-angiogenesis therapy? Nat Rev Cancer 2(11):826–835
34. Psaila B and Lyden D (2009) The metastatic niche: adapting the foreign soil. Nat Rev Cancer 9(4):285–293
35. Lyden D, Hattori K, Dias S, Costa C, Blaikie P, Butros L, Chadburn A, Heissig B et al (2001) Impaired recruitment of bone-marrow-derived endothelial and hematopoietic precursor cells blocks tumor angiogenesis and growth. Nat Med 7(11):1194–1201
36. Gao D, Nolan DJ, Mellick AS, Bambino K, McDonnell K, Mittal V (2008) Endothelial progenitor cells control the angiogenic switch in mouse lung metastasis. Science 319(5860):195–198
37. Hiratsuka S, Nakamura K, Iwai S, Murakami M, Itoh T, Kijima H, Shipley JM, Senior RM et al (2002) MMP9 induction by vascular endothelial growth factor receptor-1 is involved in lung-specific metastasis. Cancer Cell 2(4):289–300
38. Ghosh AK, Secreto CR, Knox TR, Ding W, Mukhopadhyay D, Kay NE (2010) Circulating microvesicles in B-cell chronic lymphocytic leukemia can stimulate marrow stromal cells: implications for disease progression. Blood 115(9):1755–1764
39. Rafii DC, Psaila B, Butler J, Jin DK, Lyden D (2008) Regulation of vasculogenesis by platelet-mediated recruitment of bone marrow-derived cells. Arterioscler Thromb Vasc Biol 28(2):217–222
40. Aliotta JM, Pereira M, Johnson KW, de Paz N, Dooner MS, Puente N, Ayala C, Brilliant K et al (2010) Microvesicle entry into marrow cells mediates tissue-specific changes in mRNA by direct delivery of mRNA and induction of transcription. Exp Hematol 38 (3):233–245
41. Renzulli JF, 2nd, Del Tatto M, Dooner G, Aliotta J, Goldstein L, Dooner M, Colvin G, Chatterjee D et al (2010) Microvesicle induction of prostate specific gene expression in normal human bone marrow cells. J Urol 184(5):2165–2171
42. Peinado H, Alečković M, Lavotshkin S, Matei I, Costa-Silva B, Moreno-Bueno G, Hergueta-Redondo M, Williams C, García-Santos G, Ghajar CM, Nitadori-Hoshino A, Hoffman C, Badal K, Garcia BA, Callahan MK, Yuan J, Martins VR, Skog J, Kaplan RN, Brady MS, Wolchok JD, Chapman PB, Kang Y, Bromberg J, and Lyden D (2012) Melanoma exosomes educate bone marrow progenitor cells toward a pro-metastatic phenotype through MET. Nat Med 18(6):833–891
43. Nguyen DX, Bos PD, Massague J (2009) Metastasis: from dissemination to organ-specific colonization. Nat Rev Cancer 9(4):274–284
44. Joyce JA, Pollard JW (2009) Microenvironmental regulation of metastasis. Nat Rev Cancer 9(4):239–252
45. Hanahan D, Weinberg RA (2011) Hallmarks of cancer: the next generation. Cell 144(5):646–674
46. Peinado H, Olmeda D, and Cano A (2007) Snail, Zeb and bHLH factors in tumour progression: an alliance against the epithelial phenotype? Nat Rev Cancer 7(6):415–428

47. Kalluri R (2009) EMT: when epithelial cells decide to become mesenchymal-like cells. J Clin Invest 119(6):1417–1419
48. Oskarsson T, Acharyya S, Zhang XH, Vanharanta S, Tavazoie SF, Morris PG, Downey RJ, Manova-Todorova K et al (2011) Breast cancer cells produce tenascin C as a metastatic niche component to colonize the lungs. Nat Med 17(7):867–874
49. Gao D and Mittal V (2009) The role of bone-marrow-derived cells in tumor growth, metastasis initiation and progression. Trends Mol Med 15(8):333–343
50. Gao D, Nolan D, McDonnell K, Vahdat L, Benezra R, Altorki N, Mittal V (2009) Bone marrow-derived endothelial progenitor cells contribute to the angiogenic switch in tumor growth and metastatic progression. Biochim Biophys Acta 1796(1):33–40
51. Kalluri R, Zeisberg M (2006) Fibroblasts in cancer. Nat Rev Cancer 6(5):392–401
52. Antonyak MA, Li B, Boroughs LK, Johnson JL, Druso JE, Bryant KL, Holowka DA, Cerione RA (2011) Cancer cell-derived microvesicles induce transformation by transferring tissue transglutaminase and fibronectin to recipient cells. Proc Natl Acad Sci U S A 108(12):4852–4857
53. Webber J, Steadman R, Mason MD, Tabi Z, Clayton A (2010) Cancer exosomes trigger fibroblast to myofibroblast differentiation. Cancer Res 70(23):9621–9630
54. Wysoczynski M, Ratajczak MZ (2009) Lung cancer secreted microvesicles: underappreciated modulators of microenvironment in expanding tumors. Int J Cancer 125(7):1595–1603
55. Sidhu SS, Mengistab AT, Tauscher AN, LaVail J, Basbaum C (2004) The microvesicle as a vehicle for EMMPRIN in tumor-stromal interactions. Oncogene 23(4):956–963
56. Castellana D, Zobairi F, Martinez MC, Panaro MA, Mitolo V, Freyssinet JM, Kunzelmann C (2009) Membrane microvesicles as actors in the establishment of a favorable prostatic tumoral niche: a role for activated fibroblasts and CX3CL1-CX3CR1 axis. Cancer Res 69(3):785–793
57. Dias S, Choy M, Rafii S (2001) The role of CXC chemokines in the regulation of tumor angiogenesis. Cancer Invest 19(7):732–738
58. Schorey JS, Bhatnagar S (2008) Exosome function: from tumor immunology to pathogen biology. Traffic 9(6):871–881
59. Kim CW, Lee HM, Lee TH, Kang C, Kleinman HK, Gho YS (2002) Extracellular membrane vesicles from tumor cells promote angiogenesis via sphingomyelin. Cancer Res 62(21):6312–6317
60. Dahiya R, Boyle B, Goldberg BC, Yoon WH, Konety B, Chen K, Yen TS, Blumenfeld W et al (1992) Metastasis-associated alterations in phospholipids and fatty acids of human prostatic adenocarcinoma cell lines. Biochem Cell Biol 70(7):548–554
61. Zoller M (2009) Tetraspanins: push and pull in suppressing and promoting metastasis. Nat Rev Cancer 9(1):40–55
62. Claas C, Seiter S, Claas A, Savelyeva L, Schwab M, Zoller M (1998) Association between the rat homologue of CO-029, a metastasis-associated tetraspanin molecule and consumption coagulopathy. J Cell Biol 141(1):267–280
63. Nazarenko I, Rana S, Baumann A, McAlear J, Hellwig A, Trendelenburg M, Lochnit G, Preissner KT et al (2010) Cell surface tetraspanin Tspan8 contributes to molecular pathways of exosome-induced endothelial cell activation. Cancer Res 70(4):1668–1678
64. Murdoch C, Muthana M, Coffelt SB, Lewis CE (2008) The role of myeloid cells in the promotion of tumour angiogenesis. Nat Rev Cancer 8(8):618–631
65. Ostrand-Rosenberg S (2010) Myeloid-derived suppressor cells: more mechanisms for inhibiting antitumor immunity. Cancer Immunol Immunother 59(10):1593–1600
66. Serafini P, Borrello I, Bronte V (2006) Myeloid suppressor cells in cancer: recruitment, phenotype, properties, and mechanisms of immune suppression. Semin Cancer Biol 16(1):53–65
67. Yu S, Liu C, Su K, Wang J, Liu Y, Zhang L, Li C, Cong Y et al (2007) Tumor exosomes inhibit differentiation of bone marrow dendritic cells. J Immunol 178(11):6867–6875
68. Xiang X, Poliakov A, Liu C, Liu Y, Deng ZB, Wang J, Cheng Z, Shah SV et al (2009) Induction of myeloid-derived suppressor cells by tumor exosomes. Int J Cancer 124(11):2621–2633

69. Chalmin F, Ladoire S, Mignot G, Vincent J, Bruchard M, Remy-Martin JP, Boireau W, Rouleau A et al (2010) Membrane-associated Hsp72 from tumor-derived exosomes mediates STAT3-dependent immunosuppressive function of mouse and human myeloid-derived suppressor cells. J Clin Invest 120(2):457–471
70. Xiang X, Liu Y, Zhuang X, Zhang S, Michalek S, Taylor DD, Grizzle W, Zhang HG (2010) TLR2-mediated expansion of MDSCs is dependent on the source of tumor exosomes. Am J Pathol 177(4):1606–1610
71. Yang L, DeBusk LM, Fukuda K, Fingleton B, Green-Jarvis B, Shyr Y, Matrisian LM, Carbone DP et al (2004) Expansion of myeloid immune suppressor Gr+CD11b+cells in tumor-bearing host directly promotes tumor angiogenesis. Cancer Cell 6(4):409–421
72. Yang L, Huang J, Ren X, Gorska AE, Chytil A, Aakre M, Carbone DP, Matrisian LM et al (2008) Abrogation of TGF beta signaling in mammary carcinomas recruits Gr-1+CD11b+ myeloid cells that promote metastasis. Cancer Cell 13(1):23–35
73. Ye XZ, Yu SC, Bian XW (2010) Contribution of myeloid-derived suppressor cells to tumor-induced immune suppression, angiogenesis, invasion and metastasis. J Genet Genomics 37(7):423–430
74. Asahara T, Murohara T, Sullivan A, Silver M, van der Zee R, Li T, Witzenbichler B, Schatteman G et al (1997) Isolation of putative progenitor endothelial cells for angiogenesis. Science 275(5302):964–967
75. Kopp HG, Rafii S (2007) Thrombopoietic cells and the bone marrow vascular niche. Ann N Y Acad Sci 1106:175–179
76. Mellick AS, Plummer PN, Nolan DJ, Gao D, Bambino K, Hahn M, Catena R, Turner V et al (2010) Using the transcription factor inhibitor of DNA binding 1 to selectively target endothelial progenitor cells offers novel strategies to inhibit tumor angiogenesis and growth. Cancer Res 70(18):7273–7282
77. Nolan DJ, Ciarrocchi A, Mellick AS, Jaggi JS, Bambino K, Gupta S, Heikamp E, McDevitt MR et al (2007) Bone marrow-derived endothelial progenitor cells are a major determinant of nascent tumor neovascularization. Genes Dev 21(12):1546–1558
78. Mathias RA, Wang B, Ji H, Kapp EA, Moritz RL, Zhu HJ, Simpson RJ (2009) Secretome-based proteomic profiling of Ras-transformed MDCK cells reveals extracellular modulators of epithelial-mesenchymal transition. J Proteome Res 8(6):2827–2837
79. Svensson KJ, Kucharzewska P, Christianson HC, Skold S, Lofstedt T, Johansson MC, Morgelin M, Bengzon J et al (2011) Hypoxia triggers a proangiogenic pathway involving cancer cell microvesicles and PAR-2-mediated heparin-binding EGF signaling in endothelial cells. Proc Natl Acad Sci U S A 108(32):13147–13152
80. Deng Z, Cheng Z, Xiang X, Yan J, Zhuang X, Liu C, Jiang H, Ju S et al (2011) Tumor cell cross talk with tumor-associated leukocytes leads to induction of tumor exosomal fibronectin and promotes tumor progression. Am J Pathol
81. Raman D, Baugher PJ, Thu YM, Richmond A (2007) Role of chemokines in tumor growth. Cancer Lett 256(2):137–165
82. Singh S, Sadanandam A, Singh RK (2007) Chemokines in tumor angiogenesis and metastasis. Cancer Metastasis Rev 26(3–4):453–467
83. Ratajczak J, Miekus K, Kucia M, Zhang J, Reca R, Dvorak P, Ratajczak MZ(2006) Embryonic stem cell-derived microvesicles reprogram hematopoietic progenitors: evidence for horizontal transfer of mRNA and protein delivery. Leukemia 20(5):847–856
84. Janowska-Wieczorek A, Wysoczynski M, Kijowski J, Marquez-Curtis L, Machalinski B, Ratajczak J, Ratajczak MZ (2005) Microvesicles derived from activated platelets induce metastasis and angiogenesis in lung cancer. Int J Cancer 113(5):752–760
85. Yang M, Chen J, Su F, Yu B, Su F, Lin L, Liu Y, Huang JD et al (2011) Microvesicles secreted by macrophages shuttle invasion-potentiating microRNAs into breast cancer cells. Mol Cancer 10:117
86. Lim PK, Bliss SA, Patel SA, Taborga M, Dave MA, Gregory LA, Greco SJ, Bryan M et al (2011) Gap junction-mediated import of microRNA from bone marrow stromal cells can elicit cell cycle quiescence in breast cancer cells. Cancer Res 71(5):1550–1560

87. Corcoran C, Friel AM, Duffy MJ, Crown J, O'Driscoll L (2011) Intracellular and extracellular microRNAs in breast cancer. Clin Chem 57(1):18–32
88. Zernecke A, Bidzhekov K, Noels H, Shagdarsuren E, Gan L, Denecke B, Hristov M, Koppel T et al (2009) Delivery of microRNA-126 by apoptotic bodies induces CXCL12-dependent vascular protection. Sci Signal 2(100):ra81
89. Fish JE, Santoro MM, Morton SU, Yu S, Yeh RF, Wythe JD, Ivey KN, Bruneau BG et al (2008) miR-126 regulates angiogenic signaling and vascular integrity. Dev Cell 15(2):272–284
90. Resnick KE, Alder H, Hagan JP, Richardson DL, Croce CM, Cohn DE (2009) The detection of differentially expressed microRNAs from the serum of ovarian cancer patients using a novel real-time PCR platform. Gynecol Oncol 112(1):55–59
91. Hanke M, Hoefig K, Merz H, Feller AC, Kausch I, Jocham D, Warnecke JM, Sczakiel G (2010) A robust methodology to study urine microRNA as tumor marker: microRNA-126 and microRNA-182 are related to urinary bladder cancer. Urol Oncol 28(6):655–661
92. Benameur T, Tual-Chalot S, Andriantsitohaina R, Martinez MC (2010) PPARalpha is essential for microparticle-induced differentiation of mouse bone marrow-derived endothelial progenitor cells and angiogenesis. PLoS ONE 5(8):e12392
93. Mause SF, Ritzel E, Liehn EA, Hristov M, Bidzhekov K, Muller-Newen G, Soehnlein O, Weber C (2010) Platelet microparticles enhance the vasoregenerative potential of angiogenic early outgrowth cells after vascular injury. Circulation 122(5):495–506
94. Hristov M, Erl W, Linder S, Weber PC (2004) Apoptotic bodies from endothelial cells enhance the number and initiate the differentiation of human endothelial progenitor cells in vitro. Blood 104(9):2761–2766
95. Leroyer AS, Ebrahimian TG, Cochain C, Recalde A, Blanc-Brude O, Mees B, Vilar J, Tedgui A et al (2009) Microparticles from ischemic muscle promotes postnatal vasculogenesis. Circulation 119(21):2808–2817
96. Lacroix R, Sabatier F, Mialhe A, Basire A, Pannell R, Borghi H, Robert S, Lamy E et al (2007) Activation of plasminogen into plasmin at the surface of endothelial microparticles: a mechanism that modulates angiogenic properties of endothelial progenitor cells in vitro. Blood 110(7):2432–2439
97. Sahoo S, Klychko E, Thorne T, Misener S, Schultz KM, Millay M, Ito A, Liu T et al (2011) Exosomes from human CD34(+) stem cells mediate their proangiogenic paracrine activity. Circ Res 109(7):724–728
98. Vrijsen KR, Sluijter JP, Schuchardt MW, van Balkom BW, Noort WA, Chamuleau SA, and Doevendans PA (2010) Cardiomyocyte progenitor cell-derived exosomes stimulate migration of endothelial cells. J Cell Mol Med 14(5):1064–1070
99. Dignat-George F, Boulanger CM (2011) The many faces of endothelial microparticles. Arterioscler Thromb Vasc Biol 31(1):27–33
100. Zhan R, Leng X, Liu X, Wang X, Gong J, Yan L, Wang L, Wang Y et al (2009) Heat shock protein 70 is secreted from endothelial cells by a non-classical pathway involving exosomes. Biochem Biophys Res Commun 387(2):229–233
101. Aharon A, Tamari T, Brenner B (2008) Monocyte-derived microparticles and exosomes induce procoagulant and apoptotic effects on endothelial cells. Thromb Haemost 100(5):878–885
102. Al-Nedawi K, Szemraj J, Cierniewski CS (2005) Mast cell-derived exosomes activate endothelial cells to secrete plasminogen activator inhibitor type 1. Arterioscler Thromb Vasc Biol 25(8):1744–1749
103. Gay LJ, Felding-Habermann B (2011) Contribution of platelets to tumour metastasis. Nat Rev Cancer 11(2):123–134
104. Aharon A and Brenner B (2009) Microparticles, thrombosis and cancer. Best Pract Res Clin Haematol 22(1):61–69

Chapter 10
Microvesicles and Their Emerging Role in Cellular Therapies for Organ and Tissue Regeneration

Mariusz Z. Ratajczak, Kasia Mierzejewska, Magda Kucia, Nick Greco and Janina Ratajczak

Abstract Microvesicles (MVs) are small, spherical membrane fragments shed from the cell surface or secreted from the endosomal compartment. MVs released from cells employed in regenerative medicine to rescue damaged tissues seem to play an important and underappreciated role in improving the function of damaged organs. A growing body of evidence suggests that MVs secreted from hematopoietic stem progenitor cells (HSPCs), multipotent stroma cells (MSCs), or cardiac stem cells (CSCs) employed in various treatment strategies in regenerative medicine may: (1) inhibit apoptosis of cells residing in the damaged tissues, (2) stimulate proliferation of cells that survive organ injury, and (3) stimulate vascularization of affected tissues. These proregenerative effects mediated by MVs are explained by the fact that these small, spherical membrane fragments: (1) are enriched in bioactive lipids (e.g., sphingosine-1-phosphate [S1P]), (2) may express antiapoptoic and prostimulatory growth factors or cytokines (e.g., vascular endothelial growth factor [VEGF], stem cell factor [SCF], or stromal derived factor-1 [SDF-1]) on their surface, and (3) may deliver mRNA, regulatory miRNA, and proteins to the damaged tissues that improve overall cell function. Based on these observations, the potential use of MVs, instead of whole cells, has become an exciting new concept in regenerative medicine.

10.1 Introduction

Both prokaryotic and eukaryotic cells communicate and exchange information by employing cell–cell contact mechanisms. Cells may communicate by: (1) secreted growth factors, cytokines, chemokines, and small molecule mediators (e.g., nucleotides or nitric oxide ions) [1, 2], (2) bioactive lipids (e.g., sphingosine-1-phosphate or eicosanoids), (3) interactions mediated by sets of specialized adhesion molecules and their ligands (e.g., integrins and their corresponding specific ligands

M. Z. Ratajczak (✉) · K. Mierzejewska · M. Kucia · N. Greco · J. Ratajczak
Stem Cell Institute at the James Graham Brown Cancer Center, University of Louisville, 500 South Floyd Street, Room 107, Louisville, 40202 KY, USA
e-mail: mzrata01@louisville.edu

Cleveland Cord Blood Center, Cleveland, OH, USA

H.-G. Zhang (ed.), *Emerging Concepts of Tumor Exosome–Mediated Cell–Cell Communication*, DOI 10.1007/978-1-4614-3697-3_10, expressed on interacting cells), and, as reported recently, (4) exchanging information by means of tunneling nanotubules [3, 4]. However, growing attention is now being focused on cell-to-cell communication that involves spherical membrane fragments called microvesicles (MVs) [5–8], a mechanism that has been largely overlooked for many years.

MVs are shed from the cell surface of normal healthy or damaged cells during membrane blebbing in a calcium flux- and calpain-dependent manner and "hijack" both membrane components and the engulfed cytoplasmic contents [5–8]. Shedding of membrane-derived MVs is a physiological phenomenon that accompanies cell activation and growth. Interestingly, rapidly growing cell lines tend to secrete more MVs than slowly growing ones. Generally, the number of MVs shed from cells increases upon: (1) cell activation, (2) hypoxia or irradiation, (3) oxidative injury, (4) exposure to proteins from the activated complement cascade, and (5) exposure to shearing stress [9–16]. As mentioned above, there is mounting evidence that MV shedding depends on an increase in cytosolic Ca^{2+} and degradation of the membrane skeleton in a calpain-dependent manner [9–12].

Another source of MVs is the intracellular endosomal membrane compartment. These MVs, termed exosomes, are usually released from cells as secretory granules during the process of exocytosis [5–8]. While MVs released from the surface membranes during membrane blebbing are relatively large (0.1–1 mm), exosomes are much smaller and appear more homogeneous in size (30–100 nm). MVs are often called by different names in the literature. While, for example, MVs released from the basolateral membrane of morphogen-synthesizing cells during tissue development are called argosomes [17], MVs released by activated human polymorphonuclear leukocytes are sometimes described as ectosomes [5–8]. Other terms for MVs are microparticles or signalosomes, and this latter term reflects the biological role of MVs as a kind of signaling complex.

There is no doubt that the biological significance of MVs has been largely overlooked for many years and MVs have been regarded, like apoptotic bodies, as mere cellular fragments or debris. However, compelling evidence has accumulated that these tiny spherical membrane fragments orchestrate several biological processes [5–8]. On the one hand, it is already acknowledged that MVs are secreted or shed by healthy and not dying cells, are much smaller in size than apoptotic bodies, and do not contain fragments of nuclei loaded with nuclear DNA, which is a common feature of apoptotic bodies. On the other hand, MVs contain numerous structural proteins (e.g., receptors and adhesion molecules), as well as lipids, similar to those present in the membranes of the cells from which they originate.

MVs not only contain numerous proteins and lipids similar to those present in the membranes of the cells from which they originate, but since MV membranes engulf some cytoplasm during membrane blebbing, they may also contain intracellular proteins [18], mRNA [18], regulatory miRNA [19], and even intact organelles, such as mitochondria [20]. In this transfer of mRNA or proteins, MVs act as a kind of naturally engineered liposome. Interestingly, it has been demonstrated that mRNA and miRNA molecules are somehow preferentially enriched within MVs by a mechanism that involves docking proteins [5–8, 18].

At the same time, mounting evidence suggests that several cell types employed in regenerative medicine for the therapy of damaged organs, such as bone marrow-derived mononuclear cells (BMMNCs), hematopoietic stem/progenitor cells (HSPCs), mesenchymal stem cells or multipotent stroma cells (MSCs), skeletal muscle myoblasts, cardiac stem cells (CSCs), and neural stem cells (NSCs), are a rich source of cell surface membrane- and endosomal membrane-derived MVs. Later on in this chapter, we will discuss new data showing that MVs can replace intact cells in regenerative medicine approaches.

10.2 Regenerative Medicine Is Searching for its Holy Grail

The field of regenerative medicine is searching for a source of stem and progenitor cells that can be safely and efficiently employed for regeneration of damaged organs (e.g., heart, liver, kidney, or neural tissue) [12–24]. In experimental animal models of organ damage (e.g., heart infarct, liver damage, ischemic kidney failure, or stroke), various types of stem cells isolated from adult tissues have been employed, including, as mentioned above, BMMNCs, HSCs, MSCs, skeletal muscle myoblasts, and CSCs [21–24]. Similar types of cells are employed in the clinic to treat patients, while clinical trials are ongoing or have already been completed using these cells to improve the function of damaged organs [2, 25–27]. In parallel, another promising source of stem cells for regenerative medicine has emerged: induced pluripotent stem cells (iPSCs), which are also a rich source of MVs.

Interestingly, while some beneficial effects have been reported following cell-based therapies, there is no solid evidence that the cells employed to regenerate damaged tissues truly give rise to organ-specific cell populations (e.g., functional cardiomyocytes in heart, hepatocytes in liver, or tubular epithelium cells in kidney). Overall, the concept proposed a few years ago that stem cells, for example, HSPCs or MSCs, are plastic and may transdifferentiate into cells from different germ layers (e.g., cardiomyocytes, neural cells, or hepatocytes) lacks solid experimental support and was never confirmed in experiments performed in several independent laboratories [21–27]. Therefore, the concept of stem cell plasticity or transdedifferentiation has been challenged and some positive effects of stem cell therapy have been explained by other mechanisms. These alternative possibilities are listed in Table 10.1.

First, it is possible that some of the stem cell plasticity data could be explained simply by the phenomenon of cell fusion [24]. Accordingly, the cells observed in damaged tissues that express markers of both the donor cells employed in treatment (e.g., HSCs or MSCs) and cells typical of the damaged organ (e.g., cardiomyocytes), could be heterokaryons, the result of the fusion of therapeutic cells with somatic host cells in the damaged organ (e.g., myocardium). However, cell fusion is an extremely rare event that cannot fully account for the extensive positive transdedifferentiation/plasticity data published in several reports [21, 23].

Some of the benefits observed in organ and tissue regeneration after infusion of therapeutic cells might be explained by paracrine effects. It is well known that stem

Table 10.1 Alternative explanations for the phenomenon of transdedifferentiation or plasticity of stem cells

Cell fusion	The relatively rare phenomenon by which cells employed for therapy may fuse with cells in damaged tissues and form heterokaryons. Created in this way, heterokaryons express markers of both donor and recipient cells (pseudochimerism)
Paracrine stimulation	Cells employed for therapy, particularly HSPCs or MSCs, are a source of trophic and angiopoietic factors that may promote tissue/organ repair
Heterologous population of stem cells in BM	Cells employed for therapy (e.g., BM cells) may contain various stem cell populations. Regeneration could be explained by the presence of endothelial progenitors that promote neovasculogenesis and also by the presence of other stem cells, including pluripotent very small embryonic-like stem cells (VSELs)
Microvesicle (MV)-mediated transfer (shuttling) of molecules	Some of the plasticity data could be explained by a transient modification of cell phenotype and stimulation of cells in damaged organs by MVs derived from cells employed for therapy. MVs may transfer (shuttle) between cells proteins, mRNA, and miRNA that inhibit apoptosis in damaged cells and stimulate angiogenesis

cells, for example, HSPCs or MSCs, are a source of several trophic cytokines and growth factors and that these factors, if released from these cells, could promote tissue repair and vascularization [23]. It has been reported that HSPCs secrete stem cell factor (SCF), HGF, and vascular endothelial growth factor (VEGF) [1, 2], while MSCs secrete stromal derived factor-1 (SDF-1) and IL-8 [22].

Next, cells employed for therapy in regenerative medicine (e.g., BMMNCs) may, from the beginning, contain heterogeneous populations of stem cells. This simple fact was, unfortunately, not taken into careful consideration in the first optimistic reports demonstrating stem cell plasticity [21, 22]. We argued from the beginning that studies that did not address this possibility when employing donor-derived bone marrow (BM), mobilized peripheral blood (mPB), or umbilical cord blood (UCB) cells for the regeneration of nonhematopoietic tissues (e.g., cardiomycoytes or neural cells) may have made misleading interpretations [22, 27]. It is known that BMMNCs are enriched in several types of nonhematopoietic stem/progenitor cells, including endothelial progenitor cells (EPCs), as well as a population of pluripotent very small embryonic-like stem cells (VSELs) [28].

10.3 A Novel Role for MVs in Preventing Apoptosis and Promoting Proliferation of Cells in Damaged Tissues

There is accumulating evidence that cells may transiently modify the phenotype of neighboring cells by transferring surface receptors, intracellular proteins, mRNA, and miRNA through mechanisms that involve MVs [6, 18, 29]. Overall, as all cell types secrete MVs, some of the biologically active components of MVs (e.g., membrane receptors, adhesion molecules, peptides, mRNA, and miRNA hijacked from parental cells) depend on the cell of origin. Components that are more commonly shared between cells, for example, bioactive lipids that are components of cell membranes, S1P, ceramide-1-phosphate (C1P), and lysophosphatidylic acid (LPA), are highly enriched in all types of MVs. It is well known that S1P, for example, inhibits cell apoptosis and stimulates angiogenesis [30]. Thus, the biological effects of MVs secreted from cells employed in therapy could also explain the beneficial effects of stem cell therapy in regenerative medicine (Table 10.1).

Horizontal transfer of mRNA and miRNA by microparticles plays an important role in understanding their proregenerative properties. From an historical point of view, the phenomenon of horizontal transfer of mRNA between cells was first demonstrated in a model of expansion of murine HSPCs in the presence of murine embryonic stem cell (ESC)-derived MVs [18]. Since the maintenance of pluripotency and undifferentiated propagation of ESCs in vitro requires tight cell-to-cell contacts and effective intercellular signaling, it has been hypothesized that MVs derived from ESCs express stem cell-specific molecules that also support self-renewal and expansion of adult stem cells, including HSPCs. To address this hypothesis, murine HSPCs were expanded in the presence of ESC-derived MVs. As reported, murine embryonic stem cell line D3 (ES-D3)-derived MVs in serum-free cultures: (1) significantly enhanced survival and improved expansion of murine HSPCs, (2) upregulated the expression of early pluripotency markers (Oct-4, Nanog, and Rex-1) and early hematopoietic stem cells markers (Scl, HoxB4, and GATA 2) in these cells, and (3) induced phosphorylation of MAPK p42/44 and serine-threonine kinase AKT. Molecular analysis revealed that ESC-derived MVs are selectively highly enriched in mRNA for several pluripotency transcription factors compared with parental ESCs and that these mRNAs could be delivered/shuttled by ESC-derived MVs to HSPCs and translated into the corresponding proteins. These positive biological effects of ESC-derived MVs on HSPCs were inhibited after heat inactivation or pretreatment with RNase, indicating a major involvement of protein and mRNA components in the observed phenomena. Thus, this study first clearly demonstrated that ESC-derived MVs efficiently expanded murine HSPCs ex vivo and induced pluripotency of these cells after horizontal transfer of ESC-derived mRNA and miRNA species [18].

This mechanism of modifying the phenotype of target cells by MVs has been subsequently confirmed in several other elegant studies. For example, one study showed a change of phenotype in rat BM-derived HSPCs in cocultures with murine lung tissue-derived MVs [18]. This transfer of MVs was demonstrated in vitro in a coculture system in which BM cells and lung cells were separated by a cell-impermeable membrane (0.4-μm pore size), and murine mRNAs for pulmonary epithelial cell-specific surfactants B and C were detected in the rat cells using species-specific primers. Similar results showing transfer of albumin mRNA after coculture with rat or murine liver cells were reported [32]. Furthermore, similar mechanisms of transfer of organ-specific mRNA into BM cells has been demonstrated for MVs derived from brain, heart, and liver, suggesting that MV-mediated cellular phenotype change is a universal phenomenon [7]. On the other hand, it has been reported that lung-derived MVs may transfer miRNA that negatively modulates pulmonary epithelial cell-specific mRNA levels in cocultured BM cells [7, 18]. Thus, transfer of both mRNA and miRNA by MVs has an obvious impact on the phenotype of target cells.

Based on this mechanism, MV-mediated exchange of receptors, proteins, mRNA, and miRNA between therapeutic cells (e.g., BMMNCs, HSPCs, or MSCs) and cells in damaged organs that are targets for MV delivery could temporarily modify the phenotypes of these cells [7]. Thus, transient modification of the phenotype of damaged cells by MVs derived from cells employed in therapy may give the false impression of transdedifferentiation, plasticity, or chimerism [23]. At the same time, this mechanism may have a beneficial effect on recovery and regeneration of damaged tissues, for example, where BMMNCs are employed to treat damaged lung tissue. In this setting, paracrine effects of transplanted BMMNCs in combination with BMMNC-derived MVs: (1) improve the function of damaged tissue and (2) transiently change the phenotype of the damaged cells due to transfer of BMMNC receptors, mRNA, miRNA, and proteins to pneumocytes, which may give the impression of true chimerism or stem cell plasticity.

Thus, MVs are a source of several antiapoptotic, proliferation-stimulating, and angiopoietic factors that may have a beneficial effect on improving the function of damaged tissues (e.g., myocardium after heart infarct, kidney after ischemic insult, or brain after stroke). As shown in Fig. 10.1, this underappreciated mechanism may explain the beneficial effects of stem cell therapies currently employed in cardiology, nephrology, or neurology. It is well known that very similar beneficial effects were reported, for example, in experiments and clinical trials aimed to improve the function of myocardium or neural tissues, regardless of the cell types employed for treatment (e.g., HSPCs or MSCs). These similar effects that do not depend so much on the source of cells strongly suggest the involvement of common mechanisms. Based on this reasoning, we postulate that these beneficial effects rely mostly on paracrine secretion of growth factors, cytokines, and chemokines, and, most importantly, on secretion of proregenerative signals delivered by MVs derived from cells employed for therapy (Fig. 10.1).

While unrecognized for many years, the paracrine effects of cells employed in stem cell therapy (and the involvement in this phenomenon of cell-derived MVs) is emerging as an important mechanism that explains some of the beneficial effects that have been observed. Moreover, based on this realization, a new potential application of MVs in which they replace whole cells is a promising concept in regenerative medicine. In support of this possibility, in recent elegant studies, MSC-derived MVs were found to have the same beneficial effect on regeneration of kidneys as intact MSCs [8, 19]. As reported, MSC-derived MVs were able to protect the kidney against ischemia-reperfusion-induced acute and chronic kidney injury. MVs isolated from MSCs were injected intravenously into rats immediately after monolateral nephrectomy and transient occlusion of the renal artery and vein to the remaining kidney. The authors observed that a single administration of MVs immediately after ischemia reperfusion injury protected experimental animals from acute kidney injury by inhibiting apoptosis and stimulating tubular epithelial cell proliferation [19]. This phenomenon was strongly dependent on mRNA present in these MVs, as pretreatment of MVs with RNase abrogated this effect. Based on previous work from this

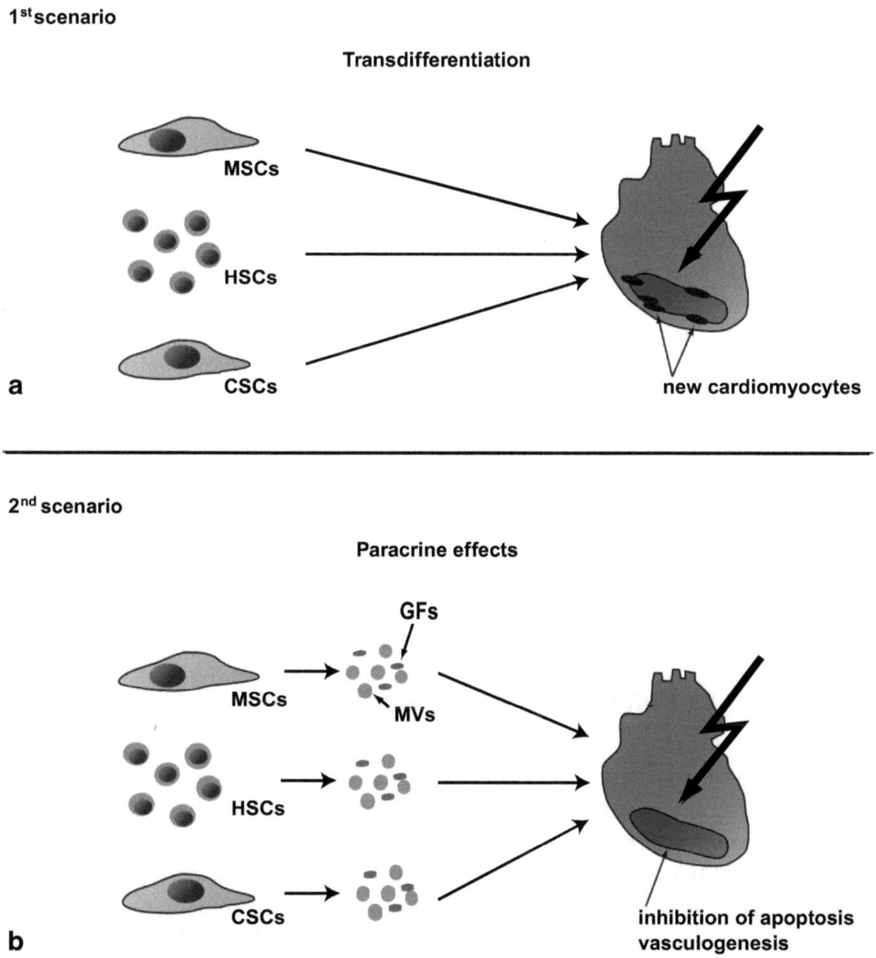

Fig. 10.1 Using a heart infarct model, two possible scenarios illustrating the beneficial effects of stem cell therapies in regenerative medicine. **a** (*1st scenario*) Cells employed for therapy (e.g., HSCs or MSCs) may theoretically transdifferentiate into cardiomyocytes. However, if this occurs at all, it is a very rare and random phenomenon and is not well substantiated by current experimental data. **b** (*2nd scenario*) Cells employed for therapy (e.g., HSCs or MSCs) do not transdifferentiate into cardiomyocytes but secrete several paracrine factors and shed MVs that inhibit apoptosis in damaged cardiomyocytes, promote their proliferation, and stimulate angiogenesis. Evidence is accumulating that this is a major effect in currently employed stem cell therapies

group in which a similar protective effect was observed after infusion of MSCs, this recent report further suggests that MSCs and MSC-derived MVs show comparable therapeutic effects [19].

In another study, it was demonstrated that human liver stem cell-derived MVs were able to accelerate regeneration of liver in partially hepatectomized rats in which $\sim 70\ \%$ of the liver parenchyma had been removed [33]. Based on in vitro coculture studies, human liver stem cell-derived MVs stimulated proliferation and induced apoptotic resistance in human and rat hepatocytes, which required internalization of MVs in an α4-integrin-dependent manner. Interestingly, as reported in the same study, MVs pretreated with RNase did not show any beneficial effects, which suggests the importance of an mRNA- and miRNA-dependent transfer mechanism to overexpress appropriate antiapoptotic genes in the target cells. In strong support of this interpretation, both microarray analysis and quantitative RT-PCR demonstrated that human liver stem cell-derived MVs are able to transfer mRNA species for genes involved in the control of transcription, translation, proliferation, and apoptosis. In the next part of this elegant study, MVs administered in vivo into hepatectomized rats accelerated the morphological and functional recovery of liver. This proregenerative effect was again associated with an increase in hepatocyte proliferation and was abolished when MVs were pretreated with RNase [33].

All these experiments imply that MVs and their mRNA and miRNA cargo are important and long-underappreciated players in cellular therapies involving MSCs. This also shows that MVs are an important source of several mRNA- and miRNA-encoded proproliferation and antiapoptotic factors. Furthermore, the mechanisms that govern the selectivity and enrichment of MVs for appropriate mRNA and miRNA species require further study.

10.4 MVs as Potent Proangiopoietic Factors

Improvement of vascularization in the damaged organ (e.g., myocardium after heart infarct) is an important step in the regenerative process. Damaged cells secrete several potential proangiopoietic factors, such as SDF-1 and VEGF, which are upregulated in the damaged tissues in an hypoxia inducible factor-1 alpha (HIF-1α)-dependent manner [23]. In addition, several factors, such as bioactive lipids (e.g., S1P or C1P) that possess strong proangiopoietic potential, are released from the damaged cells. All these factors together may promote sprouting of endothelial cells from existing vessels, as well as chemoattract EPCs, which circulate in PB and are mobilized in response to organ injury.

It is well known that during organ and tissue damage MVs and apoptotic bodies are released from damaged and dying cells, respectively. On the other hand, MVs derived from activated blood platelets, leukocytes, and endothelial cells continuously circulate in peripheral blood under steady-state conditions, with their numbers increasing during stress situations (e.g., heart infarct or stroke) [27, 34]. It has already been shown that MVs are endowed with strong proangiopoietic properties (Fig. 10.2). This could be explained by the fact that MVs are highly enriched in bioactive lipids, including S1P and C1P, that possess direct and robust vasculogenic potential [35]. Therefore, MVs, regardless from which cells they are derived, are very strong chemoattractants for endothelial cells (Fig. 10.3) and promote tube formation by these cells [36]. These stimulatory effects of MVs on angiogenesis are supported

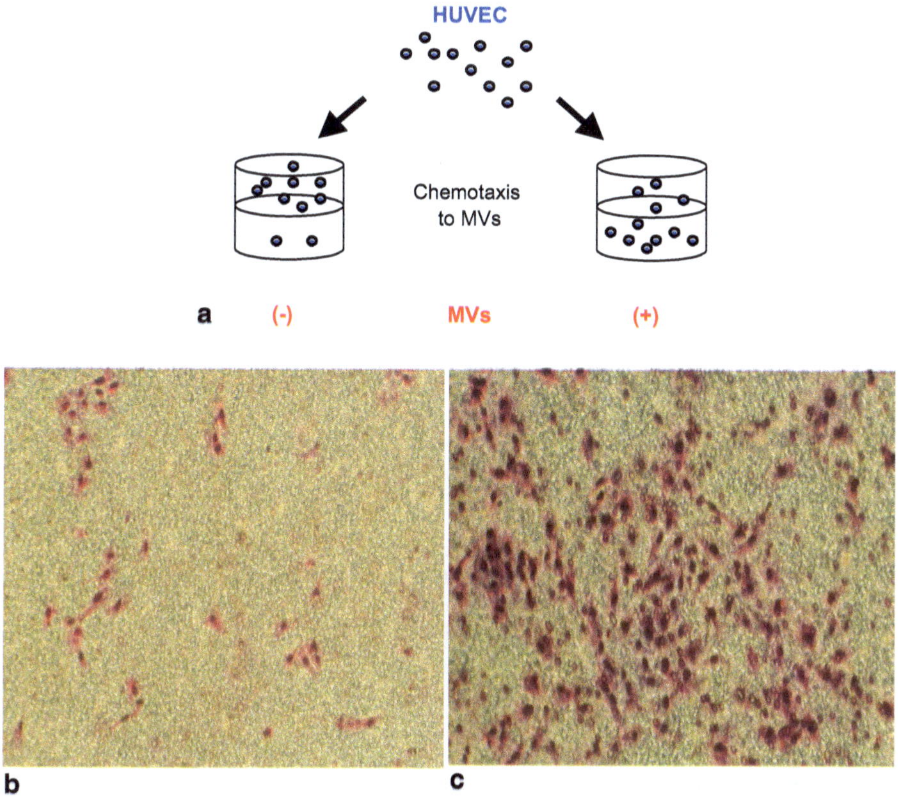

Fig. 10.2 Proangiopoietic properties of MVs. **a** Chemotaxis assay across transwell membranes employing human umbilical cord blood cells (HUVEC). **b** Chemotaxis in response to control medium. **c** Chemotaxis in response to MVs (10 μg/ml). Following the chemotaxis assay, cells were stained on the lower surface of the transwell membranes

by the observed activation of several signaling pathways in endothelial cells (e.g., MAPKp42/44 or AKT).

As mentioned above, in addition to MVs released from damaged cells, platelets, and leukocytes, several cell types employed for therapy are abundant sources of MVs. Local delivery of these cells to the damaged organ leads not only to secretion of soluble growth factors, cytokines, and chemokines but also triggers secretion of MVs by these cells. All this together may explain the robust proangiopoietic potential of transplanted cells employed to rescue damaged tissues.

Finally, since MVs are enriched for mRNA and regulatory miRNA, further studies are needed to identify the presence of appropriate mRNA and miRNA species that promote effective angiogenesis.

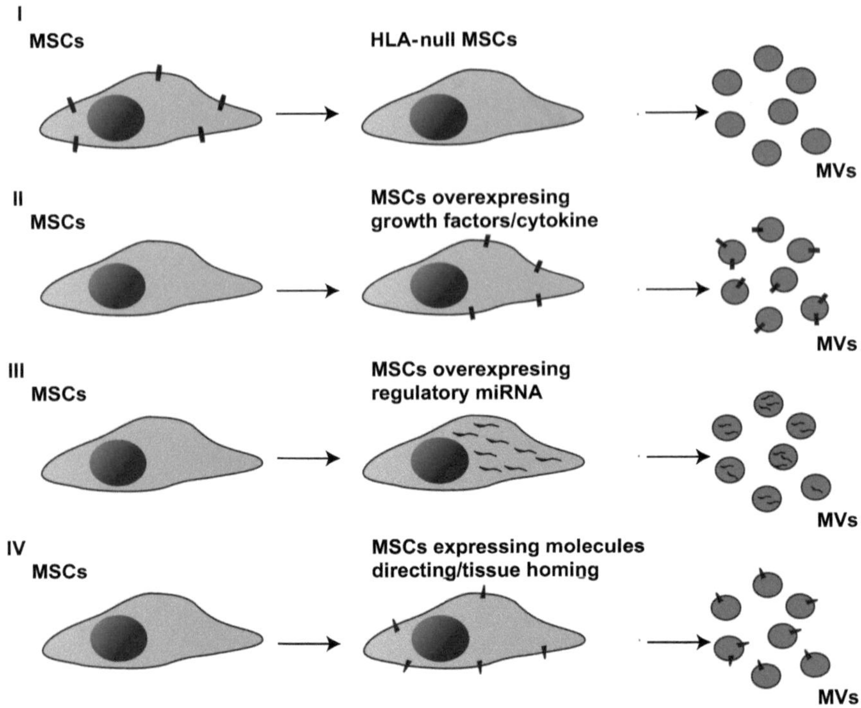

Fig. 10.3 Different approaches to generating more efficient proregenerative MVs in vitro. MVs could be harvested from in vitro cultures of MV-producing cell lines. Such cell lines may be modified to obtain MVs that: (*I*) do not express HLA antigens, (*II*) are enriched for growth factors, cytokines, and chemokines that promote regeneration of damaged organs, (*III*) are enriched in mRNA and regulatory miRNA facilitating regeneration of damaged tissues and/or promoting angiogenesis, and (*IV*) express molecules that direct them to, and subsequently be retained in, damaged tissues

10.5 Toward Development of Engineered MVs for Regenerative Medicine

Based on the fact that MVs have similar beneficial effects in regenerative medicine therapy as the intact cells from which they are derived [23], one can speculate that it is possible to engineer or modify MVs for the purposes of regenerative medicine. Several possibilities for how to make this approach more efficient are shown in Fig. 10.3. Overall, it would be tempting to modify MVs to ensure their most efficient involvement in therapy. For example, MVs could be isolated for potential application in regenerative medicine from a large-scale ex vivo expansion of cells (e.g., MSCs) cultured in appropriate generators. These MV-generating cells could be genetically modified in order to produce custom-engineered MVs more suitable for therapy. This challenging approach awaits verification in appropriate animal models.

First, as depicted in Fig. 10.3, it should be possible to expand MV-producing cell lines that lack genes encoding histocompatibility antigens. This approach would minimalize the possibility of cross-immunization with donor HLA antigens. Second, a better understanding of mechanisms that govern, and factors that enhance, secretion of MVs will allow the engineering of high-MV-producing cell lines. We speculate that MVs derived from cells in hypoxic conditions would be enriched in mRNAs and miRNAs that promote angiogenesis. Third, MV producer cell lines could be transduced with genes that overexpress on their surface: (1) peptides that protect target cells in damaged organs from apoptosis and stimulate proliferation of residual remaining cell populations (e.g., SCF or Notch ligands) or (2) factors that effectively induce angiogenesis (e.g., VEGF, FGF-2, or SDF-1). Fourth, producer cell lines could be enriched for mRNA and regulatory miRNA species that, after delivery to the damaged tissues, promote regeneration. Finally, we envision that MV producer cell lines could be enriched for molecules that facilitate their tropism to the damaged organ and subsequently promote retention of MVs in the damaged tissues. In the example mentioned above, human liver stem cell-derived MVs stimulated proliferation and induced apoptotic resistance in human and rat hepatocytes, which required their internalization in an α4-integrin-dependent manner [33].

MV-based therapies also open up new possibilities for clinical applications of iPSCs. Since *in vivo* application of iPSCs is still limited by the high risk of teratoma formation by these cells [23], MVs from patient-derived iPSCs could be employed as a novel generation of therapeutics to rescue damaged organs and tissues. Based on this possibility, we envision that patient-derived iPSCs could be employed as MV-producing cells. Moreover, taking advantage of the recently proposed epigenetic memory of cells employed for generation of iPSCs, one can also envision that, for example, MVs from keratinocyte-derived iPSCs would preferentially affect regeneration of damaged skin (e.g., after burns), MVs isolated from supernatants of cardiomyocyte-derived iPSCs would have advantages in regeneration of damaged myocardium.

10.6 Conclusions

MVs have become a focus of contemporary medicine for several reasons. Evidence is accumulating that MVs: (1) play an important role in cell–cell communication, (2) directly stimulate target cells by ligands that are expressed on their surface, (3) shuttle mRNA, regulatory miRNA, proteins, and organelles between cells, (4) serve as vehicles in the infection of cells by viruses (e.g., HIV) and prions, and (5) are central to a new biotechnological platform being developed that employs the molecular signatures (e.g., mRNA, miRNA, and protein compositions) of MVs circulating in peripheral blood in the diagnosis of different disorders [18].

Strong evidence is accumulating that MVs are abundantly secreted by stem cells infused locally or systemically to rescue damaged tissues [18, 19, 31–33]. In several elegant studies, it has been demonstrated that infusion of MVs has the same proregenerative potential as infusion of intact cells that are the source of these MVs. Based on recent data, the rescue of damaged organ or tissue following infusion of cells, which was previously attributed to stem cell plasticity or transdedifferentiation, is better explained by the paracrine effects of growth factors, cytokines, and chemokines secreted from the infused or locally delivered stem cells, as well as by secretion of proregenerative MVs by these cells [34–36].

These remarkable properties of MVs should have an impact in the development of new strategies in regenerative medicine in which MVs would be harvested from large-scale in vitro cultures of MV-producing cells engineered to overexpress appropriate growth factors, cytokines, surface molecules, mRNA, and miRNA that inhibit apoptosis of target cells and promote neovascularization of damaged tissues. Such custom-engineered "super MVs" could become a new class of cell-derived therapeutics in regenerative medicine.

Acknowledgments This work was supported by NIH grant R01 DK074720, the Stella and Henry Endowment, and the European Union structural funds (Innovative Economy Operational Program POIG.01.01.02-00-109/09-00) to MZR.

References

1. Majka M, Janowska-Wieczorek A, Ratajczak J, Ehrenman K, Pietrzkowski Z, Kowalska MA et al (2001) Numerous growth factors, cytokines, and chemokines are secreted by human CD34(+) cells, myeloblasts, erythroblasts, and megakaryoblasts and regulate normal hematopoiesis in an autocrine/paracrine manner. Blood 97:3075–3085
2. Janowska-Wieczorek A, Majka M, Ratajczak J et al (2001) Autocrine/paracrine mechanisms in human hematopoiesis. Stem Cells 19:99–107
3. Rustom A, Saffrich R, Markovic I, Walther P, Gerdes HH (2004) Nanotubular highways for intercellular organelle transport. Science 303:1007–1010
4. Vidulescu C, Clejan S, O'Connor K C (2004) Vesicle traffic through intercellular bridges in DU 145 human prostate cancer cells. J Cell Mol Med 8:388–396
5. George JN, Thoi LL, McManus LM et al (1982) Isolation of human platelet membrane microparticles from plasma and serum. Blood 60:834–839
6. Ratajczak J, Wysoczynski M, Hayek F et al (2006) Membrane-derived microvesicles (MV): important and underappreciated mediators of cell to cell communication. Leukemia 20:1487–1495
7. Quesenberry PJ, Dooner MS, Aliotta JM (2010) Stem cell plasticity revisited: the continuum marrow model and phenotypic changes mediated by microvesicles. Exp Hematol 38:581–592
8. Camussi G, Deregibus MC, Tetta C (2010) Paracrine/endocrine mechanism of stem cells on kidney repair: role of microvesicle-mediated transfer of genetic information. Curr Opin Nephrol Hypertens 19: 7–12
9. Beaudoin AR, Grondin G (1991) Shedding of vesicular material from the cell surface of eukaryotic cells: different cellular phenomena. Biochim Biophys Acta 1071:203
10. Fevrier B, Raposo G (2004) Exosomes: endosomal-derived vesicles shipping extracellular messages. Curr Opin Cell Biol Aug 16(4):415–421
11. Greenwalt TJ (2006) The how and why of exocytic vesicles. Transfusion 46(1):143–152
12. Hugel B, Martinez MC, Kunzelmann C, Freyssinet JM (2005) Membrane microparticles: two sides of the coin. Physiology (Bethesda) 20:22–27

13. Barry OP, FitzGerald GA (1999) Mechanisms of cellular activation by platelet microparticles. Thromb Haemost 82:794–800
14. Barry OP, Pratico D, Savani RC, FitzGerald GA (1998) Modulation of monocyte-endothelial cell interaction by platelet microparticles. J Clin Invest 102:136–144
15. VanWijk MJ, VanBavel E, Sturk A, Nieuwland R (2003) Microparticles in cardiovascular diseases. Cardiovasc Res 59(2):277–287
16. Horstman LL, Jy W, Jimenez JJ, Bidot C, Ahn YS (2004) New horizons in the analysis of circulating cell-derived microparticles. Keio J Med 53(4):210–230
17. Greco V, Hannus M, Eaton S (2001) Argosomes: a potential vehicle for the spread of morphogens through epithelia. Cell 106:633–645
18. Ratajczak J, Miekus K, Kucia M et al (2006) Embryonic stem cell-derived microvesicles reprogram hematopoietic progenitors: evidence for horizontal transfer of mRNA and protein delivery. Leukemia 20:847–856
19. Gatti S, Bruno S, Deregibus MC et al (2011) Microvesicles derived from human adult mesenchymal stem cells protect against ischemia-reperfusion-induced acute and chronic kidney injury. Nephrol Dialysis Transplant 26:1474–1483
20. Spees JL, Olson SD, Whitney MJ et al (2006) Mitochondrial transfer between cells can rescue aerobic respiration. Proc Natl Acad Sci U S A 103:1283–1288
21. Friedman RS, Krause DS (2009) Regeneration and repair: new findings in stem cell research and aging. Ann N Y Acad Sci 1172:88–94
22. Joyce N, Annett G, Wirthlin L et al (2010) Mesenchymal stem cells for the treatment of neurodegenerative disease. Regen Med 5:933–946
23. Ratajczak MZ, Zuba-Surma EK, Wysoczynski M et al (2008) Hunt for pluripotent stem cell—regenerative medicine search for almighty cell. J Autoimmun 30:151–162
24. O'Malley K, Scott EW (2004) Stem cell fusion confusion. Exp Hematol 32:131–134
25. Ratajczak MZ, Zuba-Surma E, Ratajczak J et al (2008) Very small embryonic like (VSEL) stem cells—characterization, developmental origin and biological significance. Exp Hematol 36:742–751
26. Tang XL, Rokosh DG, Guo Y et al (2010) Cardiac progenitor cells and bone marrow-derived very small embryonic-like stem cells for cardiac repair after myocardial infarction. Circ J 74:390–404
27. Tendera M, Wojakowski W, Ruzyłło W et al (2009) Intracoronary infusion of bone marrow-derived selected CD34+CXCR4+ cells and non-selected mononuclear cells in patients with acute STEMI and reduced left ventricular ejection fraction: results of randomized, multicentre myocardial regeneration by intracoronary infusion of selected population of stem cells in acute myocardial infarction (REGENT) trial. Eur Heart J 30:1313–1321
28. Kucia M, Reca R, Campbell FR et al (2006) A population of very small embryonic-like (VSEL) CXCR4(+)SSEA-1(+)Oct-4+ stem cells identified in adult bone marrow. Leukemia 20:857–869
29. Collino F, Deregibus MC, Bruno S et al (2010) Microvesicles derived from adult human bone marrow and tissue specific mesenchymal stem cells shuttle selected pattern of miRNAs. PLoS ONE 5:e11803
30. Ratajczak MZ, Lee H, Wysoczynski M et al (2010) Novel insight into stem cell mobilization-plasma sphingosine-1-phosphate is a major chemoattractant that directs the egress of hematopoietic stem progenitor cells from the bone marrow and its level in peripheral blood increases during mobilization due to activation of complement cascade/membrane attack complex. Leukemia 24:976–985
31. Aliotta JM, Sanchez-Guijo FM, Dooner GJ et al (2007) Alteration of marrow cell gene expression, protein production, and engraftment into lung by lung-derived microvesicles: a novel mechanism for phenotype modulation. Stem Cells 25:2245–2256
32. Del Tatto M, Ng T, Aliotta JM et al (2011) Marrow cell genetic phenotype change induced by human lung cancer cells. Exp Hematol 39:1072–1080
33. Herrera MB, Fonsato V, Gatti S et al (2010) Human liver stem cell-derived microvesicles accelerate hepatic regeneration in hepatectomized rats. J Cell Mol Med 14(6B):1605–1618

34. Paczkowska E, Kucia M, Koziarska D et al (2009) Clinical evidence that very small embryonic-like stem cells are mobilized into peripheral blood in patients after stroke. Stroke 40:1237–1244
35. Ratajczak MZ, Kim CH, Abdel-Latif A et al (2012) A novel perspective on stem cell homing and mobilization—review on bioactive lipids as potent chemoattractants and cationic peptides as underappreciated modulators of responsiveness to SDF-1 gradients. Leukemia 26:63–72
36. Ratajczak J, Kijowski J, Majka M et al (2003) Biological significance of the different erythropoietic factors secreted by normal human early erythroid cells. Leuk Lymphoma 44:767–774

Index

H.-G. Zhang (ed.), *Emerging Concepts of Tumor Exosome–Mediated Cell–Cell Communication,* DOI 10.1007/978-1-4614-3697-3,

MIX
Papier aus verantwortungsvollen Quellen
Paper from responsible sources
FSC® C105338

If you have any concerns about our products,
you can contact us on
ProductSafety@springernature.com

In case Publisher is established outside the EU,
the EU authorized representative is:
Springer Nature Customer Service Center GmbH
Europaplatz 3, 69115 Heidelberg, Germany

Printed by Libri Plureos GmbH
in Hamburg, Germany